T0371819

ALGORITHMS FOR NEXT-GENERATION SEQUENCING

Chapman & Hall/CRC Mathematical and Computational Biology Series

ALGORITHMS FOR NEXT-GENERATION SEQUENCING

Wing-Kin Sung

CRC Press
Taylor & Francis Group
Boca Raton London New York

CRC Press is an imprint of the
Taylor & Francis Group, an **informa** business

A CHAPMAN & HALL BOOK

CRC Press
Taylor & Francis Group
6000 Broken Sound Parkway NW, Suite 300
Boca Raton, FL 33487-2742

© 2017 by Taylor & Francis Group, LLC
CRC Press is an imprint of Taylor & Francis Group, an Informa business

No claim to original U.S. Government works

Printed on acid-free paper
Version Date: 20170421

International Standard Book Number-13: 978-1-4665-6550-0 (Hardback)

Visit the Taylor & Francis Web site at
http://www.taylorandfrancis.com

and the CRC Press Web site at
http://www.crcpress.com

Printed and bound in the United States of America by
Edwards Brothers Malloy on sustainably sourced paper

Contents

Preface

Next-generation sequencing (NGS) is a recently developed technology enabling us to generate hundreds of billions of DNA bases from the samples. We can use NGS to reconstruct the genome, understand genomic variations, recover transcriptomes, and identify the transcription factor binding sites or the epigenetic marks.

The NGS technology radically changes how we collect genomic data from the samples. Instead of studying a particular gene or a particular genomic region, NGS technologies enable us to perform genome-wide study unbiasedly. Although more raw data can be obtained from sequencing machines, we face computational challenges in analyzing such a big dataset. Hence, it is important to develop efficient and accurate computational methods to analyze or process such datasets. This book is intended to give an in-depth introduction to such algorithmic techniques.

The primary audiences of this book include advanced undergraduate students and graduate students who are from mathematics or computer science departments. We assume that readers have some training in college-level biology, statistics, discrete mathematics and algorithms.

This book was developed partly from the teaching material for the course on Combinatorial Methods in Bioinformatics, which I taught at the National University of Singapore, Singapore. The chapters in this book are classified based on the application domains of the NGS technologies. In each chapter, a brief introduction to the technology is first given. Then, different methods or algorithms for analyzing such NGS datasets are described. To illustrate each algorithm, detailed examples are given. At the end of each chapter, exercises are given to help readers understand the topics.

Chapter 1 introduces the next-generation sequencing technologies. We cover the three generations of sequencing, starting from Sanger sequencing (first generation). Then, we cover second-generation sequencing, which includes Illumina Solexa sequencing. Finally, we describe the third-generation sequencing technologies which include PacBio sequencing and nanopore sequencing.

Chapter 2 introduces a few NGS file formats, which facilitate downstream analysis and data transfer. They include fasta, fastq, SAM, BAM, BED, VCF and WIG formats. Fasta and fastq are file formats for describing the raw sequencing reads generated by the sequencers. SAM and BAM are file formats for describing the alignments of the NGS reads on the reference genome. BED, VCF and WIG formats are annotation formats.

To develop methods for processing NGS data, we need efficient algorithms and data structures. Chapter 3 is devoted to briefly describing these techniques.

Chapter 4 studies read mappers. Read mappers align the NGS reads on the reference genome. The input is a set of raw reads in fasta or fastq files. The read mapper will align each raw read on the reference genome, that is, identify the region in the reference genome which is highly similar to the read. Then, the read mapper will output all these alignments in a SAM or BAM file. This is the basic step for many NGS applications. (It is the first step for the methods in Chapters 6–9.)

Chapter 5 studies the de novo assembly problem. Given a set of raw reads extracted from whole genome sequencing of some sample genome, de novo assembly aims to stitch the raw reads together to reconstruct the genome. It enables us to reconstruct novel genomes like plants and bacteria. De novo assembly involves a few steps: error correction, contig assembly (de Bruijn graph approach or base-by-base extension approach), scaffolding and gap filling. This chapter describes techniques developed for these steps.

Chapter 6 discusses the problem of identifying single nucleotide variations (SNVs) and small insertions/deletions (indels) in an individual genome. The genome of every individual is highly similar to the reference human genome. However, each genome is still different from the reference genome. On average, there is 1 single nucleotide variation in every 3000 bases and 1 small indel in every 1000 bases. To discover these variations, we can first perform whole genome sequencing or exome sequencing of the individual genome to obtain a set of raw reads. After aligning the raw reads on the reference genome, we use SNV callers and indel callers to call SNVs and small indels. This chapter is devoted to discussing techniques used in SNV callers and indel callers.

Apart from SNVs and small indels, copy number variations (CNVs) and structural variations (SVs) are the other types of variations that appear in our genome. CNVs and SVs are not as frequent as SNVs and indels. Moreover, they are more prone to change the phenotype. Hence, it is important to understand them. Chapter 7 is devoted to studying techniques used in CNV callers and SV callers.

All above technologies are related to genome sequencing. We can also sequence RNA. This technology is known as RNA-seq. Chapter 8 studies methods for analyzing RNA-seq. By applying computational methods on RNA-seq, we can recover the transcriptome. More precisely, RNA-seq enables us to identify exons and split junctions. Then, we can predict the isoforms of the genes. We can also determine the expression of each transcript and each gene.

By combining Chromatin immunoprecipitation and next-generation sequencing, we can sequence genome regions that are bound by some transcription factors or with epigenetic marks. Such technology is known as ChIP-seq. The computational methods that identify those binding sites are known as ChIP-seq peak callers. Chapter 9 is devoted to discussing computational methods for such purpose.

As stated earlier, NGS data is huge; and the NGS data files are usually big. It is difficult to store and transfer NGS files. One solution is to compress the NGS data files. Nowadays, a number of compression methods have been developed and some of the compression formats are used frequently in the literatures like BAM, bigBed and bigWig. Chapter 10 aims to describe these compression techniques. We also describe techniques that enable us to randomly access the compressed NGS data files.

Supplementary material can be found at

http://www.comp.nus.edu.sg/~ksung/algo_in_ngs/.

I would like to thank my PhD supervisors Tak-Wah Lam and Hing-Fung Ting and my collaborators Francis Y. L. Chin, Kwok Pui Choi, Edwin Cheung, Axel Hillmer, Wing Kai Hon, Jansson Jesper, Ming-Yang Kao, Caroline Lee, Nikki Lee, Hon Wai Leong, Alexander Lezhava, John Luk, See-Kiong Ng, Franco P. Preparata, Yijun Ruan, Kunihiko Sadakane, Chialin Wei, Limsoon Wong, Siu-Ming Yiu, and Louxin Zhang. My knowledge of NGS and bioinformatics was enriched through numerous discussions with them. I would like to thank Ramesh Rajaby, Kunihiko Sadakane, Chandana Tennakoon, Hugo Willy, and Han Xu for helping to proofread some of the chapters. I would also like to thank my parents Kang Fai Sung and Siu King Wong, my three brothers Wing Hong Sung, Wing Keung Sung, and Wing Fu Sung, my wife Lily Or, and my three kids Kelly, Kathleen and Kayden for their support.

Finally, if you have any suggestions for improvement or if you identify any errors in the book, please send an email to me at ksung@comp.nus.edu.sg. I thank you in advance for your helpful comments in improving the book.

Wing-Kin Sung

Chapter 1

Introduction

DNA stands for deoxyribonucleic acid. It was first discovered in 1869 by Friedrich Miescher [58]. However, it was not until 1944 that Avery, MacLeod and McCarty [12] demonstrated that DNA is the major carrier of genetic information, not protein. In 1953, James Watson and Francis Crick discovered the basic structure of DNA, which is a double helix [310]. After that, people started to work on DNA intensively.

DNA sequencing sprang to life in 1972, when Frederick Sanger (at the University of Cambridge, England) began to work on the genome sequence using a variation of the recombinant DNA method. The full DNA sequence of a viral genome (bacteriophage ϕX174) was completed by Sanger in 1977 [259, 260]. Based on the power of sequencing, Sanger established genomics,[1] which is the study of the entirety of an organism's hereditary information, encoded in DNA (or RNA for certain viruses). Note that it is different from molecular biology or genetics, whose primary focus is to investigate the roles and functions of single genes.

During the last decades, DNA sequencing has improved rapidly. We can sequence the whole human genome within a day and compare multiple individual human genomes. This book is devoted to understanding the bioinformatics issues related to DNA sequencing. In this introduction, we briefly review DNA, RNA and protein. Then, we describe various sequencing technologies. Lastly, we describe the applications of sequencing technologies.

1.1 DNA, RNA, protein and cells

Deoxyribonucleic acid (DNA) is used as the genetic material (with the exception that certain viruses use RNA as the genetic material). The basic building block of DNA is the DNA nucleotide. There are 4 types of DNA nucleotides: adenine (A), guanine (G), cytosine (C) and thymine (T). The DNA

---

[1] The actual term "genomics" is thought to have been coined by Dr. Tom Roderick, a geneticist at the Jackson Laboratory (Bar Harbor, ME) at a meeting held in Maryland on the mapping of the human genome in 1986.

5′ −	A	C	G	T	A	G	C	T	−3′
	‖	‖‖	‖‖	‖	‖	‖‖	‖‖	‖	
3′ −	T	G	C	A	T	C	G	A	−5′

FIGURE 1.1: The double-stranded DNA. The two strands show a complementary base pairing.

nucleotides can be chained together to form a strand of DNA. Each strand of DNA is asymmetric. It begins from 5′ end and ends at 3′ end.

When two opposing DNA strands satisfy the Watson-Crick rule, they can be interwoven together by hydrogen bonds and form a double-stranded DNA. The Watson-Crick rule (or complementary base pairing rule) requires that the two nucleotides in opposing strands be a complementary base pair, that is, they must be an (A, T) pair or a (C, G) pair. (Note that A = T and C ≡ G are bound with the help of two and three hydrogen bonds, respectively.) Figure 1.1 gives an example double-stranded DNA. One strand is ACGTAGCT while the other strand is its reverse complement, i.e., AGCTACGT.

The double-stranded DNAs are located in the nucleus (and mitochondria) of every cell. A cell can contain multiple pieces of double-stranded DNAs, each is called a chromosome. As a whole, the collection of chromosomes is called a genome; the human genome consists of 23 pairs of chromosomes, and its total length is roughly 3 billion base pairs.

The genome provides the instructions for the cell to perform daily life functions. Through the process of transcription, the machine RNA polymerase transcribes genes (the basic functional units) in our genome into transcripts (or RNA molecules). This process is known as gene expression. The complete set of transcripts in a cell is denoted as its transcriptome.

Each transcript is a chain of 4 different ribonucleic acid (RNA) nucleotides: adenine (A), guanine (G), cytosine (C) and uracil (U). The main difference between the DNA nucleotide and the RNA nucleotide is that the RNA nucleotide has an extra OH group. This extra OH group enables the RNA nucleotide to form more hydrogen bonds. Transcripts are usually single stranded instead of double stranded.

There are two types of transcripts: non-coding RNA (ncRNA) and message RNA (mRNA). ncRNAs are transcripts that do not translate into proteins. They can be classified into transfer RNAs (tRNAs), ribosomal RNAs (rRNAs), short ncRNAs (of length < 30 bp, includes miRNA, siRNA and piRNA) and long ncRNAs (of length > 200 bp, example includes Xist, and HOTAIR).

mRNA is the intermediate between DNA and protein. Each mRNA consists of three parts: a 5' untranslated region (a 5' UTR), a coding region and a 3' untranslated region (3' UTR). The length of the coding region is of a multiple of 3. It is a sequence of triplets of nucleotides called codons. Each codon corresponds to an amino acid.

Through translation, the machine ribosome translates each mRNA into a protein, which is the sequence of amino acids corresponding to the sequence of codons in the mRNA. Protein forms complex 3D structures. Each protein is a biological nanomachine that performs a specialized function. For example, enzymes are proteins that work as catalysts to promote chemical reactions for generating energy or digesting food. Other proteins, called transcription factors, interact with the genome to turn on or off the transcriptions. Through the interaction among DNA, RNA and protein, our genome dictates which cells should grow, when cells should die, how cells should be structured, and creates various body parts.

All cells in our body are developed from a single cell through cell division. When a cell divides, the double helix genome is separated into single-stranded DNA molecules. An enzyme called DNA polymerase uses each single-stranded DNA molecule as the template to replicate the genome into two identical double helixes. By this replication process, all cells within the same individual will have the same genome. However, due to errors in copying, some variations (called mutations) might happen in some cells. Those variations or mutations may cause diseases such as cancer.

Different individuals have similar genomes, but they also have genome variations that contribute to different phenotypes. For example, the color of our hairs and our eyes are controlled by the differences in our genomes. By studying and comparing genomes of different individuals, researchers develop an understanding of the factors that cause different phenotypes and diseases. Such knowledge ultimately helps to gain insights into the mystery of life and contributes to improving human health.

1.2 Sequencing technologies

DNA sequencing is a process that determines the order of the nucleotide bases. It translates the DNA of a specific organism into a format that is decipherable by researchers and scientists. DNA sequencing has allowed scientists to better understand genes and their roles within our body. Such knowledge has become indispensable for understanding biological processes, as well as in application fields such as diagnostic or forensic research. The advent of DNA sequencing has significantly accelerated biological research and discovery.

To facilitate the genomics study, we need to sequence the genomes of different species or different individuals. A number of sequencing technologies have been developed during the last decades. Roughly speaking, the development of the sequencing technologies consists of three phases:

- First-generation sequencing: Sequencing based on chemical degradation and gel electrophoresis.

- Second-generation sequencing: Sequencing many DNA fragments in parallel. It has higher yield, lower cost, but shorter reads.

- Third-generation sequencing: Sequencing a single DNA molecule without the need to halt between read steps.

In this section, we will discuss the three phases in detail.

1.3 First-generation sequencing

Sanger and Coulson proposed the first-generation sequencing in 1975 [259, 260]. It enables us to sequence a DNA template of length $500 - 1000$ within a few hours. The detailed steps are as follows (see Figure 1.3).

1. Amplify the DNA template by cloning.

2. Generate all possible prefixes of the DNA template.

3. Separation by electrophoresis.

4. Readout with fluorescent tags.

Step 1 amplifies the DNA template. The DNA template is inserted into the plasmid vector; then the plasmid vector is inserted into the host cells for cloning. By growing the host cells, we obtain many copies of the same DNA template.

Step 2 generates all possible prefixes of the DNA template. Two techniques have been proposed for this step: (1) the Maxam-Gilbert technique [194] and (2) the chain termination methodology (Sanger method) [259, 260]. The Maxam-Gilbert technique relies on the cleaving of nucleotides by chemical. Four different chemicals are used and generate all sequences ending with A, C, G and T, respectively. This allows us to generate all possible prefixes of the template. This technique is most efficient for short DNA sequences. However, it is considered unsafe because of the extensive use of toxic chemicals.

The chain termination methodology (Sanger method) is a better alternative. Given a single-stranded DNA template, the method performs DNA polymerase-dependent synthesis in the presence of (1) natural deoxynucleotides (dNTPs) and (2) dideoxynucleotides (ddNTPs). ddNTPs serve as non-reversible synthesis terminators (see Figure 1.2(a,b)). The DNA synthesis reaction is randomly terminated whenever a ddNTP is added to the growing oligonucleotide chain, resulting in truncated products of varying lengths with an appropriate ddNTP at their 3' terminus.

After we obtain all possible prefixes of the DNA template, the product is a mixture of DNA fragments of different lengths. We can separate these DNA

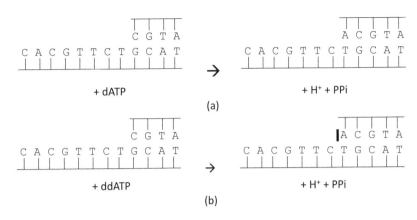

FIGURE 1.2: (a) The chemical reaction for the incorporation of dATP into the growing DNA strand. (b) The chemical reaction for the incorporation of ddATP into the growing DNA strand. The vertical bar behind A indicates that the extension of the DNA strand is terminated.

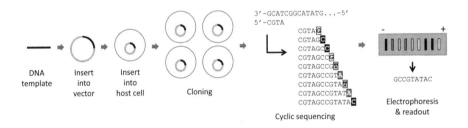

FIGURE 1.3: The steps of Sanger sequencing.

fragments by their lengths using gel electrophoresis (Step 3). Gel electrophoresis is based on the fact that DNA is negatively charged. When an electrical field is applied to a mixture of DNA on a gel, the DNA fragments will move from the negative pole to the positive pole. Due to friction, short DNA fragments travel faster than long DNA fragments. Hence, the gel electrophoresis separates the mixture into bands, each containing DNA molecules of the same length.

Using the fluorescent tags attached to the terminal ddNTPs (we have 4 different colors for the 4 different ddNTPs), the DNA fragments ending with different nucleotides will be labeled with different fluorescent dyes. By detecting the light emitted from different bands, the DNA sequence of the template will be revealed (Step 4).

In summary, the Sanger method can generate sequences of length ∼800 bp. The process can be fully automated and hence it was a popular DNA sequencing method in 1970 − 2000. However, it is expensive and the throughput is slow. It can only process a limited number of DNA fragments per unit of time.

1.4   Second-generation sequencing

Second-generation sequencing can generate hundreds of millions of short reads per instrument run. When compared with first-generation sequencing, it has the following advantages: (1) it uses clone-free amplification, and (2) it can sequence many reads in parallel. Some commercially available technologies include Roche/454, Illumina, ABI SOLiD, Ion Torrent, Helicos BioSciences and Complete Genomics.

In general, second-generation sequencing involves the following two main steps: (1) Template preparation and (2) base calling in parallel. The following Section 1.4.1 describes Step 1 while Section 1.4.2 describes Step 2.

1.4.1   Template preparation

Given a set of DNA fragments, the template preparation step first generates a DNA template for each DNA fragment. The DNA template is created by ligating adaptor sequences to the two ends of the target DNA fragment (see Figure 1.4(a)). Then, the templates are amplified using PCR. There are two common methods for amplifying the templates: (1) emulsion PCR (emPCR) and (2) solid-phase amplification (Bridge PCR).

emPCR amplifies each DNA template by a bead. First of all, one piece of DNA template and a bead are inserted within a water drop in oil. The surface of every bead is coated with a primer corresponding to one type of adaptor. The DNA template will hybridize with one primer on the surface of the bead. Then, it is PCR amplified within a water drop in oil. Figure 1.4(b) illustrates the emPCR. emPCR is used by 454, Ion Torrent and SOLiD.

For bridge PCR, the amplification is done on a flat surface (say, glass), which is coated with two types of primers, corresponding to the adaptors. Each DNA template is first hybridized to one primer on the flat surface. Amplification proceeds in cycles, with one end of each bridge tethered to the surface. Figure 1.4(c) illustrates the bridge PCR process. Bridge PCR is used by Illumina.

Although PCR can amplify DNA templates, there is amplification bias. Experiments revealed that templates that are AT-rich or GC-rich have a lower amplification efficient. This limitation creates uneven sequencing of the DNA templates in the sample.

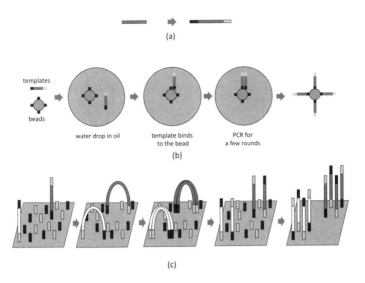

FIGURE 1.4: (a) From the DNA fragments, DNA template is created by attaching the two ends with adaptor sequences. (b) Amplifying the template by emPCR. (c) Amplifying the template by bridge PCR.

1.4.2 Base calling

Now we have many PCR clones of amplified templates (see Figure 1.5(a)). This step aims to read the DNA sequences from the amplified templates in parallel. This method is called the cyclic-array method. There are two approaches: the polymerase-mediated method (also called sequencing by synthesis) and the ligase-mediated method (also called sequencing by ligation). The polymerase-mediated method is further divided into methods based on reversible terminator nucleotides and methods based on unmodified nucleotides. Below, we will discuss these approaches.

1.4.3 Polymerase-mediated methods based on reversible terminator nucleotides

A reversible terminator nucleotide is a modified nucleotide. Similar to ddNTPs, during the DNA polymerase-dependent synthesis, if a reversible terminator nucleotide is incorporated onto the DNA template, the DNA synthesis is terminated. Moreover, we can reverse the termination and restart the DNA synthesis.

Figure 1.5(b) demonstrates how we use reversible terminator nucleotides for sequencing. First, we hybridize the primer on the adaptor of the template. Then, by DNA polymerase, a reversible terminator nucleotide is incorporated onto the template. After that, we scan the signal of the dye attached to the

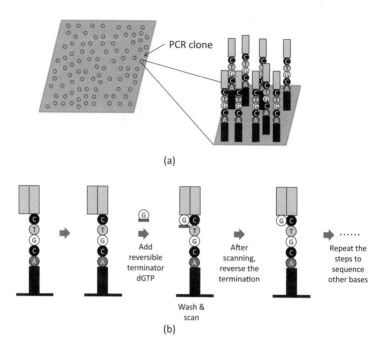

FIGURE 1.5: Polymerase-mediated sequencing methods based on reversible terminator nucleotides. (a) PCR clones of the DNA templates are evenly distributed on a flat surface. Each PCR clone contains many DNA templates of the same type. (b) The steps of polymerase-mediated methods are based on reversible terminator nucleotides.

reversible terminator nucleotide by imaging. After imaging, the termination is reversed by cleaving the dye-nucleotide linker. By repeating the steps, we can sequence the complete DNA template.

Two commercial sequencers use this approach. They are Illumina and Helicos BioSciences.

The Illumina sequencer amplifies the DNA templates by bridge PCR. Then, all PCR clones are distributed on the glass plate. By using the four-color cyclic reversible termination (CRT) cycle (see Figure 1.6(b)), we can sequence all the DNA templates in parallel.

The major error of Illumina sequencing is substitution error, with a higher portion of errors occurring when the previous incorporated nucleotide is a base G.

Another major error of Illumina sequencing is that the accuracy decreases with increasing nucleotide addition steps. The errors accumulate due to the failure in cleaving off the fluorescent tags or due to errors in controlling the

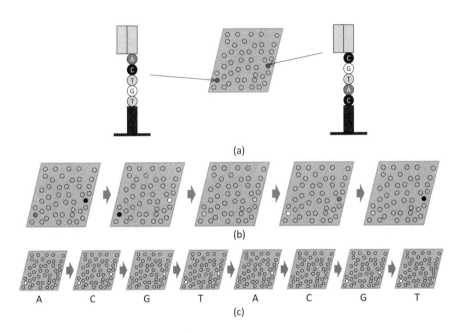

(a)

(b)

A   C   G   T   A   C   G   T

(c)

FIGURE 1.6: Polymerase-mediated sequencing methods based on reversible terminator nucleotides. (a) A flat surface with many PCR clones. In particular, we show the DNA templates for two clones. (b) Four-color cyclic reversible termination (CRT) cycle. Within each cycle, we extend the template of each PCR clone by one base. The color indicates the extended base. Precisely, the four colors, dark gray, black, white and light gray, correspond to the four nucleotides A, C, G and T, respectively. (c) One-color cyclic reversible termination (CRT) cycle. Each cycle tries to extend the template of each PCR clone by one particular base. If the extension is successful, the white color is lighted up.

reversible terminator nucleotides. Then, bases fail to get incorporated to the template strand or extra bases might get incorporated [190].

Helicos BioSciences does not perform PCR amplification. It is a single molecular sequencing method. It first immobilizes the DNA template on the flat surface. Then, all DNA templates on the surface are sequenced in parallel by using a one-color cyclic reversible termination (CRT) cycle (see Figure 1.6(c)). Note that this technology can also be used to sequence RNA directly by using reverse transcriptase instead of DNA polymerase. However, the reads generated by Helicos BioSciences are very short ($\sim$25 bp). It is also slow and expensive.

1.4.4 Polymerase-mediated methods based on unmodified nucleotides

The previous methods require the use of modified nucleotides. Actually, we can sequence the DNA templates using unmodified nucleotides. The basic observation is that the incorporation of a deoxyribonucleotide triphosphate (dNTP) into a growing DNA strand involves the formation of a covalent bond and the release of pyrophosphate and a positively charged hydrogen ion (see Figure 1.2). Hence, it is possible to sequence the DNA template by detecting the concentration change of pyrophosphate or hydrogen ion. Roche 454 and Ion Torrent are two sequencers which take advantage of this principle.

The Roche 454 sequencer performs sequencing by detecting pyrophosphates. It is called pyrosequencing. First, the 454 sequencer uses emPCR to amplify the templates. Then, amplified beads are loaded into an array of wells. (Each well contains one amplified bead which corresponds to one DNA template.) In each iteration, a single type of dNTP flows across the wells. If the dNTP is complementary to the template in a well, polymerase will extend by one base and relax pyrophosphate. With the help of enzymes sulfurylase and luciferase, the pyrophosphate is converted into visual light. The CDC camera detects the light signal from all wells in parallel. For each well, the light intensity generated is recorded as a flowgram. For example, if the DNA template in a well is TCGGTAAAAAACAGTTTCCT, Figure 1.7 is the corresponding flowgram. Precisely, the light signal can be detected only when the dNTP that flows across the well is complementary to the template. If the template has a homopolymer of length k, the light intensity detected is k-fold higher. By interpreting the flowgram, we can recover the DNA sequence.

However, when the homopolymer is long (say longer than 6), the detector is not sensitive enough to report the correct length of the homopolymer. Therefore, the Roche 454 sequencer gives higher rate of indel errors.

Ion Torrent was created by the person as Roche 454. It is the first semiconductor sequencing chip available on the commercial market. Instead of detecting pyrophosphate, it performs sequencing by detecting hydrogen ions. The basic method of Ion Torrent is the same as that of Roche 454. It also uses emPCR to amplify the templates and the amplified beads are also loaded into

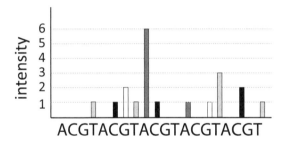

FIGURE 1.7: The flowgram for the DNA sequence TCG-GTAAAAAACAGTTTCCT.

a high-density array of wells, and each well contains one template. In each iteration, a single type of dNTP flows across the wells. If the dNTP is complementary to the template, polymerase will extend by one base and relax H+. The relaxation of H+ changes the pH of the solution in the well and an ISFET sensor at the bottom of the well measures the pH change and converts it into electric signals [251]. The sensor avoids the use of optical measurements, which require a complicated camera and laser. This is the main difference between Ion Torrent sequencing and 454 sequencing. The unattached dNTP molecules are washed out before the next iteration. By interpreting the flowgram obtained from the ISFET sensor, we can recover the sequences of the templates.

Since the method used by Ion Torrent is similar to that of Roche 454, it also has the disadvantage that it cannot distinguish long homopolymers.

1.4.5 Ligase-mediated method

Instead of extending the template base by base using polymerase, ligase-mediated methods use probes to check the bases on the template. ABI SOLiD is the commercial sequencer that uses this approach. In SOLiD, the templates are first amplified by emPCR. After that, millions of templates are placed on a plate. SOLiD then tries to probe the bases of all templates in parallel. In every iteration, SOLiD probes two adjacent bases of each template, i.e., it uses two-base color encoding. The color coding scheme is shown in the following table. For example, for the DNA template $ATGGA$, it is coded as $A3102$.

	A	C	G	T
A	0	1	2	3
C	1	0	3	2
G	2	3	0	1
T	3	2	1	0

The primary advantage of the two-base color encoding is that it improves the single nucleotide variation (SNV) calling. Since every base is covered by two color bases, it reduces the error rate for calling SNVs. However, conversion from color bases to nucleotide bases is not simple. Errors may be generated during the conversion process.

In summary, second-generation sequencing enables us to generate hundreds of billions of bases per run. However, each run takes days to finished due to a large number of scanning and washing cycles. Adding of a base per cycle is not 100% correct. This causes sequencing errors. Furthermore, base extensions of some strands may be lag behind or lead forward. Hence, errors accumulate as the reads get long. This is the reason why second-generation sequencing cannot get very long read. Furthermore, due to the PCR amplification bias, this approach may miss some templates with high or low GC content.

1.5   Third-generation sequencing

Although many of us are still using second-generation sequencing, third-generation sequencing is coming. There is no fixed definition for third-generation sequencing yet. Here, we define it as a single molecule sequencing (SMS) technology without the need to halt between read steps (whether enzymatic or otherwise). A number of third-generation sequencing methods have been proposed. They include:

- Single-molecule real-time sequencing

- Nanopore-sequencing technologies

- Direct imaging of individual DNA molecules using advanced microscopy techniques

1.5.1   Single-molecule real-time sequencing

Pacific BioSciences released their PacBio RS sequencing platform [71]. Their approach is called single-molecule real-time (SMRT) sequencing. It mimics what happens in our body as cells divide and copies their DNA with the DNA polymerase machine. Precisely, PacBio RS immobilizes DNA polymerase molecules on an array slide. When the DNA template gets in touch with the DNA polymerase, DNA synthesis happens with four fluorescently labeled nucleotides. By detecting the light emitted, PacBio RS reconstructs the DNA sequences. Figure 1.8 illustrates the SMRT sequencing approach.

PacBio RS sequencing requires no prior amplification of the DNA template. Hence, it has no PCR bias. It can achieve more uniform coverage and lower GC bias when compared with Illumina sequencing [79]. It can read long sequences

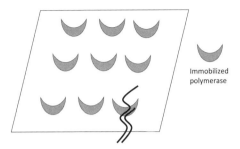

FIGURE 1.8: The illustration of PacBio sequencing. On an array slide, there are a number of immobilized DNA polymerase molecules. When a DNA template gets in touch with the DNA polymerase (see the polymerase at the lower bottom right), DNA synthesis happens with the fluorescently labeled nucleotides. By detecting the emitted light signal, we can reconstruct the DNA sequence.

of length up to 20,000 bp, with an average read length of about 10,000 bp. Another advantage of PacBio RS is that it can sequence methylation status simultaneously.

However, PacBio sequencing is more costly. It is about $3 - 4$ times more expensive than short read sequencing. Also, PacBio RS has a high error rate, up to 17.9% errors [46]. The majority of the errors are indel errors [71]. Luckily, the error rate is unbiased and almost constant throughout the entire read length [146]. By repeatedly sequencing the same DNA template, we can reduce the error rate.

1.5.2 Nanopore sequencing method

A nanopore is a pore of nano size on a thin membrane. When a voltage is applied across the membrane, charged molecules that are small enough can move from the negative well to the positive well. Moreover, molecules with different structures will have different efficiencies in passing through the pore and affect the electrical conductivity. By studying the electrical conductivity, we can determine the molecules that pass through the pore.

This idea has been used in a number of methods for sequencing DNA. These methods are called the nanopore sequencing method. Since nanopore methods use unmodified DNA, it requires an extremely small amount of input material. They also have the potential to sequence long DNA reads efficiently at low cost. There are a number of companies working on the nanopore sequencing method. They include (1) Oxford Nanopore, (2) IBM Transistor-mediated DNA sequencing, (3) Genia and (4) NABsys.

Oxford nanopore technology detects nucleotides by measuring the ionic current flowing through the pore. It allows the single-strand DNA sequence to

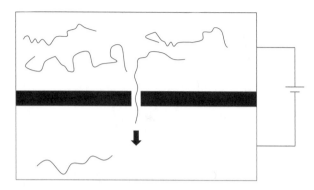

FIGURE 1.9: An illustration of the sequencing technique of Oxford nanopore.

flow through the pore continuously. As illustrated in Figure 1.9, DNA material is placed in the top chamber. The positive charge draws a strand of DNA moving from the top chamber to the bottom chamber flowing through the nanopore. By detecting the difference in the electrical conductivity in the pore, the DNA sequence is decoded. (Note that IBM's DNA transistor is a prototype which uses a similar idea.)

The approach has difficulty in calling the individual base accurately. Instead, Oxford nanopore technology will read the signal of k (say 5) bases in each round. Then, using a hidden Markov model, the DNA base can be decoded base by base.

Oxford nanopore technology has announced two sequencers: MiniION and GridION. MiniION is a disposable USB-key sequencer. GridION is an expandable sequencer. Oxford nanopore technology claimed that GridION can sequence 30x coverage of a human genome in 6 days at US$2200 − $3600. It has the potential to decode a DNA fragment of length $100,000$ bp. Its cost is about US$25 − $40 per gigabyte. Although it is not expensive, the error rate is about 17.8% (4.9% insertion error, 7.8% deletion error and 5.1% substitution error) [115].

Unlike Oxford nanopore technology, Genia suggested combining nanopore and the DNA polymerase to sequence a single-strand DNA template. In Genia, the DNA polymerase is tethered with a biological nanopore. When a DNA template gets in touch with the DNA polymerase, DNA synthesis happens with four engineered nucleotides for A, C, G and T, each attached with a different short tag. When a nucleotide is incorporated into the DNA template, the tag is cleaved and it will travel through the biological nanopore and an electric signal is measured. Since different nucleotides have different tags, we can reconstruct the DNA template by measuring the electric signals.

NABsys is another nanopore sequencer. It first chops the genome into DNA fragments of length $100,000$ bp. The DNA fragments are hybridized with a particular probe so that specific short DNA sequences on the DNA fragments

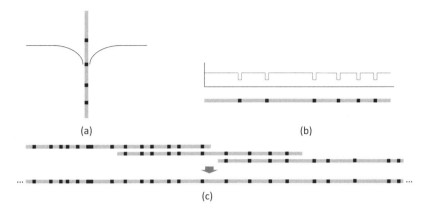

(a)   (b)

(c)

FIGURE 1.10: Consider a DNA fragment hybridized with a particular probe. After it passes through the nanopore (see (a)), an electrical signal profile is obtained (see (b)). By aligning the electrical signal profiles generated from a set of DNA fragments, we obtain the probe map for a genome (see (c)).

are bounded by the probes. Those DNA fragments with bound probes are driven through a nanopore (see Figure 1.10(a)), creating a current-versus-time tracing. The trace gives the position of the probes on the fragment. (See Figure 1.10(b).) We can align the fragments based on their inter-probe distance; then, we obtain a probe map for the genome (see Figure 1.10(c)). We can obtain the probe maps for different probes. By aligning all of them, we obtain the whole genome.

Unlike Genia, Oxford nanopore technology and the IBM DNA transistor, NABsys does not require a very accurate current measurement from the nanopore. The company claims that this method is cheap, and that read length is long and fast. Furthermore, it is accurate!

1.5.3 Direct imaging of DNA using electron microscopy

Another choice is to use direct imaging. ZS genetics is developing methods based on transmission electron microscopy (TEM). Reveo is developing a technology based on scanning tunneling microscope (STM) tips. DNA is placed on a conductive surface for detecting bases electronically using STM tips and tunneling current measurements. Both approaches have the potential to sequence very long reads (in millions) at low cost. However, they are still in the development phase. No sequencing machine is available yet.

TABLE 1.1: Comparison of the three generations of sequencing

	First generation	Second generation	Third generation
Amplification	In-vivo cloning and amplification	In-vitro PCR	Single molecule
Sequencing	Electrophoresis	Cyclic array sequencing	Nanopore, electronic microscopy or real-time monitoring of PCR
Starting material	More	Less ($< 1\mu g$)	Even less
Cost	Expensive	Cheap	Very cheap
Time	Very slow	Fast	Very fast
Read length	About 800bp	Short	Very long
Accuracy	$< 1\%$ error	$< 1\%$ error (mismatch or homopolmer error)	High error rate

1.6   Comparison of the three generations of sequencing

We have discussed the technologies of the three generations of sequencing. Table 1.1 summarizes their key features. Currently, we are in the late phase of second-generation sequencing and at the early phase of third-generation sequencing. We can already see a dramatic drop in sequencing cost. Figure 1.11 shows the sequence cost over time. Cost per genome is calculated based on 6-fold coverage for Sanger sequencing, 10-fold coverage for 454 sequencing and 30-fold coverage for Illumina (or SOLiD) sequencing. As a matter of fact, the sequencing cost does not include the data management cost and the bioinformatics analysis cost. Note that there was a sudden reduction in sequencing cost in January 2008, which is due to the introduction of second-generation sequencing. In the future, the sequencing cost is expected to drop further.

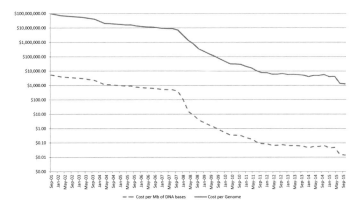

FIGURE 1.11: The sequencing cost over time. There are two curves. The blue curve shows the sequencing cost per million of sequencing bases while the red curve shows the sequencing cost per human genome. (Data is obtained from *http://www.genome.gov/sequencingcosts.*)

1.7 Applications of sequencing

The previous section describes three generations of sequencing. This section describes their applications.

Genome assembly: Genome assembly aims to reconstruct the genome of some species. Since our genome is long, we still cannot read the whole chromosome in one step. The current solution is to sequence the fragments of the genome one by one using a sequencer. Then, by overlapping the fragments computationally, the complete genome is reconstructed.

Many genome assembly projects have been finished. We have obtained the reference genomes for many species, including human, mouse, rice, etc. The human genome project is properly the most important genome assembly project. This project started in 1984 and declared complete in 2003. The project cost was more than 3 billion US$. Although it is expensive, the project enables us to have a better understanding of the human genome. Given the human reference genome, researchers can examine the list of genes in humans. We know that the number of protein coding genes in humans is about 20,000, which covers only 3% of the whole genome. Subsequently, we can also understand the differences among individuals and understand the differences between cancerous and healthy human genomes.

The project also improves the genome assembly process. It leads to a whole genome shotgun approach, which is the most common assembly approach nowadays. By coupling the whole genome shotgun approach and next-generation sequencing, we obtain the reference genomes of many species. (See Chapter 5 for methods to reconstruct a genome.)

Genome variations finding: The genome of each individual is different from that of the reference human genome. Roughly speaking, there are four types of genome variations: single nucleotide variations (SNVs), short indels, copy number variations (CNVs) and structural variations (SVs). Figure 6.1 illustrates these four types of variations. Genome variations can cause cancer. For example, in chronic myelogenous leukemia (CML), a translocation exists between chromosome 9 and chromosome 22, which fuses the ABL1 and BCR genes together to form a fusion gene, BCL-ABL1. Such a translocation is known to be present in 95 percent of patients with CML. Another example occurs with a deletion in chromosome 21 that fuses the ERG and TMPRSS2 genes. The TMPRSS2-ERG fusion is seen in approximately 50 percent of prostate cancers, and researchers have found that this fusion enhances the invasiveness of prostate cancer. Genome sequencing of cancers enables us to identify the variation of each individual. Apart from genome variations in cancers, many novel disease-causing variations have been discovered for childhood diseases and neurological diseases. In the future, we expect everyone will perform genome sequencing. Depending on the variations, different personalized therapeutics can be applied to different patients. This is known as personalized medicine or stratified medicine. (See Chapters 7 and 6 for methods to call genome variations.)

Reconstructing the transcriptome: Although every human cell has the same human genome, human cells in different tissues express different sets of genes at different times. The set of genes expressed in a particular cell type is called its transcriptome. In the past, the transcriptome was extracted using technologies like microarray. However, microarray can only report the expression of known genes. They fail to discover novel splice variants and novel genes. Due to the advance in sequencing technologies, we can use RNA-seq to solve these problems. We can not only measure gene expression more accurately, but can also discover novel genes and novel splice variants. (See Chapter 8 for methods to analyze RNA-seq data.)

Decoding the transcriptional regulation: Some proteins called transcription factors (TFs) bind on the genome and regulate the expression of genes. If a TF fails to bind on the genome, the corresponding target gene will fail to express and the cell cannot function properly. For example, one type of breast cancer is ER+ cancer cells. In ER+ cancer, ER, GATA3 and FoxA1 form a functional enhanceosome that regulates a set of genes and drives the core ERα function. It is important to understand how they work together.

To know the binding sites of each TF, we can apply ChIP-seq. ChIP-seq is a sequencing protocol that enables us to identify the binding sites of each TF on a genome-wide scale. By studying the ChIP-seq data, we can understand how TFs work together, the relationship between TFs and transcriptomes, etc. (See Chapter 9 for methods to analyze ChIP-seq data.)

Many other applications: Apart from the above applications, sequencing has been applied to many other research areas, including metagenomics, 3D modeling of the genome, etc.

1.8   Summary and further reading

This chapter summarizes the three generations of sequencing. It also briefly describes their applications. There are a number of good surveys of second generation-sequencing. Please refer to [200]. For more detail on third-generation sequencing, please refer to [263].

1.9   Exercises

1. Consider the DNA sequence 5'-ACTCAGTTCG-3'. What is its reverse complement? The SOLiD sequencer will output color-based sequences. What is the expected color-based sequence for the above DNA sequence and its reverse complement? Do you observe an interesting property?

2. Should we always use second- or third- generation sequencing instead of first-generation sequencing? If not, when should we use Sanger sequencing?

Chapter 2

NGS file formats

2.1 Introduction

NGS technologies are widely used now. To facilitate NGS data analysis and NGS data transfer, a few NGS file formats are defined. This chapter gives an overview of these commonly used file formats.

First, we briefly describe the NGS data analysis process. After a sample is sequenced using a NGS sequencer, some reads (i.e., raw DNA sequences) are generated. These raw reads are stored in raw read files, which are in the fasta, fastq, fasta.gz or fastq.gz format. Then, these raw reads are aligned on the reference genome (such as the human genome). The alignments of the reads are stored in alignment files, which are in the SAM or BAM format.

From these alignment files, downstream analysis can be performed to understand the sample. For example, if the alignment files are used to call mutations (like single nucleotide variants), we will obtain variant files, which are in the VCF or BCF format. If the alignment files are used to obtain the read density (like copy number of each genomic region), we will obtain density files, which are in the Wiggle, BedGraph, BigWig or cWig format. If the alignment files are used to define regions with read coverage (like regions with RNA transcripts or regions with TF binding sites), we will obtain annotation files, which are in the bed or bigBed format.

Figure 2.1 illustrates the relationships among these NGS file formats. In the rest of the chapter, we detail these file formats.

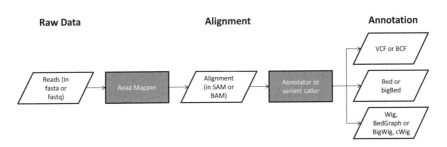

FIGURE 2.1: The relationships among different file formats.

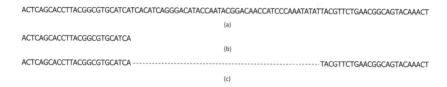

FIGURE 2.2: (a) An example DNA fragment. (b) A single-end read extracted from the DNA fragment in (a). (c) A paired-end read extracted from the DNA fragment in (a).

2.2   Raw data files: fasta and fastq

From DNA fragments, NGS technologies extract reads from them. Figure 2.2 demonstrates the process. Given a DNA fragment in Figure 2.2(a), NGS technologies enable us to sequence either one end or both ends of the fragment. Then, we obtain a single-end read or a paired-end read (see Figure 2.2(b,c)).

To store the raw sequencing reads generated by a sequencing machine, two standard file formats, fasta and fastq, are used.

The fasta file has the simplest format. Each read is described by (1) a read identifier (a line describing the identifier of a read) and (2) the read itself (one or more lines). Figure 2.3(a) gives an example consisting of two reads. The line describing the read identifier must start with >. There are two standard formats for the read identifier: (a) the Sanger standard and (b) the Illumina standard. The Sanger standard uses free text to describe the read identifier. The Illumina standard uses a fixed read identifier format (see Figure 2.4 for an example).

Sequencing machines are not 100% accurate. A number of sequencers (like the Illumina sequencer) can estimate the base calling error probability P of each called base. People usually convert the error probability P into the PHRED quality score Q (proposed by Ewing and Green [73]) which is computed by the following equation:

$$Q = -10 * \log_{10} P.$$

Q is truncated and limited to only integers in the range 0..93. Intuitively, if Q is big, we have high confidence that the base called is correct. To store Q, the value Q is offset by 33 to make it in the range from 33 to 126 which are ASCII codes of printable characters. This number is called the Q-score.

The file format fastq stores both the DNA bases and the corresponding quality scores. Figure 2.3(b) gives an example. In a fastq file, each read is described by 4 parts: (1) a line that starts with @ which contains the read identifier, (2) the DNA read itself (one or more lines), (3) a line that starts

```
>seq1                              @seq1
ACTCAGCACCTTACGGCGTGCATCA          ACTCAGCACCTTACGGCGTGCATCA
>seq2                              +seq1
CCGTACCGTTGACAGATGGTTTACA          !''**()%A541;djgp0i345adn
                                   @seq2
                                   CCGTACCGTTGACAGATGGTTTACA
                                   +seq2
                                   #$SG12j;askjpqo2i3nz!;lak

            (a)                                (b)
```

FIGURE 2.3: (a) is an example fasta file while (b) is an example fastq file. Both (a) and (b) describe the same DNA sequences. Moreover, the fastq file also stores the quality scores. Note that seq1 is the read obtained from the DNA fragment in Figure 2.2(b).

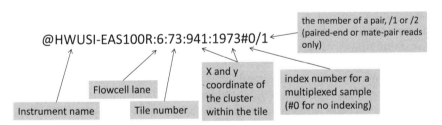

FIGURE 2.4: An example of the read identifier in the fasta or fastq file.

with + and (4) the sequence of quality scores for the read (which is of the same length as the read).

The above description states the format for storing the single-end reads. For paired-end reads, we use two fastq files (suffixed with _1.fastq and _2.fastq). The two files have the same number of reads. The ith paired-end read is formed by the ith read in the first file and the ith read in the second file. Note that the ith read in the second file is the reverse complement. For example, for the paired-end read in Figure 2.2(c), the sequence in the first file is ACTCAGCACCTTACGGCGTGCATCA while the sequence in the second file is AGTTTGTACTGCCGTTCAGAACGTA (which is the reverse complement of TACGTTCTGAACGGCAGTACAAACT).

Given the raw read data, we usually need to perform a quality check prior to further analysis. The quality check not only can tell us the quality of the sequencing library, it also enables us to filter reads that have low complexity and to trim some positions of a read that have a high chance of sequencing error. A number of tools can be used to check quality. They include SolexaQA[57], FastQC (http://www.bioinformatics.bbsrc.ac.uk/projects/fastqc/) and PRINSEQ[265].

2.3   Alignment files: SAM and BAM

The raw reads (in fasta or fastq format) can be aligned on some reference genome. For example, Figure 2.5(a) shows the alignments of a set of four reads $r1, r2, r3$ and $r4$. $r1$ is a paired-end read while $r2, r3$ and $r4$ are single-end reads. The single-end read $r3$ has two alignments. Chapter 4 discusses different mapping algorithms for aligning reads.

The read alignments are usually stored using SAM or BAM format [169]. The SAM (Sequence Alignment/Map) format is a generic format for storing alignments of NGS reads. The BAM (Binary Alignment/Map) format is the binary equivalent of SAM. It compresses the SAM file using the BGZF library (a generic library developed by Li et al. [169]). BGZF divides the SAM file into blocks and compresses each block using the standard gzip. It provides good compression while allowing fast random access.

The detailed description of SAM and BAM formats can be found in https://samtools.github.io/hts-specs/SAMv1.pdf. Briefly, SAM is a tab-delimited text file. It consists of a header section (optional) and an alignment section.

The alignment section stores the alignments of the reads. Each row corresponds to one alignment of a read. A read with c alignments will occupy c rows. For a paired-end read, the alignments of its two reads will appear in different rows. Each row has 11 mandatory fields: QNAME, FLAG, RNAME, POS, MAPQ, CIGAR, MRNM, MPOS, TLEN, SEQ and QUAL. Optional

```
                  1          2          3          4
         1234567   8901234567890123456789012345678901234567
ref:     ATCGAAC**TGACTGGACTAGAACCGTGAATCCACTGATCTAGACTCGA
+r1/1    TCG-ACGGTGACTG
+r2           GAAC-GTGACaactg
-r3               GCCTGGttcac
-r1/2                         CCGTGAATC
+r3                           GTGAAccaggc
+r4                                ATCC..................AGACTCGA
```

(a)

```
@HD VN:1.3 SO:coordinate
@SQ SN:ref LN:47
r1      99  ref    2  30  3M1D2M2I6M  =   22   29  TCGACGGTGACTG    *
r2       0  ref    4  30  4M1P1I4M5S  *    0    0  GAACGTGACAACTG   *
r3      16  ref    9  30  6M5S        *    0    0  GCCTGGTTCAC      *    NM:i:1
r1     147  ref   22  30  9M          =    2  -29  CCGTGAATC        *
r3    2048  ref   24  30  5M6H        *    2    0  GTGAA            *    NM:i:0
r4       0  ref   28  30  4M8N8M      *    0    0  ATCCAGACTCGA     *
        ↑   ↑      ↑  ↑   ↑           ↑    ↑    ↑  ↑                ↑    ↑
QNAME FLAG RNAME POS MAPQ CIGAR     MRNM MPOS TLEN   SEQ          QUAL TAG:TYPE:VAL
```

(b)

FIGURE 2.5: (a) An example of alignments of a few short reads. (b) The SAM file for the corresponding alignments is shown in (a).

fields can be appended after these mandatory fields. For each field, if its value is not available, we set it to *. We briefly describe the 11 mandatory fields here.

- QNAME is the name of the aligned read. (In Figure 2.5, the read names are $r1$, $r2$, $r3$, and $r4$. Note that $r3$ has two alignments. Hence, there are two rows with $r3$. $r1$ is a paired-end read. Hence, there are two rows for the two reads of $r1$.)

- FLAG is a 16-bit integer describing the characteristics of the alignment of the read (see Section 2.3.1 for detail).

- RNAME is the name of the reference sequence to which the read is aligned. (In Figure 2.5, the reference is **ref**.)

- POS is the leftmost position of the alignment.

- MAPQ is the mapping quality of the alignment (in Phred-scaled). (In Figure 2.5, the MAPQ scores of all alignments are 30.)

- CIGAR is a string that describes the alignment between the read and the reference sequence (see Section 2.3.2 for detail).

- MRNM is the name of the reference sequence of its mate read ("=" if it is the same as RNAME). For single-end read, its value is "*" by default.

- MPOS is the leftmost position of its mate read. For a single-end read, its value is zero by default.

- TLEN is the inferred signed observed template length (i.e., the inferred insert size for paired-end read). For a single-end read, its value is zero by default.

- SEQ is the query sequence on the same strand as the reference.

- QUAL is the quality score of the query sequence.

Figure 2.5(b) gives the SAM file for the example alignments in Figure 2.5(a). The following subsections give more information for FLAG and CIGAR.

Note that SAM and BAM use two different coordinate systems. SAM uses the 1-based coordinate system. The first base of a sequence is 1. A region is specified by a close interval. For example, the region between the 3rd base and the 6th base inclusive is $[3, 6]$. BAM uses the 0-based coordinate system. The first base of a sequence is 0. A region is specified by a half-closed, half-open interval. For example, the region between the 3rd base and the 6th base inclusive is $[3, 7)$.

To manipulate the information from SAM and BAM files, we use samtools [169] and bamtools [16]. Samtools is a set of tools that allows us to interactively view the alignments in both SAM and BAM files. It also allows us to post-process and extracts information from both SAM and BAM files. Bamtools provides a set of C++ API for us to access and manage the alignments in the BAM file.

2.3.1 FLAG

The bitwise FLAG is a 16-bit integer describing the characteristics of the alignment. Its binary format is $b_{15}b_{14}b_{13}b_{12}b_{11}b_{10}b_9b_8b_7b_6b_5b_4b_3b_2b_1b_0$. Only bits $b_0, \ldots, b_{11}$ are used by SAM. The meaning of each bit b_i is described in Table 2.1.

For example, for the first read of r1 in Figure 2.5(b), its flag is $99 = 0000000001100011$. b_0, b_1, b_5 and b_6 are ones. This means that it has multiple segments (b_0), the segment is properly aligned (b_1), the next segment maps on the reverse complement (b_5), and this is the first read in the pair (b_6).

2.3.2 CIGAR string

The CIGAR string is used to describe the alignment between the read and the reference genome. The CIGAR operations are given in Table 2.2.

For example, the CIGAR string for the first read of r1 is 3M1D2M2I6M (see Figure 2.5), which means that the alignment between the read and the

TABLE 2.1: The meaning of the twelve bits in the FLAG field.

Bit	Description
$b_0 = 1$	if the read is paired
$b_1 = 1$	if the segment is a proper pair
$b_2 = 1$	if the segment is unmapped
$b_3 = 1$	if the mate is unmapped
$b_4 = 1$	if the read maps on the reverse complement
$b_5 = 1$	if the mate maps on the reverse complement.
$b_6 = 1$	if this is the first read in the pair
$b_7 = 1$	if this is the second read in the pair
$b_8 = 1$	if this is a secondary alignment (i.e., the alternative alignment when multiple alignments exist)
$b_9 = 1$	if this alignment does not pass the quality control
$b_{10} = 1$	if it is a PCR duplicate
$b_{11} = 1$	if it is a supplementary alignment (part of a chimeric alignment)

reference genome has 3 alignment matches, 1 deletion, 2 alignment matches, 2 insertions and 6 alignment matches. (Note that the alignment match can be a base match/mismatch.)

2.4 Bed format

The bed format is a flexible way to represent and annotate a set of genomic regions. It can be used to annotate repeat regions in the genome, open regions or transcription factor binding sites in the genome. We can also use it to annotate genes with different isoforms. For example, Figure 2.6 shows the bed file for representing the two isoforms of the gene VHL.

The bed format is a row-based format, where each row represents an annotated item. Each annotated item consists of some fields separated by tabs. The first three fields of each row are compulsory. They are chromosome, chromosome start and chromosome end. These three fields are used to indicate the genomic regions. (Similar to BAM, bed also uses the 0-based coordinate system. Any genomic region is specified by a half-closed, half-open interval.) There are 9 additional optional fields to annotate each genomic region: name, score, strand, thickStart, thickEnd, itemRgb, blockCount, blockSizes, and block-Starts. The bed format allows user to include more fields for each annotated item.

As an illustration, the two isoforms of the gene VHL are represented using two rows in the bed format. The bottom of Figure 2.6 shows the bed file for the two isoforms. Since both isoforms are in chr3 : 10183318 − 10195354,

TABLE 2.2: Definitions of Different CIGAR Operations.

Op	Description
M	alignment match (can be a sequence match or mismatch)
I	insertion to the reference
D	deletion from the reference
N	skipped region from the reference
S	soft clipping (clipped sequences present in SEQ)
H	hard clipping (clipped sequences NOT present in SEQ)
P	padding (silent deletion from padded reference)
=	sequence match
X	sequence mismatch

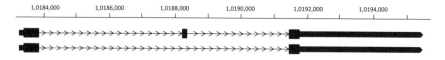

Bed file:

chr3	10183318	10195354	VHL	1	+	10183531	10191649	0,0,0	3	490,123,3884,	0,4879,8152,
chr3	10183318	10195354	VHL	1	+	10183531	10191649	0,0,0	2	490,3884,	0,8152,

FIGURE 2.6: The gene VHL has two splicing variants. One of them has 3 exons while the other one has 2 exons (is missing the middle exon). The solid bars are the exons. The thick solid bars are the coding regions of VHL. The bed file corresponding to these two isoforms is shown at the bottom of the figure.

chromosome, chromosome start, and chromosome end are set to be chr3, 10183318 and 10195354, respectively. Since the coding regions are in chr3 : 10183531 − 10191649, thickStart and thickEnd are set to be 10183531 and 10191649, respectively. The two isoforms are in black color, itemRgb is set to be $0, 0, 0$. In these two isoforms, three exons A, B and C are involved. The sizes of the three exons are 490, 123 and 3884, respectively. Relative to the start of the gene VHL, the starting positions of the three exons are 0, 4879 and 8152, respectively. The first isoform consists of all three exons A, B and C while the second isoform consists of exons A and C. So, for the first isoform, the blockCount is 3, blockSizes are $490, 123$ and 3884 and blockStarts are $0, 4879$ and 8152. For the second isoform, the blockCount is 2, blockSizes are 490 and 3884 and blockStarts are 0 and 8152.

Bed files are uncompressed files. They can be huge. To reduce the file size, we can use the compressed version of the bed format, which is the bigBed format. BigBed partitions the data into blocks of size 512 bp. Each block is compressed using a gzip-like algorithm. Efficient querying for bigBed is important since users want to extract information without scanning the whole file. BigBed includes an R-tree-like data structure that enables us to list, for any interval, all annotated items in the bed file. BigBed can also report summary statistics, like mean coverage, and count the number of annotated items within any genomic interval (with the help of the command bigBedSummary).

To manipulate the information in bed files, we use bedtools [237]. To convert between bed format and bigBed format, we can use bedToBigbed and bigBedToBed. To obtain a bam file from a bed file, we can use bedToBam.

2.5 Variant Call Format (VCF)

VCF [60] is the format specially designed to store the genomic variations, which are generated by variant callers (see Chapter 6). It is the standardized format used in the 1000 Genome Project. Information stored in a VCF file includes genomic coordinates, reference allele and alternate allele, etc. Since VCF may be big, Binary Variant Call Format (BCF) was also developed. BCF is a binary version of VCF, which is compressed and indexed for efficient analysis of various calling results.

Figure 2.7 illustrates the VCF file. It consists of a header section and a data section. The header section contains arbitrary number of meta-information lines (begin with characters ##) followed by a TAB-delimited field definition line (begins with the character #). The meta-information lines describe the tags and annotations used in the data section. The field definition line defines the columns in the data section. There are 8 mandatory fields: CHROM, POS, ID, REF, ALT, QUAL, FILTER and INFO. CHROM and POS define the position of each genomic variation. REF gives the reference allele while ALT

(a) `Chr1`
```
                        1111111111222222 2222333333333334444444444455555555556666666666777777
             1234567890123456789012345 678901234567890123456789012345678901234567890123456789012345
   REF:      ACGTACAGACAGACTTAGGACAGAT--CGTCACACTCGGACTGACCGTCACAACGGTCATCACCGGACTTACAATCG

 Sample1:    GTACACACAGAC        CAGATAACGTCAC   CGGACTGACCGTCA AACGGT--------------CAATCG
             ACACACAGACTT
             CACACAGACTTA

 Sample2: ACGTACAGACAG       GACAGATAACGTC   TCGGACT---CG  ACAACGGT--------------CAAT
          CGTACAGACAGA       GGACAGATT-CGT                       CAACGGT--------------CAATC
                             AGGACAGATT-CGT
```

(b)
```
##fileformat=VCFv4.2
##fileDate=20110705
##source=VCFtools
##reference=NCBI36
##ALT=<ID=DEL,Description="Deletion">
##FILTER=<ID=q10,Description="Quality below 10">
##INFO=<ID=SVTYPE,umber=1,Type=String,Description="Type of structural variant">
##INFO=<ID=END,Number=1,Type=Integer,Description="End position of the variant">
##FORMAT=<ID=GQ,Number=1,Type=Integer,Description="Genotype Quality (phred score)">
##FORMAT=<ID=GT,Number=1,Type=String,Description="Genotype">
##FORMAT=<ID=DP,Number=1,Type=Integer,Description="Read Depth">
#CHROM POS ID REF  ALT      QUAL FILTER INFO            FORMAT Sample1 Sample2
1      8   .  G    C        .    PASS   .               GT:DP  1/1:3   0/0:2
1      25  .  T    TAA,TT   .    q10    .               GT:DP  1/1:1   1/2:3
1      40  .  TGAC T        .    PASS   .               GT:GQ  1/1:50  0/0:70
1      55  .  T    <DEL>    .    PASS   SVTYPE=DEL;END=69 GT    1/1     1/1
```

FIGURE 2.7: Consider two samples. Each sample has a set of length-12 reads. Sample (a) shows the alignment of those reads on the reference genome. There are 4 regions with genomic variations. At position 8, an SNV appears in sample 1. At position 25, two different insertions appear in both samples. At position 40, a deletion appears in sample 2. At position 55, a deletion appears in both samples. Sample (b) shows the VCF file that describes the genomic variations in the 4 loci.

gives a list of alternative non-reference alleles (separated by semi-colon). ID is the unique identifier of the variant, QUAL is the PHRED-scaled quality score, FILTER stores the site filtering informations and INFO stores the annotation of the variant.

If n samples are included in the VCF file, we have an additional field FORMAT and n other fields, each representing a sample. The FORMAT field is a colon-separated list of tags, and each tag defines some attribute of a genotype. For example, GT:GQ:DP corresponds to a format with three attributes: genotype, genotype quality, and read depth of each variant. Read depth (DP) is the number of reads covered by the locus. Genotype quality (GQ) is the phred-scaled quality score of the genotype in the sample. Genotype (GT) is encoded as allele values separated by either / or |. The allele values can be 0 for the reference allele (in REF field), 1 for the first alternative allele (in ALT field), 2 for the second alternative allele, and so on. For example, 0/1 is a diploid call that consists of the reference allele and the first alternative allele. The separators / and | mean unphased and phased genotypes, respectively.

To manipulate the information in VCF and BCF files, we use vcftools [60]

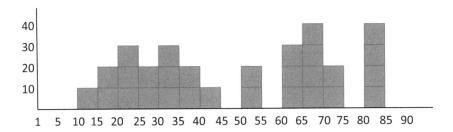

FIGURE 2.8: An example of a density curve in chr1 : 1 − 90.

and bcftools [163]. (Note that VCF files use the 1-based coordinate system while BCF files use the 0-based coordinate system.)

2.6   Format for representing density data

Read density measures the number of reads covering each base. It is formed by aligning the NGS reads on a reference genome; then, the number of reads covering each base can be counted.

Density data can describe the transcript expression in RNA-seq (see Section 8), the peak intensity in ChIP-seq (see Chapter 9), the open chromatin profile (see Chapter 9), the copy number variation in whole genome sequencing (see Chapter 7), etc.

Density data can be represented as a list of r_i, where r_i is the density value for position i. Figure 2.8 shows an example list of density values. In the example, $r_5 = 0, r_{10} = 10, r_{17} = 20, r_{20} = 30$. To represent the list of r_i, we can use the bedGraph or wiggle (wig) format.

The wiggle format is a row-based format. It consists of one track definition line (defining the format of the lines) and a few lines of data points. There are two options: variableStep and fixedStep. Variable step data describe the densities of data points in irregular intervals. It has an optional parameter span s, which is the number of bases spanned with the same density for each data point. To describe variable step data, the first line is the track definition line, which contains information about the chromosome and the span s. Then, it is followed with some lines, where each line contains two numbers: a position and a density value for that position.

Fixed step data describe the densities of data points in regular interval. To describe fixed step data, the first line is the track definition line, which contains the chromosome, the start b, the step δ and the span s. Then, it follows with some lines, where the ith line contains the density value for position $b + \delta i$, which spans s bases. The following box gives the wiggle file for Figure 2.8. (Note that the wiggle file uses a 1-based coordinate system.)

```
fixedStep chrom=chr1 start=10 step=5 span=5
10
20
30
20
30
20
10
variableStep chrom=chr1 span=5
50 20
60 30
65 40
70 20
80 40
```

BedGraph is in a bed-like format where each line represents a genomic region with the same density value. Each line consists of 4 fields: chromosome, chromosome start, chromosome end, and density. (Note that bedGraph uses 0-based coordinate format. The coordinate is described in a half-closed, half-open interval.) The following box gives the bedGraph file for Figure 2.8.

```
chr1 9 14 10
chr1 14 19 20
chr1 19 24 30
chr1 24 29 20
chr1 29 34 30
chr1 34 39 20
chr1 39 44 10
chr1 49 54 20
chr1 59 64 30
chr1 64 69 40
chr1 69 74 20
chr1 79 84 40
```

However, wiggle and bedGraph are uncompressed text formats and they are usually huge. Hence, when we want to represent a genome-wide density profile, we usually use the compressed version, which is in the bigWig format [135] or cWig format [112]. Both bigWig and cWig support efficient queries over any selected interval. Precisely, they support 4 operations. Let the density values of the N bases be $r_1, \ldots, r_N$. For any chromosome interval $p..q$, the four operations are:

- $coverage(p, q)$: Proportion of bases that contain non-zero density values in $p..q$, that is, $\frac{N}{q-p+1}$.

- $mean(p, q)$: The mean of the density values in $p..q$, that is, $\frac{1}{N}\sum_{i=1}^{N} r_i$.

- $minVal(p, q)$ and $maxVal(p, q)$: The minimum and maximum of the density values in $p..q$, that is, $\min_{i=1..N} r_i$ and $\max_{i=1..N} r_i$.

- $stdev(p, q)$: The standard derivation of the density values in $p..q$, that is, $\sqrt{\frac{1}{N}\sum_{i=1}^{N}(r_k - mean(p, q))^2}$.

To manipulate the information in bigWig and cWig files, we use bwtool [234] and cWig tools [112].

2.7 Exercises

1. Suppose chromosome 2 is the following sequence ACACGACTAA....

   - For the genomic region in chromosome 2 containing the DNA sequence ACGAC, if we describe it using bed format, what are the chromosome, start, and end?

   - In SAM, what is the alignment position of the DNA sequence ACGAC?

   - In BAM, what is the alignment position of the DNA sequence ACGAC?

2. Please perform the following conversions.

   (a) Convert the following set of intervals from the 0-based coordinate format to the 1-based coordinate format: 3..100, 0..89 and 1000..2000.

   (b) Convert the following set of intervals from the 1-based coordinate format to the 0-based coordinate format: 3..100, 1..89 and 1000..2000.

3. Given a BAM file `input.bam`, we want to find all alignments with $maQ > 0$ using samtools. What should be the command?

4. Given two BED files `input1.bed` and `input2.bed`, we want to find all genomic regions in `input1.bed` that overlap with some genomic regions in `input2.bed`. What should be the command?

5. For the following wiggle file, can you compute $coverage(3,8)$, $mean(3,8)$, $minVal(3,8)$, $maxVal(3,8)$ and $stdev(3,8)$?

```
fixedStep chrom=chr1 start=1 step=1 span=1
20
10
15
30
20
25
30
20
10
30
```

6. Can you propose a script to convert a BAM file into a bigWig file?

Chapter 3

Related algorithms and data structures

3.1 Introduction

This chapter discusses various algorithmic techniques and data structures used in this book. They include:

- Recursion and dynamic programming (Section 3.2)

- Parameter estimation and the EM algorithm (Section 3.3)

- Hash data structures (Section 3.4)

- Full-text index (Section 3.5)

- Data compression techniques (Section 3.6)

3.2 Recursion and dynamic programming

Recursion (also called divide-and-conquer) and dynamic programming are computational techniques that solve a computational problem by partitioning it into subproblems. The development of recursion and dynamic programming algorithms involves three stages.

1. **Characterize the structure of the computational problem**: Identify the subproblems of the computational problem.

2. **Formulate the recursive equation**: Define an equation that describes the relationship between the problem and its subproblems.

3. **Develop an algorithm**: Based on the recursive equation, develop either a recursive algorithm or a dynamic programming algorithm.

Below, we illustrate these three steps using two computational problems: (1) a key searching problem and (2) an edit-distance problem.

Algorithm **search**$(i..j, v)$
1: **if** $i = j$ **then**
2:   **if** $a_i = v$ **then**
3:     Return *true*;
4:   **else**
5:     Return *false*;
6:   **end if**
7: **end if**
8: Return $search(i + 1..j, v)$;

FIGURE 3.1: First recursive algorithm to check if v appears in $a_i, \ldots, a_j$.

3.2.1 Key searching problem

Given a sorted list of integers $a_1, \ldots, a_n$ (i.e., $a_1 < a_2 < \ldots < a_n$), the key searching problem asks if a key v appears in $a_1, \ldots, a_n$.

For example, suppose the list of integers is $3, 7, 12, 14, 18, 23, 37, 45, 87, 92$. If the key is 15, *false* is reported since 15 is not in the list of integers. If the key is 18, *true* is reported.

We use the above three stages to design a recursive algorithm that solves the key searching problem. First, we characterize the key searching problem and identify subproblems. The full problem is to check if v appears in $a_1, \ldots, a_n$. The subproblem can be defined as checking if v appears in $a_i, \ldots, a_j$ for $1 \le i \le j \le n$. Precisely, we define $search(i..j, v)$ to be *true* if $v \in \{a_k \mid i \le k \le j\}$; and *false* otherwise. Our aim is to compute $search(1..n, v)$.

The second stage formulates the recursive equation. We need to define the base case and the recursive case. The base case is the case where the answer is known. In the key searching problem, the base case occurs when the interval is of length 1, i.e., $i = j$. We have

$$search(i..i, v) = \begin{cases} true & \text{if } a_i = v \\ false & \text{otherwise.} \end{cases}$$

The recursive case occurs when $i < j$. We can reduce the problem into subproblems by checking if $a_i = v$. We have

$$search(i..j, v) = \begin{cases} true & \text{if } a_i = v \\ search(i + 1..j, v) & \text{otherwise.} \end{cases} \tag{3.1}$$

Based on this recursive formula, the third stage formulates the recursive algorithm. We develop an algorithm that handles both the base case and the recursive case. Here, we try to develop a recursive algorithm which is detailed in Figure 3.1. Steps 1–5 handle the base case while Step 6 handles the recursive case. The algorithm basically checks if $a_i = v$ for i from 1 to n. Its running time is $O(n)$.

Algorithm **search**$(i..j, v)$
1: **if** $i = j$ and $a_i \neq v$ **then**
2:    Return *false*;
3: **else if** $a_i = v$ **then**
4:    Return *true*;
5: **end if**
6: Set $\kappa = \lfloor \frac{i+j}{2} \rfloor$;
7: **if** $v \leq a_\kappa$ **then**
8:    Return $search(i..\kappa, v)$;
9: **else**
10:    Return $search(\kappa + 1..j, v)$;
11: **end if**

FIGURE 3.2: Second recursive algorithm to check if v appears in $a_i, \ldots, a_j$.

The above algorithm is not efficient. By defining a different recursive formula in the second stage, we can develop a more efficient algorithm. Precisely, for Equation 3.1, instead of comparing v to a_i, we compare v to a_κ, where $\kappa = \lfloor \frac{i+j}{2} \rfloor$. Then, we obtain the following recursive equation.

$$search(i..j, v) = \begin{cases} search(i..\kappa, v) & \text{if } v \leq a_\kappa \\ search(\kappa + 1..j, v) & \text{otherwise} \end{cases}$$

Based on this recursive formula, $search(1..n, v)$ can be computed using the recursive algorithm in Figure 3.2. Interestingly, this algorithm will not check every a_i. It performs at most $\log n$ comparisons between a_i's and v and hence, it runs in $O(\log n)$ time. It is more efficient than the previous solution.

3.2.2 Edit-distance problem

The second example is the edit-distance problem. Given two strings $S[1..n]$ and $T[1..m]$, the edit-distance problem aims to find $edit(S, T)$, which is the minimum number of edit operations (i.e., insertion, deletion and substitution) to transform S to T. For example, suppose $S = ants$ and $T = bent$. S can be transformed to T by inserting "b", substituting "a" with "e", and deleting "s". So, $edit(S, T) = 3$.

We use the above three stages to develop an algorithm to solve the edit-distance problem. First, we characterize the problem. The full problem finds the edit distance between $S[1..n]$ and $T[1..m]$. The subproblem can be defined to compute the edit distance between the prefixes of S and T. Precisely, for $0 \leq i \leq n$ and $0 \leq j \leq m$, we define $V(i, j)$ be $edit(S[1..i], T[1..j])$.

Second, we need to define the recursive formula. For the base case (either $i = 0$ or $j = 0$), we have

$$V(0, i) = V(i, 0) = i.$$

For the recursive case (both $i > 0$ and $j > 0$), we aim to find the minimum number of edit operations to transform $S[1..i]$ to $T[1..j]$. There are three subcases depending on the edit operation on the rightmost character: (1) substitution (substitute $S[i]$ by $T[j]$), (2) insertion (insert $T[j]$) or (3) deletion (delete $S[i]$). For subcase (1), $V(i,j) = V(i-1,j-1) + \delta(S[i],T[j])$, where $\delta(a,b) = 0$ if $a = b$; and 1, otherwise. For subcase (2), $V(i,j) = V(i,j-1)+1$. For subcase (3), $V(i,j) = V(i-1,j)+1$. Hence, we have the following recursive formula.

$$V(i,j) = \min \begin{cases} V(i-1,j-1) + \delta(S[i],T[j]) & \text{substitution} \\ V(i,j-1) + 1 & \text{insertion} \\ V(i-1,j) + 1 & \text{deletion} \end{cases}$$

Given the recursive formula, the third stage constructs either a recursive algorithm or a dynamic programming algorithm. We can compute edit distance by recursion. However, such a method runs in $O(3^{n+m})$ time. The inefficiency of the recursive algorithm is due to the recomputation of $V(i,j)$. (Precisely, the value $V(i,j)$ will be recomputed when we need to compute the values $V(i+1,j+1), V(i+1,j)$ and $V(i,j+1)$.) If we store $V(i,j)$ after it is computed, we can avoid the recomputation and reduce the running time. Such an approach is called dynamic programming. Dynamic programming generates the dependency graph of all values $V(i,j)$. Then, it first computes the values $V(i,j)$ for base cases. Afterward, for recursive cases, the values $V(i,j)$ are computed in bottom-up order according to the dependency graph. Figure 3.3 details the algorithm to compute $V(i,j)$ for $0 \le i \le n$ and $0 \le j \le m$. Steps 1−6 are for the base cases while Steps 7−11 are for the recursive cases. This algorithm runs in $O(nm)$ time. (Note: Apart from bottom-up dynamic programming, we can also use the technique of "memoization". Initialize all table entries $V(i,j)$ to be -1 and run the recursive algorithm to compute $V(n,m)$. When the recursive algorithm computes $V(i,j)$, if its value is -1, the algorithm applies the recursive formula and computes recursively. When the value of the entry is ≥ 0, we just return the value of the entry. This solution also runs in $O(nm)$ time.)

3.3   Parameter estimation

Consider a set of n observed data $X = \{X_1, \ldots, X_n\}$. Suppose each observation X_i can be modeled by a distribution with an unknown parameter θ. We aim to find θ. For example, suppose we flip a coin. Let θ be the (unknown) chance that the outcome is a head. Suppose we flip the coin n times. Let $X = \{X_1, \ldots, X_n\}$ be the list of n outcomes, where $X_i = 1$ if the ith outcome is a head; and 0 otherwise. Our aim is to estimate θ.

Algorithm **EditDist**$(S[1..n], T[1..m])$

1: **for** $i = 0$ to n **do**
2:     Set $V(i, 0) = i$;
3: **end for**
4: **for** $j = 0$ to m **do**
5:     Set $V(0, j) = j$;
6: **end for**
7: **for** $i = 1$ to n **do**
8:     **for** $j = 1$ to m **do**
9:        $V(i, j) = \min\{V(i - 1, j - 1) + \delta(S[i], T[j]), V(i, j - 1) + 1, V(i - 1, j) + 1\}$;
10:     **end for**
11: **end for**
12: Return $V(n, m)$;

FIGURE 3.3: The dynamic programming algorithm to compute the edit distance between $S[1..n]$ and $T[1..m]$.

3.3.1 Maximum likelihood

The parameter θ can be estimated by maximum likelihood. Assume all observations in X are independent. Then, the probability of observing X given the parameter θ is

$$L(\theta|X) = Pr(X|\theta) = \prod_{i=1}^{n} Pr(X_i|\theta).$$

This function is called the likelihood function. The log likelihood function is defined to be $\log L(\theta|X) = \sum_{i=1}^{n} \log Pr(X_i|\theta)$. The maximum likelihood problem aims to estimate the θ that maximizes $L(\theta|X)$ or $\log L(\theta|X)$.

For the coin-flipping example, its likelihood function and log likelihood function are

$$L(\theta|X) = Pr(X|\theta) = \prod_{i=1}^{n} Pr(X_i|\theta) = \prod_{i=1}^{n} \theta^{X_i}(1 - \theta)^{(1-X_i)} \qquad (3.2)$$

$$\log L(\theta|X) = \sum_{i=1}^{n} (X_i \log \theta + (1 - X_i) \log(1 - \theta)). \qquad (3.3)$$

By differentiating $L(\theta|X)$ (or $\log L(\theta|X)$) with respect to θ, we can show that $\theta = \frac{\sum_{i=1}^{n} X_i}{n}$ maximizes $L(\theta|X)$.

For example, suppose we toss the coin 10 times and the observed outcome is $(X_1 = 1, X_2 = 0, X_3 = 0, X_4 = 1, X_5 = 1, X_6 = 1, X_7 = 1, X_8 = 0, X_9 = 1, X_{10} = 1)$. Then, we estimate $\theta = \frac{\sum_{i=1}^{n} X_i}{n} = 0.7$.

3.3.2   Unobserved variable and EM algorithm

This section introduces the problem of parameter estimation when there are unobserved variables. Before we introduce unobserved variables, we first discuss an example. Consider two coins A and B. Let θ_A and θ_B be the chances of getting heads when we toss coins A and B, respectively. We perform n experiments. In the ith experiment, either coin A or coin B is chosen and it is tossed five times. Let λ be the chance that coin A is chosen in each experiment. Suppose $n = 5$. The following table shows the possible outcome.

i	Coin chosen (Z_i)	Outcomes of the five tosses	Number of heads (X_i)
1	A	THHHH	4
2	B	THTTT	1
3	A	HHTHH	4
4	A	HHHTH	4
5	B	TTTTT	0

In the above table, Z_i is the coin chosen in the ith experiment and X_i is the number of heads in the five tosses. Given $\{(X_1, Z_1), \ldots, (X_n, Z_n)\}$, we aim to estimate θ_A, θ_B and λ. This problem can be solved by maximum likelihood. For $i = 1, \ldots, n$, we have:

$$
\begin{aligned}
Pr(X_i, Z_i | \theta) &= Pr(Z_i | \theta) Pr(X_i | Z_i, \theta) \\
&= \begin{cases} \lambda \theta_A^{X_i} (1 - \theta_A)^{(5 - X_i)} & Z_i = A \\ (1 - \lambda) \theta_B^{X_i} (1 - \theta_B)^{(5 - X_i)} & Z_i = B \end{cases}
\end{aligned} \tag{3.4}
$$

Hence, for the above dataset, the complete likelihood function is

$$
\begin{aligned}
L(\theta | X, Z) &= Pr(X, Z | \theta) \\
&= \prod_{i=1}^{5} Pr(X_i, Z_i | \theta) \\
&= Pr(4, A | \theta) Pr(1, B | \theta) Pr(4, A | \theta) Pr(4, A | \theta) Pr(0, B | \theta) \\
&= \lambda^3 \theta_A^{4+4+4} (1 - \theta_A)^{1+1+1} (1 - \lambda)^2 \theta_B^{1+0} (1 - \theta_B)^{4+5}.
\end{aligned}
$$

By differentiation, we can show that $\lambda = 0.6$, $\theta_A = 0.8$ and $\theta_B = 0.1$.

The above example assumes the full data is known. In real life, some data may not be observable. For example, suppose we don't know whether coin A or coin B is chosen. Then, Z_i is an unobserved variable while X_i is an observed variable. In such case, the observed data is $\{X_1, \ldots, X_n\}$. The marginal likelihood function of the observed data is

$$
L(\theta | X) = \prod_{i=1}^{n} \left(\sum_{Z_i \in \{A, B\}} Pr(X_i, Z_i | \theta) \right).
$$

We aim to estimate θ by maximizing $L(\theta|X)$. However, the marginal likelihood function may be difficult to optimize since the unobserved data is unknown. In such case, we can apply the Expectation Maximization (EM) algorithm.

The EM algorithm iteratively improves the estimation of θ, that is, it tries to iteratively update θ so that $L(\theta|X)$ is getting bigger and bigger. Starting from an initial guess of θ, each iteration of the EM algorithm involves two steps: the expectation step (E-step) and the maximization step (M-step). In the E-step, given θ, it estimates the values of the unobserved variables Z_i. Precisely, we estimate the chance that $Z_i = z$ given the parameter θ, which is

$$t_i(z) = Pr(Z_i = z|X_i, \theta) = \frac{Pr(X_i, Z_i = z|\theta)}{Pr(X_i|\theta)} = \frac{Pr(X_i, Z_i = z|\theta)}{\sum_{Z_i} Pr(X_i, Z_i|\theta)}. \quad (3.5)$$

Given the estimated distribution of Z_i (i.e., $t_i(z)$), the M-step tries to improve θ to θ' so that $Pr(X|\theta') > Pr(X|\theta)$ where

$$Pr(X|\theta) = \prod_{i=1}^{n} \sum_{z} t_i(z) Pr(X_i, Z_i = z|\theta). \quad (3.6)$$

However, the above equation is sometime time consuming to evaluate. It is suggested that the above equation can be transformed to $Q(\theta|t_i)$, which is as follows.

$$Q(\theta|t_i) = \prod_{i=1}^{n} \prod_{z} Pr(X_i, Z_i = z|\theta)^{t_i(z)} \quad (3.7)$$

The equation $Q(\theta|t_i)$ is expected to be easier to maximize. More importantly, Lemma 3.1 showed that, when θ' maximizes $Q(\theta'|t_i)$, we guarantee $Pr(X|\theta') \geq Pr(X|\theta)$. Hence, we can iteratively apply the E-step and the M-step to find θ, which improves the likelihood function $L(\theta|X)$. This process is repeated until $L(\theta|X)$ is converged. Figure 3.4 details the EM algorithm.

Lemma 3.1 $\frac{Pr(X|\theta')}{Pr(X|\theta)} \geq \frac{Q(\theta'|t_i)}{Q(\theta|t_i)}$.

Proof See Exercise 3. ∎

To illustrate the usage of the EM algorithm, we use the coin-tossing example. In the E-step, by Equations 3.5 and 3.4, we have: $t_i(B) = 1 - t_i(A)$ and

$$
\begin{aligned}
t_i(A) &= Pr(Z_i = A|X_i, \theta) \\
&= \frac{Pr(X_i, Z_i = A|\theta)}{Pr(X_i, Z_i = A|\theta) + Pr(X_i, Z_i = B|\theta)} \\
&= \frac{\lambda \theta_A^{X_i}(1-\theta_A)^{(5-X_i)}}{\lambda \theta_A^{X_i}(1-\theta_A)^{(5-X_i)} + (1-\lambda)\theta_B^{X_i}(1-\theta_B)^{(5-X_i)}}. \quad (3.8)
\end{aligned}
$$

EM algorithm

Require: The observed data X

Ensure: The parameter θ that maximizes $L(\theta|X) = Pr(X|\theta) = \sum_Z Pr(X, Z|\theta)$.

1: Initialize θ (randomly or using some simple heuristics);
2: **repeat**
3:     E-step: Compute $t_i(z) = Pr(Z_i = z|X_i, \theta)$ for $i \in 1, \ldots, m$;
4:     M-step: Compute $\theta = argmax_{\theta'} Q(\theta'|t_i)$.
5: **until** $L(\theta|X)$ is converged;
6: **return** θ;

FIGURE 3.4: The standard EM algorithm.

Given t_i's, the M-step aims to find $\theta' = (\theta'_A, \theta'_B, \lambda')$ that maximizes $Q(\theta'|t_i) = \prod_{i=1}^{n} \prod_{z \in \{A,B\}} Pr(X_i, Z_i = z|\theta')^{t_i(z)}$ where $t_i(A)$ satisfies Equation 3.8, $t_i(B) = 1 - t_i(A)$ and $Pr(X_i, Z_i|\theta')$ satisfies Equation 3.4. By differentiating $Q(\theta'|t_i)$ with respect to θ'_A, θ'_B and λ', we can show that $\theta'_A = \frac{\sum_{i=1}^{n} t_i(A)X_i}{n \sum_{i=1}^{n} t_i(A)}$, $\theta'_B = \frac{\sum_{i=1}^{n} t_i(B)X_i}{n \sum_{i=1}^{n} t_i(B)}$ and $\lambda' = \frac{\sum_{i=1}^{n} t_i(A)}{n}$. Hence, we have the algorithm in Figure 3.5.

Given the observed data $(X_1 = 4, X_2 = 1, X_3 = 4, X_4 = 4, X_5 = 0)$, suppose we initialize $\theta_A = 0.6, \theta_B = 0.5$ and $\lambda = 0.5$. The following table shows the values of $\theta_A, \theta_B, \lambda$ after the kth iteration of the algorithm in Figure 3.5 for $k = 0, 1, 2, 3, 4$. At the 4th iteration, we observe that the values are close to the optimal solution $\theta_A = 0.8, \theta_B = 0.1, \lambda = 0.6$.

Iteration	θ_A	θ_B	λ
0	0.6	0.5	0.5
1	0.638577564	0.406249756	0.489610971
2	0.737306832	0.28673575	0.517707992
3	0.787973134	0.152522412	0.578294391
4	0.794271411	0.10301972	0.603224969

Note that the EM algorithm just iteratively improves θ. It may be trapped in the local optimal. (For example, if we initialize $\theta_A = 0.5, \theta_B = 0.5, \lambda = 0.5$, the algorithm in Figure 3.5 will be trapped in the suboptimal solution.) To obtain the global optimal, we need a good method to initialize θ or we can run the EM algorithm multiple rounds with different initial guesses of θ.

CoinTossingEM algorithm

Require: The observed data $X = (X_1, \dots, X_n)$
Ensure: The parameters $\theta_A, \theta_B, \lambda$
1: Initialize $\theta_A, \theta_B, \lambda$ (randomly or using some simple heuristics);
2: **repeat**
3:    /* E-step */
4:    **for** $i = 1, \dots, n$ **do**
5:       Set $t_i(A) = \dfrac{\lambda \theta_A^{X_i}(1-\theta_A)^{(5-X_i)}}{\lambda \theta_A^{X_i}(1-\theta_A)^{(5-X_i)} + (1-\lambda)\theta_B^{X_i}(1-\theta_B)^{(5-X_i)}}$;
6:       Set $t_i(B) = 1 - t_i(A)$;
7:    **end for**
8:    /* M-step */
9:    Set $\theta_A = \dfrac{\sum_{i=1}^{n} t_i(A)X_i}{n\sum_{i=1}^{n} t_i(A)}$;
10:    Set $\theta_B = \dfrac{\sum_{i=1}^{n} t_i(B)X_i}{n\sum_{i=1}^{n} t_i(B)}$;
11:    Set $\lambda = \dfrac{\sum_{i=1}^{n} t_i(A)}{n}$;
12: **until** the estimation of $\theta_A, \theta_B, \lambda$ has little change;
13: **return** $\theta_A, \theta_B, \lambda$;

FIGURE 3.5: The EM algorithm for the coin-tossing problem.

3.4 Hash data structures

Hashing is a frequently used technique in many domains including NGS data processing. It is a technique that maps a set of keys into a set of small values by a hash function. This section briefly covers four applications of hashing: (1) associate array, (2) maintaining a set, (3) maintaining a multiset and (4) estimating the similarity of two sets.

3.4.1 Maintain an associative array by simple hashing

Consider a collection of $(key, value)$ pairs. An associative array maintains them and supports the following three operations: (1) add a $(key, value)$ pair, (2) remove a $(key, value)$ pair and (3) look up the value associated with a key.

This problem can be solved using a data structure like a binary search tree. Although it works, the insert/remove/lookup time is $\Theta(\log n)$, where n is the number of elements in the binary search tree.

Alternatively, an associative array can be implemented using hashing. It consists of two parts: a hash function $h()$ and an array $H[0..n-1]$ (known as a hash table). The hash function $h()$ maps each key to an integer between 0 and $n-1$. The hash table $H[0..n-1]$ is for storing the $(key, value)$ pairs.

For example, suppose we have a set of individuals where each individual is represented by a $(key, value)$ pair, where the key is his/her length-k DNA string and the $value$ is his/her age. To maintain the $(key, value)$ pairs in a hash table $H[0..n-1]$, we need a hash function $h()$ to map each DNA string $key = b_{k-1} \dots b_1 \dots b_0$ into some integer between 0 and $n-1$. A possible hash function is as follows:

$$h(b_{k-1} \dots b_1 b_0) = \left(\sum_{i=0}^{k-1} 4^i c(b_i) \right) \quad \text{mod } n$$

where $c(\mathtt{A}) = 0$, $c(\mathtt{C}) = 1$, $c(\mathtt{G}) = 2$ and $c(\mathtt{T}) = 3$.

For example, assume the DNA string is of length $k = 4$ and the hash table size is $n = 13$. Then, $h(\mathtt{ACTC}) = \left(4^0 * 1 + 4^1 * 3 + 4^2 * 1 + 4^3 * 0 \right) \mod 13 = 3$.

Given the hash function $h()$, we can support the three operations in constant time. To insert a (key, age) pair into a hash table $H[0..n-1]$, we first compute $t = h(key)$; then, we assign $H[t] = (key, value)$. For example, to insert $(\mathtt{ACTC}, 20)$ into $H[0..12]$, we compute $h(\mathtt{ACTC}) = 3$ and assign $H[3] = (\mathtt{ACTC}, 20)$. Similarly, to insert $(\mathtt{GTCA}, 51)$ into $H[0..12]$, we compute $h(\mathtt{GTCA}) = 11$ and we assign $H[11] = (\mathtt{GTCA}, 51)$.

To find the value associated with the key, we just need to compute $t = h(key)$ and extract the $(key, value)$ pair from $H[t]$. To delete a $(key, value)$ pair, we compute $t = h(key)$; then, set $H[t]$ to be empty.

The above method can only work when $n \geq 4^k$. In such case, any length-k DNA string can be hashed into a unique entry in the array $H[0..n-1]$. All operations can be performed in $O(1)$ worst case time. This is called perfect hashing.

When $n < 4^k$, multiple $(key, value)$ pairs may be hashed to the same entry in the hash table $H[0..n-1]$. This is called collision. The above method cannot work. There are two ways to resolve collisions: separate chaining and open addressing. Separate chaining uses a link list to store all the $(key, value)$ pairs hashed to the same entry.

Open addressing maintains at most one $(key, value)$ pair in each hash entry. (Note: The number of $(key, value)$ pairs that can be stored by an open addressing scheme is upper bounded by the hash table size. If we want to store more pairs, we need to resize the hash table.) The simplest open addressing scheme is linear probing. To insert the $(key, value)$ into the hash table, let $h(key) = t$. If $H[t]$ is non-empty and $H[t] \neq key$, a collision occurs. Linear probing iteratively increments the index t by 1 when collision occurs until either $H[t]$ is an empty entry or the key of $H[t]$ is key. If $H[t]$ is an empty entry, we assign $H[t] = (key, value)$. Otherwise, if the key of $H[t]$ is key, this means that key is already stored in the hash table H. We can implement the other two operations (i.e., deletion and lookup) similarly using linear probing.

In the worst case, the insert/delete/lookup operation of the linear probing scheme takes $\Theta(n)$ time. However, the average performance of the linear probing scheme is not bad. Empirical study showed that the expected running time of all three operations is $O(1)$ when the load factor (i.e., the proportion of non-empty entries in the hash table) is lower than 0.7.

Linear probing is the simplest open-addressing scheme. You can also use some better scheme like Cuckoo hashing [222] to resolve collisions.

3.4.2 Maintain a set using a Bloom filter

Consider a set X of elements. We hope to maintain X and support the following operations: (1) insert an element into X and (2) query if an element w exists in X. (Note: We don't support deleting an element from X.)

The set X can be maintained using an associative array. It works well when X is of small or medium size. When the number of elements in X is huge, one solution is to use a Bloom filter proposed by Burton Howard Bloom [23] in 1970. A Bloom filter is a space-efficient probabilistic data structure to solve this problem. The insert/query time is worst case $O(1)$. Although it may give a false positive (i.e., it may incorrectly report that an element exists), it does not give a false negative.

The Bloom filter consists of (1) k hash functions $h_1, \ldots, h_k$ and (2) a length-n bit array $B[0..n-1]$. $B[0..n-1]$ is initialized as a zero-bit array. A Bloom filter is small and it uses $O(n)$ bit space.

To insert an element w into X, we set $B[h_i(w)]$ to be 1 for every $1 \leq i \leq k$. To query if an element w exists in X, we compute $h_1(w), \ldots, h_k(w)$. If $B[h_i(w)] = 0$ for some i, then $w \notin X$; otherwise, w is predicted to exist in X.

A Bloom filter may give a false positive prediction. Precisely, assume the hash function is truly uniformly random, after m elements are inserted into X, the false positive rate is $\left[1 - \left(1 - \frac{1}{n}\right)^{km}\right]^k \approx \left(1 - e^{-\frac{km}{n}}\right)^k$. Hence, when $n \gg m$, the false positive rate is low.

Given n and m, the optimal number of hash functions that minimizes the false positive rate is $k \approx \frac{n}{m} \ln 2$. For example, if $n = 8m$, the optimal $k \approx 8 \ln 2 = 5.545$. If we set $k = 6$, the false positive rate is 0.022.

We give an example to illustrate the idea. Consider a Bloom filter consists of a bit array $B[0..12]$ and three hash functions h_1, h_2 and h_3, where $h_1(w) = w \bmod 13$, $h_2(w) = w^2 \bmod 13$ and $h_3(w) = (w + w^2) \bmod 13$. We initialize $B[i] = 0$ for $0 \leq i \leq 12$ (see Figure 3.6(a)). Figure 3.6(b–e) demonstrates the steps of inserting $12, 4, 31, 27$ into the bit array B.

To query 4, we have $h_1(4) = 4$, $h_2(4) = 3$ and $h_3(4) = 7$. Since $B[4] = B[3] = B[7] = 1$, 4 is predicted to appear in X. To query 35, we have $h_1(35) = 9$, $h_2(35) = 3$ and $h_3(35) = 12$. Since $B[9]$ is 0, we predict that 35 does not appear in X. To query 40, we have $h_1(40) = 1$, $h_2(40) = 1$ and $h_3(40) = 2$. Since $B[1] = B[2] = 1$, we predict that 40 appears in X. However, this is a false positive.

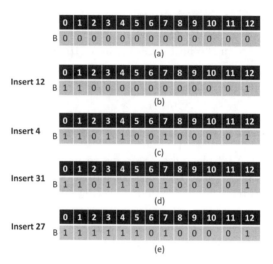

FIGURE 3.6: An example of a Bloom filter.

3.4.3   Maintain a multiset using a counting Bloom filter

Consider a multiset X of elements. We hope to maintain X and support the following operations: (1) insert an element into X, (2) delete an element from X and (3) query the number of occurrences of an element in X.

Fan et al. [74] proposed the counting Bloom filter to solve this problem. A counting Bloom filter is a generalization of the Bloom filter. It is a space-efficient probabilistic data structure that enables constant time insert/delete/query. Although it may give false positives (i.e., it may incorrectly report an element exists or over-estimate the count of the element), it does not give false negatives.

The counting Bloom filter consists of (1) k hash functions $h_1, \ldots, h_k$ and (2) a length-n integer array $B[0..n-1]$. All entries in the integer array $B[0..n-1]$ are initialized as zeros. To insert an element w into X, we increment $B[h_i(w)]$ by 1 for every $1 \le i \le k$. Similarly, to delete an element w that exists in X, we decrement $B[h_i(w)]$ by 1 for every $1 \le i \le k$.

To query if an element w belongs to X, we compute $h_1(w), \ldots, h_k(w)$. If $B[h_i(w)] = 0$ for some i, then $w \notin X$; otherwise, w is estimated to occur $\min_{i \in 1..k}\{B[h_i(w)]\}$ times in X. Note that w may be a false positive or the count may be inflated. But the chance is low when n is big enough.

To illustrate the idea, we give an example. Consider a counting Bloom filter that consists of an integer array $B[0..12]$ and three hash functions h_1, h_2 and h_3, where $h_1(w) = w \mod 13$, $h_2(w) = w^2 \mod 13$ and $h_3(w) = (w + w^2) \mod 13$. We initialize $B[i] = 0$ for $0 \le i \le 12$ (see Figure 3.7(a)). Figure 3.7(b–e) demonstrates the steps of inserting $4, 31, 4$ into the integer array B; then deleting 31.

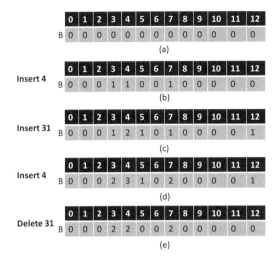

FIGURE 3.7: An example of a counting Bloom filter.

We use the example in Figure 3.7(d) to illustrate the counting query. For the element 4, since $\min\{B[h_1(4)], B[h_2(4)], B[h_3(4)]\} = \min\{B[4], B[3], B[7]\} = 2$, we conclude that 4 occurs twice. For the element 31, since $\min\{B[h_1(31)], B[h_2(31)], B[h_3(31)]\} = \min\{B[5], B[12], B[4]\} = 1$, we conclude that 31 occurs once. For the element 33, since $\min\{B[h_1(33)], B[h_2(33)], B[h_3(33)]\} = \min\{B[7], B[10], B[4]\} = 0$, we conclude that 33 does not exist.

Note that when too many elements are inserted into the counting bloom filter, the chance of giving false positives increases. To make the counting filter smaller and scalable, Bonomi et al. [27] introduced a data structure based on d-left hashing. This counting filter is known as the d-left counting Bloom filter, which is functionally equivalent to the normal counting Bloom filter but uses approximately half as much space; furthermore, once the designed capacity is exceeded, the keys will be reinserted using a new hash table of double size. This makes a d-left counting Bloom filter scalable.

3.4.4 Estimating the similarity of two sets using minHash

Let U be a universe. For any two sets $A, A' \subseteq U$, we can measure their similarity by Jaccard similarity, where Jaccard similarity is defined to be $J(A, A') = \frac{|A \cap A'|}{|A \cup A'|}$. For example, consider two sets $A = \{2, 14, 32, 41\}$ and $A' = \{5, 14, 41, 48, 72\}$. We have $J(A, A') = \frac{2}{7}$.

Suppose we have many sets and the sizes of the sets are big. It is time consuming to compute the Jaccard similarity of every pair of sets. This section aims to find a short signature for each set A that allows us to quickly estimate Jaccard similarity of any two sets. More precisely, we aim to find a similarity-preserving signature for every set such that similar sets have similar signatures while dissimilar sets are unlikely to have similar signatures. One possible signature is minHash.

First, we define the minHash function. Let $h()$ be a function that maps U to any permutation of U. For any set $A \subseteq U$, we define the minHash function $\omega_h(A) = \min\{h(x) \mid x \in A\}$. The following lemma is the key.

Lemma 3.2 *Let $h()$ be a function that maps U to a random permutation of U. For any two sets $A, A' \subseteq U$. $Pr(\omega_h(A) = \omega_h(A')) = J(A, A')$.*

Proof Note that $\omega_h(A) = \omega_h(A')$ if and only if $\min_{x \in A \cup A'} h(x) = \min_{x \in A \cap A'} h(x)$.

As $h()$ is a random permutation function, $\frac{|A \cap A'|}{|A \cup A'|}$ is the probability that $\min_{x \in A \cup A'} h(x) = \min_{x \in A \cap A'} h(x)$. Hence, the lemma follows. ∎

Based on the above lemma, we can define a length-K signature of a set A as follows. Let $h_1, \ldots, h_K$ be K different random permutation functions of U. For any set A, denote the signature of A as

$$sig(A) = (\omega_{h_1}(A), \ldots, \omega_{h_K}(A)).$$

For any two sets A and A', their similarity $sim(sig(A), sig(A'))$ is defined to be $\frac{1}{K} \sum_{i=1}^{K} \delta(\omega_{h_i}(A), \omega_{h_i}(A'))$. ($\delta(x, y) = 1$ if $x = y$; otherwise $\delta(x, y) = 0$.) We have the following lemma.

Lemma 3.3 *The expected value of $sim(sig(A), sig(A'))$ is $J(A, A')$.*

Proof See Exercise 6. ∎

By the above lemma, we know $sim(sig(A_i), sig(A_j)) \approx J(A_i, A_j)$. Below, we give an example. Suppose the universe is $\{0, 1, \ldots, 15\}$. Let $h_1(x) = x \oplus 1010_2$, $h_2(x) = x \oplus 0110_2$ and $h_3(x) = x \oplus 0100_2$ be three permutation functions. The following table shows the values of the three functions for three sets. We have $sim(A_1, A_2) = 2/3 = 0.6667$, $sim(A_1, A_3) = sim(A_2, A_3) = 0$. Note that $J(A_1, A_2) = 3/6 = 0.5$, $J(A_1, A_3) = J(A_1, A_3) = 0$. Hence, $sim()$ and $J()$ have approximately the same values.

A	$h_1()$	$h_2()$	$h_3()$
$A_1 = \{1, 4, 6, 8, 15\}$	2	0	0
$A_2 = \{4, 7, 8, 15\}$	2	1	0
$A_3 = \{3, 9, 12, 14\}$	3	5	7

3.5 Full-text index

Consider a text $T[1..n]$; its full-text index is an index that allows efficient pattern matching on T. There are three important full-text indexes: the suffix tree, suffix array, and FM-index. The following subsections briefly describe them. A more detailed description can be found in [289].

3.5.1 Suffix trie and suffix tree

Suffix tree is first proposed by Weiner in 1973 [313]. It is a fundamental data structure for pattern matching. Given a text T over an alphabet $\mathcal{A}$, we first define suffix trie τ (or τ_T for clarity), which is a variant of the suffix tree. A trie (derived from the word re*trie*val) is a rooted tree where every edge is labeled by a character in $\mathcal{A}$. It represents a set of strings formed by concatenating the characters on the unique paths from the root to the leaves of the trie. A suffix trie is simply a trie storing all possible suffixes of T. Figure 3.14(a) shows all possible suffixes of a text $T[1..7] = $ ACACAG\$, where \$ is a special symbol that indicates the end of T. For example, the 4th suffix is CAG\$ The corresponding suffix trie is shown in Figure 3.14(b). In a suffix trie, every possible suffix of T is represented as a path from the root to a leaf. This implies that every possible substring of T is represented as a path from the root to an internal node.

For any node u in the suffix trie τ for T, $Leaf(u)$ (or $Leaf_T(u)$ for clarity) is defined as the set of descendant leaves of u. Also, the path label of u, denoted as $plabel(u)$ (or $plabel_T(u)$ for clarity) is defined as the concatenation of the edge labels from the root to the node. We define $depth(u)$ (or $depth_T(u)$ for clarity) be the string depth of a node u, which is the length of the path label $plabel(u)$. For example, in Figure 3.14(b), the leaf label set of the node v is $Leaf(v) = \{1, 3\}$, the path label of the node v is $plabel(v) = $ ACA, and the string depth of the node v is $depth(v) = 3$.

A suffix trie may contain $O(n^2)$ nodes, which is too big. A suffix tree is a compact representation of a suffix trie, which is a rooted tree formed by contracting all internal nodes in the suffix trie with a single child and a single parent. Figure 3.14(c) shows the suffix tree made from the suffix trie in Figure 3.14(b). For each edge in the suffix tree, the edge label is defined as the concatenation of the characters on the edges of the suffix trie, which are merged. For example, in Figure 3.14(c), the edge label of (u, v) is CA. Given a string T of length n, its suffix tree can be stored in $O(n \log n)$ bits. It also can be constructed in $O(n)$ time [76, 104].

A suffix tree/trie enables us to perform pattern searching efficiently. Given a query pattern Q, we can determine if Q exists in T by checking if there exists a node u such that $plabel(u) = Q$. Then, $Leaf(u)$ is the set of occurrences of Q, which can be found by running a depth-first search at the node u. In total,

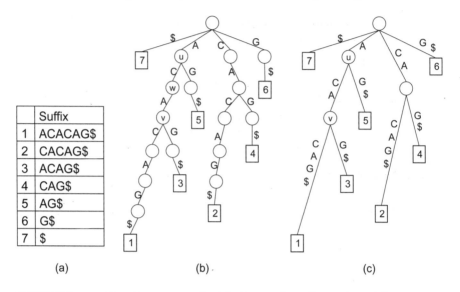

	Suffix
1	ACACAG$
2	CACAG$
3	ACAG$
4	CAG$
5	AG$
6	G$
7	$

(a)   (b)   (c)

FIGURE 3.8: For $T = $ ACACAG$, (a) shows all suffixes of T, (b) shows the suffix trie of T, and (c) shows the suffix tree of T.

the occurrences of any pattern Q of length m can be found in $O(m + occ)$ time where occ is the number of occurrences of the pattern Q in S.

For the example suffix tree/trie of $T = $ ACACAG$ in Figure 3.14(b,c), the occurrences of $Q = $ ACA can be found by traversing down the suffix tree/trie to find a path label which equals Q. Since $plabel(v) = $ ACA $= Q$, we confirm that Q occurs in T. The occurrences are $Leaf(v)$, which is $\{1, 3\}$.

Consider the case when $Q = $ ACC. When we traverse down the suffix tree/trie, we find a node w with path label AC (see Figure 3.14(b)). However, from this point onward, we cannot extend the path for the character C; thus we report that the pattern ACC does not occur in T.

3.5.2 Suffix array

A suffix tree is big since we need to maintain the tree structure. To reduce space, Manber and Myers [186] proposed a data structure called a suffix array, in 1993, which is functionally similar to a suffix tree. Let $T[1..n]$ be a text over an alphabet $\mathcal{A}$, which ends at a unique terminator $. Assume $ is lexicographically the smallest character. The suffix array $SA[1..n]$ is a permutation of $\{1, \ldots, n\}$ such that the $SA[i]$-th suffix is lexicographically smaller than the $SA[i + 1]$-th suffix for $i = 1, \ldots, n - 1$. A suffix array can be stored in $n \log n$ bits as every integer in the suffix array is less than n and can be stored in $\log n$ bits. For example, consider $T = $ ACACAG$. Its suffix array is $(7, 1, 3, 5, 2, 4, 6)$ (see Figure 3.9).

The suffix tree and suffix array are highly correlated. When the leaves of a

i	$SA[i]$	$SA[i]$-th suffix
1	7	$
2	1	ACACAG$
3	5	ACAG$
4	6	AG$
5	2	CACAG$
6	3	CAG$
7	4	G$

FIGURE 3.9: Consider a text $T = $ ACACAG$. The table has 7 rows. For the ith row, the second column is the suffix array value $SA[i]$ and the third column is the $SA[i]$-th suffix.

suffix tree are traversed in lexicographical depth-first search order, they form the suffix array of that string. (Note that the left-to-right order of the leaves in Figure 3.14 is in fact the suffix array in Figure 3.9.) Thus the suffix array of $T[1..n]$ can be constructed in $O(n)$ time by first constructing the suffix tree $\mathcal{T}$; then the suffix array can be generated by traversing $\mathcal{T}$ using lexicographical depth-first traversal.

Similar to the suffix tree, we can perform exact pattern matching using a suffix array. Consider a query string Q. Let i and j be the smallest and the largest index such that both the $SA[i]$-th suffix and the $SA[j]$-th suffix contain Q as their prefix. Then, $i..j$ is called the SA range of Q, denoted as $range(T, Q)$. Note that positions $SA[i], \ldots, SA[j]$ are all occurrences of Q in T. For example, considering the suffix array for the text $T = $ ACACAG$ (see Figure 3.9), CA occurs in $SA[5]$ and $SA[6]$. In other words, $range(T, CA) = 5..6$.

$range(T, Q)$ can be found by applying a binary search on the suffix array $SA[1..n]$. Precisely, we compare the query Q with the middle sequence in the suffix array. If Q is lexicographically smaller than the middle sequence, we recursively perform a binary search on the first half of the suffix array; otherwise, we perform a binary search on the second half. After performing at most $\log n$ comparisons, we can locate the pattern. As each comparison takes at most $O(m)$ time, the algorithm runs in $O(m \log n + occ)$ time. Manber and Myers [186] gave a modified version of the binary search algorithm, whose practical running time is $O(m + \log n + occ)$ for most cases.

3.5.3 FM-index

Although a suffix array is smaller than suffix tree, it is still too big for indexing genomes. For example, the suffix tree and suffix array of the human genome require ~40 gigabytes and ~13 gigabytes, respectively. To reduce space, Ferragina and Manzini [78] proposed the FM-index, which is much smaller. For the human genome, its FM-index takes less than 2 gigabytes, which is small enough to be stored in a personal computer.

Given a text $T[1..n]$ over an alphabet $\mathcal{A}$, the FM-index of T consists of three parts:

(1) C,

(2) the Burrow-Wheeler Transform (BWT) of the text T, and

(3) an auxiliary index $Occ(y, i)$.

For every $y \in \mathcal{A}$, $C(y)$ is defined to be the total number of characters in the text T that are lexicographically smaller than y. For example, for $T =$ ACACAG\$, we have $C(\$) = 0, C(\text{A}) = 1, C(\text{C}) = 4, C(\text{G}) = 6, C(\text{T}) = 7$.

i	Rotation	$F[i]$	$B[i]$	$LF(i)$	$SA[i]$	$SA[i]$-th suffix
1	\$ACACAG	\$	G	7	7	\$
2	ACACAG\$	A	\$	1	1	ACACAG\$
3	ACAG\$AC	A	C	5	3	ACAG\$
4	AG\$ACAC	A	C	6	5	AG\$
5	CACAG\$A	C	A	2	2	CACAG\$
6	CAG\$ACA	C	A	3	4	CAG\$
7	G\$ACACA	G	A	4	6	G\$

FIGURE 3.10: Consider a text $T =$ ACACAG\$. The table has 7 rows. For the ith row of the table, the second column is the last-to-first mapping function $LF(i)$, the third column is the BWT $B[i]$, the fourth column is the suffix array $SA[i]$, and the fifth column is the $SA[i]$-th suffix. Note that $B[i]$ is the preceding character of the $SA[i]$-th suffix.

The BWT B is an easily invertible permutation of T. We first define rotations of T. For any $i \in 1..n$, $T[i..n]T[1..i-1]$ is a rotation of T. For example, for $T =$ ACACAG\$, the second column of Figure 3.10 shows the lexicographical order of all rotations of T. Note that the lexicographical order of rotations of T is the same as the lexicographical order of suffixes of T. (That is, the $SA[i]$-th rotation is the i-th smallest rotation of T.)

The BWT is defined to be the last symbols of all sorted rotations of T. For example, for $T[1..7] =$ ACACAG\$, $B[1..7] =$ G\$CCAAA (see Figure 3.10). Mathematically, B and SA are related by the formula $B[i] = T[SA[i] - 1]$ for $SA[i] > 1$; otherwise, $B[i] = \$$. By the mathematical definition, we know that $B[i]$ is the preceding character of the $SA[i]$-th suffix (assuming the preceding character of the first suffix of T is $T[n] = \$$).

Given B, we define the function Occ such that, for any $y \in \mathcal{A}$ and $1 \le i \le n$, $Occ(y, i)$ is the number of occurrences of y in $B[1..i]$ and $Occ(y, 0) = 0$. When $|\mathcal{A}|$ is constant, [78] showed that the Occ data structure can be stored in $O(n \frac{\log \log n}{\log n})$ bits while any entry $Occ(y, i)$ can be accessed in constant time.

Algorithm BWT2Text

Require: The BWT $B[1..n]$

Ensure: The text $T[1..n]$

1: $T[n] = \$$;
2: $i = 1$ and $T[n-1] = B[i]$;
3: **for** $k = n - 2$ down to 1 **do**
4:   $i = LF(i)$;
5:   $T[k] = B[i]$;
6: **end for**
7: Return $T[1..n]$;

FIGURE 3.11: The algorithm for inverting the BWT $B[1..n]$ to the original text $T[1..n]$.

3.5.3.1   Inverting the BWT B to the original text T

The BWT B can be inverted to the original text T by utilizing the last-to-first mapping function $LF()$. Recall that $B[]$ is the array of the last symbols of all sorted rotations of T. Denote $F[]$ as the array of the first symbols of all sorted rotations of T (see Figure 3.10). The last-to-first mapping function $LF()$ maps the ith occurrence of the symbol c in $B[]$ to the ith occurrence of the symbol c in $F[]$. For example, in Figure 3.10, $B[6]$ is the second occurrence of A in $B[]$ while $F[3]$ is the second occurrence of A in $F[]$. So, we have $LF(6) = 3$.

Lemma 3.4 states the properties of $LF()$. By the first property of Lemma 3.4, $LF(i) = C(B[i]) + Occ(B[i], i)$. Using the FM-index, $LF(i)$ can be computed in $O(1)$ time.

We can recover $T[k]$ for $k = n, n - 1, \ldots, 1$ iteratively by the algorithm BWT2Text in Figure 3.11. The correctness follows from property 2 of Lemma 3.4 (see Exercise 8). Since $LF(i)$ can be computed in $O(1)$ time, the algorithm runs in $O(n)$ time.

Lemma 3.4 (BWT inversion) *LF has the following properties:*

1. $LF(i) = C(B[i]) + Occ(B[i], i)$.

2. $SA[i] = SA[LF(i)] + 1$. *(In other words, $T[SA[LF(i)]..n] = B[i]T[SA[i]..n]$.)*

Proof The following shows the proofs for the two properties:

1. Let $c = B[k]$ and $j = Occ(B[k], k)$. This means that the j-th occurrence of c in $B[1..n]$ is at position k. The array $F[1..n]$ is sorted. Hence, the j-th occurrence of c in $F[1..n]$ is at position $C(c) + j$. Thus, $LF(k) = C(c) + j = C(B[k]) + Occ(B[k], k)$.

Algorithm SA_value(i)

Require: the FM-index of T, the auxiliary data structure $D[]$ and an integer i between 1 and n

Ensure: $SA[i]$

1: $count = 0$;
2: **while** $D[i] = 0$ **do**
3: $\quad i = LF(i)$;
4: $\quad count = count + 1$;
5: **end while**
6: Return $D[i] - count$;

FIGURE 3.12: The algorithm for computing $SA[i]$.

2. By definition, $T[SA[i]..n]$ is the rank-i suffix and $B[i]$ is its preceding character. Hence, we have $T[SA[i] - 1..n] = B[i]T[SA[i]..n]$.

   Note that the rank of $B[i]T[SA[i]..n]$ is $LF(i)$. Hence, we have $SA[LF(i)] = SA[i] - 1$. ∎

3.5.3.2 Simulate a suffix array using the FM-index

We can use the FM-index to simulate a suffix array by including an auxiliary data structure. Here, we describe an $O(\frac{n \log \log n}{\log n})$-bit data structure that enables us to access every $SA[k]$ using $O(\log^2 n)$ time. We sample and store $SA[i]$ for all i such that $SA[i]$ is the multiple of κ where $\kappa = \log^2 n$. Precisely, we create an integer array $D[1..n]$ such that $D[i] = SA[i]$ if $SA[i]$ is a multiple of κ; and 0 otherwise. Then, using a packed B-tree, $D[1..n]$ can be stored using $O(\frac{n \log \log n}{\log n})$ bits while every $D[i]$ can be accessed in $O(1)$ time.

Given the FM-index and the array $D[1..n]$, we can compute $SA[i]$ using $O(\log^2 n)$ time using the algorithm in Figure 3.12. The correctness of the algorithm is due to the following lemma.

Lemma 3.5 *Given the FM-index of T and $D[]$, $SA[i]$ can be computed in $O(\log^2 n)$ time for any i.*

Proof The algorithm iteratively performs $i = LF(i)$ until $D[i] \neq 0$. Assume it performs d iterations of "$i = LF(i)$" and denote $i^{(1)}, \ldots, i^{(d)}$ as the values of i after the 1st, 2nd, ..., dth iterations, respectively. By Lemma 3.4, $SA[i] = SA[i^{(1)}] + 1 = SA[i^{(2)}] + 2 = \ldots = SA[i^{(d)}] + d = D[i^{(d)}] + d$. As $LF()$ can be computed in $O(1)$ time, the algorithm runs in $O(d)$ time.

Finally, note that we sample $SA[i]$ for all i such that $SA[i]$ is a multiple of $\log^2 n$. Hence, $d < \log^2 n$. ∎

```
Algorithm BackwardSearch(Q[1..m])
 1: Set i..j = 1..n;
 2: for k = m to 1 do
 3:     // Compute the SA range of Q[k..m]
 4:     y = Q[k];
 5:     i = C(y) + Occ(y, i − 1) + 1;
 6:     j = C(y) + Occ(y, j);
 7:     if i > j then
 8:         Return "Q does not exist";
 9:     end if
10: end for
11: Return i..j;
```

FIGURE 3.13: The algorithm for a backward search.

3.5.3.3 Pattern matching

Consider a text $T[1..n]$. For any query $Q[1..m]$, we aim to compute all occurrences of Q in T. This is equivalent to computing the SA range $range(T, Q)$. Section 3.5.2 showed that $range(T, Q)$ can be computed by a binary search of the suffix array $SA[]$, which takes $O(m \log n)$ time. We can also compute $range(T, Q)$ from the FM-index of T. This method is known as a backward search. More interestingly, a backward search is more efficient and takes $O(m)$ time.

A backward search is based on the following key lemma.

Lemma 3.6 (Backward Search) *Consider any pattern Q and any character y. Let $i..j$ be the SA range $range(T, Q)$. Then, the SA range $range(T, yQ)$ is $i'..j'$ where $i' = C(y) + Occ(y, i − 1) + 1$ and $j' = C(y) + Occ(y, j)$.*

Proof See Exercise 9. ∎

The detail of the backward search algorithm is stated in Figure 3.13. We start with an empty string, whose SA range is $1..n$ (Step 1). Iteratively, we compute $range(T, Q[k..m])$ for $k = m$ down to 1 by applying Lemma 3.6 (Steps 2−10). Since $C()$ and $Occ()$ can be computed in $O(1)$ time, the algorithm BackwardSearch($Q[1..m]$) runs in $O(m)$ time.

3.5.4 Simulate a suffix trie using the FM-index

Consider a text $T[1..n]$. The suffix trie τ of T is too big to be stored in memory. Here, we show that the suffix trie τ can be simulated using the FM-index of $\overline{T}$, where $\overline{T}$ is the reverse of T.

We aim to implement a suffix trie τ that supports the following operations:

- $child(u, c)$: Reports the child v of u such that the edge label of (u, v) is c.

- $Leaf(u)$: Reports all the leaves of τ under the node u.

First, we describe how to represent the nodes in τ. Every node u in τ is represented by a tuple $(|P|, i..j)$ where $P = plabel(u)$ and $i..j$ is $range(\overline{T}, \overline{P})$. In particular, the root is represented by $(0, 1..n)$. For example, Figure 3.14(b) gives the suffix trie τ for $T = $ ACGCG$ and Figure 3.14(a) gives the suffix array for $\overline{T}$. The node with path label CG is denoted by the tuple $(2, 5..6)$ since 2 is its depth and $5..6 = range(\overline{T}, $ GC$)$. Note that every node in τ is represented by a distinct tuple.

Below, we describe how to implement the operation $child(u, c)$. Suppose the node u is represented by $(d, i..j)$. By definition, $d = |P|$ and $i..j = range(\overline{T}, \overline{P})$, where $P = plabel(u)$. Then, the node v is $(d + 1, i'..j')$ where $i'..j' = range(\overline{T}, c\overline{P})$. Note that, given the FM-index of $\overline{T}$, $i'..j'$ can be found in $O(1)$ time by applying Lemma 3.6. Hence, we have the following lemma.

Lemma 3.7 *Consider a text T. We can simulate the suffix trie τ of T by representing every node u by $(|plabel(u)|, range(\overline{T}, \overline{plabel(u)}))$. Then, given the FM-index of $\overline{T}$, $child(u, c)$ can be computed in $O(1)$ time for every node u and every character c.*

For the implementation of $Leaf(u)$, we leave it as an exercise (see Exercise 11).

3.5.5   Bi-directional BWT

Consider a text T over an alphabet $\mathcal{A}$. Section 3.5.3 describes the FM-index that enables us to perform a backward search. Precisely, consider any pattern Q and any character $y \in \mathcal{A}$. Given $range(T, Q)$, the FM-index enables us to compute $range(T, yQ)$ in $O(1)$ time (see Lemma 3.6).

By symmetry, another operation is the forward search. For any pattern Q and any character y, given $range(T, Q)$, the forward search operation aims to compute $range(T, Qy)$. However, given the FM-index of T, the forward search operation cannot be solved efficiently.

Lam et al. [152] proposed the bi-directional BWT to solve this problem. Bi-directional BWT consists of two FM-indexes, that is, the FM-indexes for both T and $\overline{T}$. For every pattern Q, the bi-directional BWT aims to compute both $range(T, Q)$ and $range(\overline{T}, \overline{Q})$. The following is the key lemma.

Lemma 3.8 *Consider any pattern Q and any character $y \in \mathcal{A}$. Given the SA range $range(T, Q)$ and $range(\overline{T}, \overline{Q})$, we can compute*

(1) $range(T, yQ)$ and $range(\overline{T}, \overline{yQ})$

T =`ACGCG$`
$\overline{T}$ =`$GCGCA`

i	$SA_{\overline{T}}[i]$	Suffix of $\overline{T}$
1	1	`$GCGCA`
2	6	`A`
3	5	`CA`
4	3	`CGCA`
5	4	`GCA`
6	2	`GCGCA`

(a)

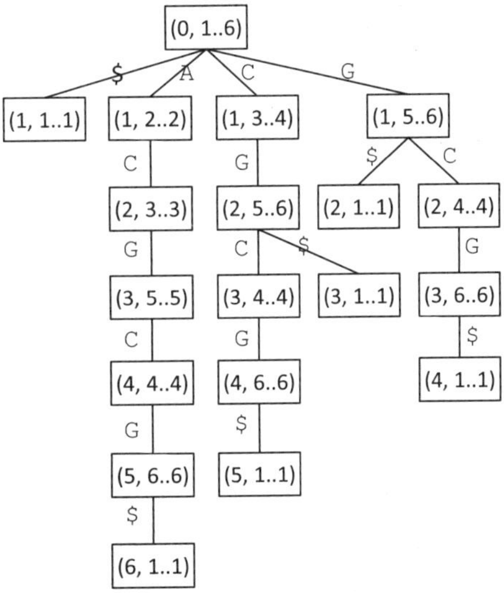

(b)

FIGURE 3.14: Consider a text T = `ACGCG$`. $\overline{T}$ = `$GCGCA` is the reverse of T. (a) is the suffix array of $\overline{T}$. (b) is the suffix trie τ of T.

(2) $range(T, Qy)$ and $range(\overline{T}, \overline{Qy})$

in $O(|\mathcal{A}|)$ time. (Note that $\overline{yQ} = \overline{Q}y$ and $\overline{Qy} = y\overline{Q}$.)

Proof We only illustrate the proof for (1). (2) can be proved similarly.

Let $i..j = range(T, Q)$ and $i'..j' = range(\overline{T}, \overline{Q})$. Suppose $\mathcal{A} = \{a_1, \ldots, a_{|\mathcal{A}|}\}$ where $a_1 < \ldots < a_{|\mathcal{A}|}$.

For every character $a_k \in \mathcal{A}$, $i_{a_k}..j_{a_k} = range(T, a_kQ)$ can be computed in $O(1)$ time by a backward search (see Lemma 3.6). Hence, using $O(|\mathcal{A}|)$ time, we compute $i_{a_k}..j_{a_k}$ for all $k = 1, \ldots, |\mathcal{A}|$.

It is easy to show that $range(\overline{T}, \overline{a_kQ}) = i'_{a_k}..j'_{a_k}$ where $j'_{a_k} = i' - 1 + \sum_{p=1}^{k}(j_{a_p} - i_{a_p} + 1)$ and $i'_{a_k} = i' + \sum_{p=1}^{k-1}(j_{a_p} - i_{a_p} + 1)$. The lemma follows. ∎

By the above lemma, if we maintain both $range(T, Q)$ and $range(\overline{T}, \overline{Q})$, we can perform a forward and a backward search in $O(|\mathcal{A}|)$ time.

By combining Lemma 3.7 and Lemma 3.8, we have the following lemma.

Lemma 3.9 *Consider a text T over an alphabet $\mathcal{A}$. We can simulate the suffix trie τ_T of T and the suffix trie $\tau_{\overline{T}}$ of $\overline{T}$ using a bi-directional BWT of T. Let Q be any substring of T. Let u and $\overline{u}$ be the nodes in τ_T and $\tau_{\overline{T}}$, respectively, such that $plabel_T(u) = Q$ and $plabel_{\overline{T}}(\overline{u}) = \overline{Q}$. Given $(u, \overline{u})$, we enable three operations:*

(1) For any character y, compute $(v, \overline{v})$ where $plabel_T(v) = Qy$ and $plabel_{\overline{T}}(\overline{v}) = \overline{Qy}$ in $O(|\mathcal{A}|)$ time.

(2) For any character y, compute $(v, \overline{v})$ where $plabel_T(v) = yQ$ and $plabel_{\overline{T}}(\overline{v}) = \overline{yQ}$ in $O(|\mathcal{A}|)$ time.

(3) Compute $Leaf_T(u)$ and $Leaf_{\overline{T}}(\overline{u})$ in $O(|Leaf_T(u)| \log^2 n)$ time.

Proof Let $Q = plabel_T(u)$. Observe that the node u in T is represented as $(|Q|, range(\overline{T}, \overline{Q}))$ while the node $\overline{u}$ in $\overline{T}$ is represented as $(|Q|, range(T, Q))$.

For (1), v is represented as $(|Q|+1, range(\overline{T}, \overline{Qy}))$ while $\overline{v}$ is represented as $(|Q|+1, range(T, Qy))$ (see Lemma 3.7). Given $range(\overline{T}, \overline{Q})$ and $range(T, Q)$, Lemma 3.8 shows that both $range(T, Qy)$ and $range(\overline{T}, \overline{Qy})$ can be computed in $O(|\mathcal{A}|)$ time.

For (2), $(v, \overline{v})$ can similarly be computed in $O(|\mathcal{A}|)$ time. For (3), see Exercise 11. ∎

3.6 Data compression techniques

This section describes some coding schemes that are used in Chapter 10.

3.6.1 Data compression and entropy

Data compression aims to reduce the total space to store a text T over some alphabet $\mathcal{A}$. Note that different characters have different numbers of occurrences. One can use fewer bits to represent frequent characters in T and more bits to represent rare characters in T. Then, we can reduce the storage size. Let $p(x)$ be the frequency for every character $x \in \mathcal{A}$ ($0 \leq p(x) \leq 1$). The optimal coding scheme encodes each character x using $-\log_2 p(x)$ bits. The entropy is defined to be the average number of bits per character when the text T is stored using the optimal coding scheme. Precisely, the entropy equals $H_0(T) = -\sum_{x \in T} p(x) \log_2 p(x)$. If the entropy is small, the data is compressible.

For example, consider a sequence $X = (c, b, d, c, b, d, a, a)$. $p(a) = p(b) = p(c) = p(d) = 2/8$. Hence, the entropy of X is $-2/8 \log_2(2/8) - 2/8 \log_2(2/8) - 2/8 \log_2(2/8) - 2/8 \log_2(2/8) = 2.0$ bits. To achieve the entropy, we represent a by 00, b by 01, c by 10 and d by 11. In other words, the sequence X is represented by $(10, 01, 10, 11, 00, 00, 01, 11)$, which requires 16 bits (or 2 bits per character).

Consider another sequence $Y = (c, b, d, a, b, a, b, b)$. $p(a) = 2/8, p(b) = 4/8$ and $p(c) = p(d) = 1/8$. The entropy of Y equals $-2/8 \log_2(2/8) -$

$4/8 \log_2(4/8) - 1/8 \log_2(1/8) - 1/8 \log_2(1/8) = 1.75$ bits. To achieve the entropy, we represent a by 01, b by 1, c by 000 and d by 001. In other words, the sequence Y is represented by $(000, 1, 001, 01, 1, 01, 1, 1)$, which requires 14 bits (or 1.75 bits per element). Hence, sequence Y is more compressible and has a smaller entropy.

Compression schemes exist that can encode a text T optimally or near optimally. The Huffman code (see Section 3.6.4) encodes the text T using at most $n(H_0(T) + 1)$ bits. The arithmetic code (see Section 3.6.5) encodes the text T optimally using $nH_0(T)$ bits.

The above discussion assumes the symbols are independent. However, the occurrence of some symbol may depend on the preceding symbols. For example, in English, after the character "q", it is highly probable that the next character is "u". Also, after the characters "th", it is likely that the next character is "e". If the coding scheme is allowed to encode the symbols using a different number of bits depending on its k preceding symbols (called context), then we can further reduce the storage space. As such, the minimum number of bits required per symbol is called the order-k entropy, which is defined as follows. If a string $s \in \mathcal{A}^k$ precedes a symbol c in T, s is called the context of c. We denote $T^{(s)}$ to be the string formed by concatenating all symbols whose contexts are s. The order-k entropy of T is defined as $H_k(T) = \frac{1}{n}\sum_{s \in \mathcal{A}^k} |T^{(s)}| H_0(T^{(s)})$. Note that $H_{k+1}(T) \le H_k(T)$ for any $k \ge 0$.

For example, for the sequence $X = (c, b, d, c, b, d, a, a)$ above, $X^{(a)} = (a), X^{(b)} = (d, d), X^{(c)} = (b, b), X^{(d)} = (c, a)$. We have $H_0(X^{(a)}) = 0, H_0(X^{(b)}) = 0, H_0(X^{(c)}) = 0$ and $H_0(X^{(d)}) = -1/2 \log_2(1/2) - 1/2 \log_2(1/2) = 1$. The order-1 entropy $H_1(X) = \frac{1}{8}(1 \cdot H_0(X^{(a)}) + 2 \cdot H_0(X^{(b)}) + 2 \cdot H_0(X^{(c)}) + 2 \cdot H_0(X^{(d)}) = 0.25$, which is smaller than $H_0(X) = 2$.

Manzini [187] showed that, for any compression scheme that encodes each symbol in T depending on the symbol itself and its length-k context, the storage of T requires at least $nH_k(T)$ bits. A number of popular compression methods like gzip and bzip can encode the text T using $nH_k(T) + o(n)$ bits.

3.6.2 Unary, gamma, and delta coding

There are many coding schemes. Here, we present 3 simple coding schemes for encoding a nature number w (i.e. $w \ge 1$): unary code, gamma code and delta code. All of them are prefix code (i.e., no code word is the prefix of another code word).

Unary code transforms w to $(w-1)$'s 1-bits followed by a single 0-bit. Thus, unary code uses w bits to represent the integer w. For example, an integer 4 is represented by 1110.

Gamma code represents w by $(v, w - 2^v)$ where $v = \lfloor \log_2 w \rfloor$. Gamma code stores (1) the unary code for $v + 1$ using $v + 1$ bits and (2) the binary code for $(w - 2^v)$ using v bits. In total, gamma code uses $2\lfloor \log_2 w \rfloor + 1$ bits. For example, an integer 14 is represented by $(v = 3, w - 2^v = 6) = 1110110$.

Delta code represents w by (1) gamma code for $v + 1$ where $v = \lfloor \log_2 w \rfloor$

(which uses $2\lfloor \log_2 v \rfloor + 1$ bits) and (2) binary code for $w - 2^v$ (using v bits). In total, delta code uses $2\lfloor \log_2(\lfloor \log_2 w \rfloor + 1) \rfloor + \lfloor \log_2 w \rfloor + 1$ bits. For example, an integer 50 is represented by $(v = 5, w - 2^v = 18) = (11010, 10010) = 1101010010$.

When the numbers are mostly small, unary code is a good compression scheme. When the numbers are big, we can use gamma code. When the numbers are very big, we can use delta code.

The following example gives encoding of a sequence $4, 9, 5, 11$ in unary code, gamma code and delta code.

- Unary code: 1110, 111111110, 11110, 11111111110

- Gamma code: 11000, 1110001, 11001, 1110011

- Delta code: 10100, 11000001, 10101, 11000011

3.6.3   Golomb code

Golomb code is also a prefix code. It is highly suitable when the input has significantly more small values than large values. (More precisely, Golomb code has the best performance when the frequency distribution of the values follows the geometric distribution.) Golomb code requires a parameter M. Then, any integer w is represented by (1) the unary code of $q + 1$ and (2) a binary code of $r = w - qM$, where $q = \lfloor w/M \rfloor$. We code r differently depending on the value of r. When $r < 2^b - M$ where $b = \lceil \log_2 M \rceil$, we code r as a plain binary using $b - 1$ bits; otherwise, we code the number $2^b + r - M$ in plain binary representation using b bits. (Note that r (i.e., the second part of the Golomb code) can be represented using either b bits or $b - 1$ bits. During decoding, if the first bit of the second part is 1, we know that r is represented in b bits; otherwise, r is represented in $b - 1$ bits.)

For example, suppose the parameter $M = 10$. Then, $b = \lfloor \log_2 M \rfloor = 4$. The integer 42 can be represented by $(q = \lfloor 42/10 \rfloor = 4, r = 42 - Mq = 2)$. Since $2 < 2^4 - 10$, 42 is encoded by the unary code of $4 + 1$ and the binary code of 2 using 3 bits, that is, 11110 010. As another example, the integer 39 is represented by $(q = 3, r = 9)$. Since $9 > 2^4 - 10$, 39 is encoded by the unary code of $3 + 1$ and the binary code of $2^4 + 9 - 10$ using 4 bits, that is, 1110 1111.

One issue of Golomb code is to set the parameter M. To obtain a small size, we set M to be the mean of all encoded integers. Note that when M equals the power of 2, the code is also called Rice code.

3.6.4   Huffman coding

Huffman coding is a prefix code that encodes high-frequency symbols with shorter codes and low-frequency symbols with longer codes. The codes of the symbols can be represented by a tree called the Huffman tree. For example, consider the sentence "apple is a health fruit." The frequencies of the alphabets

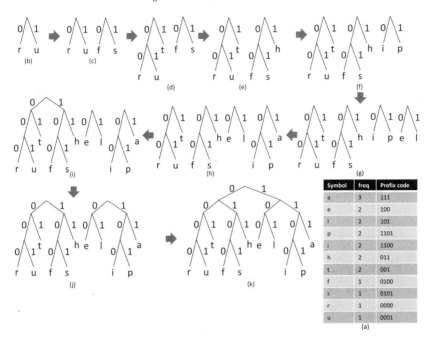

FIGURE 3.15: (a) is the table of the frequencies of the alphabets in the sentence "apple is a health fruit." (b−j) illustrate how to construct the Huffman tree. (j) is the Huffman tree of the sentence.

in the sentence are shown in Figure 3.15(a). Figure 3.15(j) gives the Huffman tree of the sentence. Every path from the root to the leaf is a prefix code of a symbol. For example, the prefix code of the symbol s is 0101. Based on the Huffman codes for the symbols, we can encode the sentence as

111 1101 1101 101 100 1100 0101 111 011 100 111 101 001 011 0100 0000 0001 1100 001

The Huffman tree of a text T can be constructed by a greedy algorithm as follows. Let $\mathcal{A}$ be the set of characters that appear in T. Let $f(x)$ be the frequency of x for every character $x \in \mathcal{A}$. To build the Huffman tree, a leaf node is created for every character in $\mathcal{A}$ and all nodes are included in the priority queue Q. The Huffman tree can be built in $|\mathcal{A}| - 1$ iterations. Every iteration performs 4 steps. First, two nodes x and y are extracted from Q which have the lowest frequencies among all nodes in Q. Second, a new node z is created and x and y are set to be its children. Third, the frequency of z is denoted as $f(z) = f(x) + f(y)$. Fourth, z is included in the priority queue. After $|\mathcal{A}| - 1$ iterations, the Huffman tree is obtained. The detail of the algorithm is shown in Figure 3.16. Figure 3.15(b−j) demonstrates the steps to build the Huffman tree for the text "apple is a health fruit."

When every symbol is encoded separately with a prefix code, Huffman coding is an optimal encoding scheme. Hence, Huffman code is popularly

---

Algorithm HuffmanTree(T)

Require: a text T

Ensure: A Huffman tree for the text T

1: Let $\mathcal{A}$ be the set of characters appear in T; Let $f(x)$ be the frequency of every character $x \in \mathcal{A}$;

2: Every $x \in \mathcal{A}$ is a leaf node and is included in a priority queue Q;

3: **while** Q contains at least two nodes **do**

4:    Extract two nodes x and y which have the lowest frequencies among all nodes in Q;

5:    Create a new node z and set x and y be its children;

6:    $f(z) = f(x) + f(y)$;

7:    Insert z into the priority queue Q;

8: **end while**

9: Let r be the remaining node in Q and return the tree rooted at r;

FIGURE 3.16: The algorithm for constructing the Huffman tree.

used for lossless data compression. However, Huffman is not optimal when we don't have the symbol-by-symbol encoding restriction. In particular, when the alphabet size is small, Huffman code can have a lot of redundancy. "Blocking," i.e., grouping multiple symbols, can reduce the redundancy. (See Exercise 15.)

3.6.5   Arithmetic code

Huffman code assigns a code to each character. This approach may waste some space. For example, consider a sequence with 3 symbols A, B, C which occur in the same frequency. By Huffman, one symbol (say A) uses 1 bit while the other two symbols (say B and C) each use 2 bits. Hence, each symbol uses $\frac{5}{3} = 1.666$ bits on average. By entropy (see Section 3.6.1), each symbol should use $\log_2 3 = 1.58496$ bits on average. This means that Huffman code may waste some space.

To reduce the wastage, we can use arithmetic code. Arithmetic code does not encode symbol by symbol. It encodes an entire sequence into a real number in $[0, 1)$.

Let the alphabet $\mathcal{A} = \{a_1, \dots, a_m\}$. Let $f(a_i)$ be the frequency of a_i for every symbol $a_i \in \mathcal{A}$. Let $F_0 = 0$ and $F_i = \sum_{j=1}^{i} f(a_i)$ for $i = 1, \dots m$. (Note that $F_m = 1$.) In arithmetic code, a text T is represented by a range $[p, q)$ where $0 \le p \le q < 1$. The range corresponds to a text that can be recursively defined as follows. The empty text is represented by the range $[0..1)$. The one symbol text a_i is represented by $[F_{i-1}, F_i)$. Recursively, if a text T is represented by a range $[p..q)$, then the text Ta_i is represented by a range $[p + (q - p)F_{i-1}, p + (q - p)F_i)$. Suppose T is represented by $[p..q)$. Then, the arithmetic code for T is the binary representation of p. Figure 3.17 details the algorithm to compute the arithmetic code for a text T.

Algorithm ArithmeticCoder(T)
Require: T is a text over an alphabet $\{a_1, \ldots, a_m\}$
Ensure: The arithmetic code for T
1: Set $p = 0$; $q = 1$;
2: Set $f(a_i)$ be the frequency of a_i in T for $i = 1, \ldots, m$;
3: Set $F_0 = 0$ and $F_i = \sum_{j=1}^{i} f(a_i)$ for $i = 1, \ldots m$;
4: **while** T is not an empty string **do**
5:    Suppose a_i is the first symbol in T;
6:    Set $[p, q) = [p + (q - p)F_{i-1}, p + (q - p)F_i)$;
7:    Delete the first symbol from T;
8: **end while**
9: Report the shortest binary sequence for v among all values $v \in [p, q)$;

FIGURE 3.17: The algorithm ArithmeticCoder(T) computes the arithmetic code for T.

For example, consider a text CAGC. Figure 3.18 shows the steps to obtain its arithmetic code. Denote $a_1 = $ A, $a_2 = $ C and $a_3 = $ G. $f(\text{A}) = f(a_1) = 0.25$, $f(\text{C}) = f(a_2) = 0.5$, $f(\text{G}) = f(a_3) = 0.25$. Then, $F_0 = 0$, $F_1 = 0.25$, $F_2 = 0.75$ and $F_3 = 1$.

- The sequence C is represented by $[0.25, 0.75)$.

- The sequence CA is represented by $[0.25 + 0.5 * 0, 0.25 + 0.5 * 0.25) = [0.25, 0.375)$.

- The sequence CAG is represented by $[0.25 + 0.125 * 0.75, 0.25 + 0.125 * 1) = [0.34375, 0.375)$.

- The sequence CAGC is represented by $[0.34375 + 0.03125 * 0.25, 0.34375 + 0.03125 * 0.75) = [0.3515625, 0.3671875)$.

Then, CAGC can be represented by the shortest binary sequence for v among all values $v \in [0.3515625, 0.3671875)$. The binary code of 0.3515625 is 0.0101101_2. The binary code of 0.3671875 is 0.0101111_2. The shortest binary sequence among all binary values in $[0.328125, 0.34375)$ is 0.010111_2. Hence, the arithmetic code for CAGC is 010111.

The decoding algorithm for arithmetic code is as follows. We first convert the arithmetic code back to a value v in $[0, 1)$. Then, the sequence can be decoded symbol by symbol iteratively. Initially, the sequence is an empty string and it corresponds to the region $[0, 1)$. The region is divided into subregions proportional to the symbol frequencies. By checking the region containing v, we can decode the first symbol. Precisely, the first symbol is a_i if $v \in [F_{i-1}, F_i)$. Then, the region for the sequence is $[F_{i-1}, F_i)$. Iteratively, for $j = 2, \ldots, n$, the jth iteration decodes the jth symbol and computes the region that corresponds to the first j characters. Precisely, let $[p, q)$ be the interval

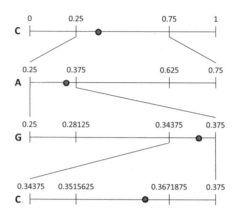

FIGURE 3.18: Illustration of the encoding and decoding of arithmetic code. The sequence is CAGC. The dot is $0.010111_2 = 0.359375$.

that corresponds to the first $j - 1$ characters. $[p, q)$ is divided into subregions proportional to the symbol frequencies. Then, the jth symbol is a_i if $v \in [p + (q - p)F_{i-1}, p + (q - p)F_i)$. The interval that corresponds to the first j characters is $[p + (q-p)F_{i-1}, p + (q-p)F_i)$. Figure 3.18 illustrates the decoding of the arithmetic code 010111.

As a final note, arithmetic code can achieve entropy with at most 2 bit difference.

3.6.6 Order-k Markov Chain

Although arithmetic coding can achieve the optimal order-0 entropy, it still has redundancy when adjacent characters have dependency. In this situation, we can use Markov encoding to improve the compression. Order-k Markov encoding encodes every character depending on its preceding k characters (called the context). Given a text T, for every length-k substring $X \in \mathcal{A}^k$ and every $x \in \mathcal{A}$, we estimate $Pr(x|X) = \frac{\#Xx}{\#X}$, where $\#Xx$ and $\#X$ are the number of occurrences of Xx and X, respectively, in T. Then, for every position i in T, $T[i]$ can be encoded using Huffman code or arithmetic code using $- \log(Pr(T[i]|T[i - k..i - 1]))$ bits. The order-k Markov encoding can achieve the optimal order-k entropy.

Below, we demonstrate how to encode $T =$ CATGTCATCTCATTACC using order-3 Markov encoding. We denote a k-mer to be a length-k DNA sequence. Figure 3.19(a) gives the frequencies of all 4-mers $b_1 b_2 b_3 b_4$ of T. It also gives $Pr(b_4|b_1 b_2 b_3)$. For example, consider $b_1 b_2 b_3 = CAT$. There are three 4-mers that begin with CAT and they are CATC, CATG and CATT. So, after CAT, it can be either C, G or T. We have $Pr(\text{C}|\text{CAT}) = \frac{\#\text{CATC}}{\#\text{CAT}} = 1/3, Pr(\text{G}|\text{CAT}) = \frac{\#\text{CATG}}{\#\text{CAT}} = 1/3, Pr(\text{T}|\text{CAT}) = \frac{\#\text{CATT}}{\#\text{CAT}} = 1/3$. Using Huffman code, we can encode C|CAT, G|CAT, T|CAT by 00, 01, 1, respectively. Another example is $b_1 b_2 b_3 =$ ATC.

$b_1b_2b_3b_4$	occ	$Pr(b_4\|b_1b_2b_3)$	Huffman code of $b_4\|b_1b_2b_3$
ATCT	1	1	ϵ
ATGT	1	1	ϵ
ATTA	1	1	ϵ
CATC	1	1/3	00
CATG	1	1/3	01
CATT	1	1/3	1
CTCA	1	1	ϵ
GTCA	1	1	ϵ
TACC	1	1	ϵ
TCAT	2	1	ϵ
TCTC	1	1	ϵ
TGTC	1	1	ϵ
TTAC	1	1	ϵ

(a)

$T =$	CAT	G	T	C	A	T	C	T	C	A	T	T	A	C	C
Encoding	010011	01					00					1			

(b)

FIGURE 3.19: (a) The frequency table of all 4-mers $b_1b_2b_3b_4$ in $T =$ CATGTCATCTCATTACC. It also shows $Pr(b_4|b_1b_2b_3)$ and the Huffman code of $b_4|b_1b_2b_3$. (b) The order-3 Markov encoding of T.

There is only one length-4 substring begin with ATC and it is ATCT. So, we have $Pr(\text{T}|\text{ATC}) = 1$. We don't need to use any bits to encode T|ATC.

To store the order-3 Markov encoding of $T =$ CATGTCATCTCATTACC, we need to store the Huffman tree for every length-3 context and the actual encoding. The actual encoding has two parts (see Figure 3.19(b)): (1) the encoding of the first 3 bases of T (i.e., CAT) is 010011 and (2) the Huffman encoding of the remaining 14 bases is 01001.

3.6.7 Run-length encoding

For some sequence, it may have many runs of the same values. One example is the BWT sequence, which is known to have many runs of the same character.

When a sequence contains many runs of the same values, we can compress the sequence using run-length encoding. In run-length encoding, each run of a value v of length ℓ is stored as a pair (ℓ, v). For example, for the following sequence, its run-length encoding is 5C17G1F2<3@6#2:,1:.

CCCCCGGGGGGGGGGGGGGGGGF<<@@@#######::,,:

3.7   Exercises

1. We aim to compute the x^n in $O(\log n)$ time. Propose your method by answering the following three questions:

   - Can you formulate the subproblem?
   - Can you give the recursive formula? (You need to give the formula for both the base case and the recursive case.)
   - Can you give an efficient algorithm to solve this problem? (Your algorithm must be either a dynamic programming algorithm or a recursive algorithm.) What is the running time of your algorithm?

2. Given an array of n integers $A[1..n]$. A sequence $A[i_1], A[i_2], \ldots, A[i_k]$ is called an increasing subsequence if $1 \le i_1 < i_2 < \ldots < i_k \le n$ and $A[i_1] \le \ldots \le A[i_k]$. We aim to find the length of the longest increasing subsequence of $A[1..n]$. Propose your method by answering the following three questions:

   - Can you formulate the subproblem?
   - Can you give the recursive formula? (You need to give the formula for both the base case and the recursive case.)
   - Can you give an efficient algorithm to solve this problem? (Your algorithm must be either a dynamic programming algorithm or a recursive algorithm.) What is the running time of your algorithm?

3. Show that $\frac{Pr(X|\theta')}{Pr(X|\theta)} \ge \frac{Q(\theta'|t_i)}{Q(\theta|t_i)}$.

4. For the algorithm in Figure 3.5, what are the values of $\theta_A, \theta_B, \lambda$ after 4 iterations assuming the following two initial conditions?

   - Initial condition (a): $\theta_A = 0.5, \theta_B = 0.5, \lambda = 0.5$.
   - Initial condition (b): $\theta_A = 0.3, \theta_B = 0.4, \lambda = 0.3$.

5. Consider a set of n heights $\{X_1, \ldots, X_n\}$. Assume these heights are measured from a mixture of children and adults. Let $Z_i = 1$ if X_i is measured from a child; and 0, otherwise. Let λ be the chance that the height is measured from a child. Suppose the height of the children follows a normal distribution $N(\mu_1, \sigma_1)$ and the height of the adults follows another normal distribution $N(\mu_0, \sigma_0)$. Can you propose an algorithm to estimate $\theta = (\mu_0, \sigma_0, \mu_1, \sigma_1, \lambda)$?

6. Show that Lemma 3.3 is correct.

7. Consider a text $T[1..n]$. Let $SA[1..n]$ be the suffix array of T. Suppose the SA range $range(T, Q) = s..e$, For any character c, show that $range(T, Qc)$ can be computed in $O(\log n)$ time.

8. Show that the algorithm BWT2Text in Figure 3.11 can correctly invert the BWT $B[1..n]$ to the original text $T[1..n]$.

9. Please prove Lemma 3.6.

10. Given the FM-index of a length-n text T and the $D[]$ array in Section 3.5.3.2. Let $\Psi(i) = SA^{-1}[SA[i] + 1]$. Can you give an algorithm that computes $\Psi(i)$ in $O(\log^2 n)$ time?

11. Consider a length-n text T over an alphabet $\mathcal{A}$. Can you propose a way to simulate the suffix trie τ of T using the FM-index? Please ensure your data structure allows the following operations:

    - For any node u in τ and any $\sigma \in \mathcal{A}$, $child(u, \sigma)$ (i.e. return the child v of u such that the edge label of (u, v) is σ) can be computed in $O(1)$ time.
    - For any node u in τ, $Leaf(u)$ can be computed in $O(|Leaf(u)| \log^2 n)$ time.

12. For a sequence T, can you show that $H_k(T) \geq H_{k+1}(T)$?

13. Can you give the unary code, gamma code, delta code and golomb code for one to twenty? (For golomb code, assume $M = 5$)

14. Consider the following sequence ACACGACTAA. What is the Huffman code for this sequence? What is the length of the encoding?

15. Refer to the above question. If every two characters are grouped in one block, what is the Huffman code for the blocks? Is its encoding shorter than the Huffman code for the original sequence?

16. Consider the following text $T = 949496497$. (1) Give the gamma code encoding of T. (2) Give the Huffman tree and the Huffman code encoding of T. Of (1) and (2), which encoding is shorter?

17. Consider a text T over an alphabet $\mathcal{A} = \{a_1, \ldots, a_k\}$. For $i = 1, \ldots, k$, suppose the frequency of a_i is $\frac{1}{2^{\delta_i}}$, where δ_i is an integer. Can you show that Huffman code is optimal (i.e. achieves zero-order entropy)?

18. Create sequences for the following cases:

    - Give a sequence such that both arithmetic code and Huffman code are of the same length.
    - Give a sequence such that Huffman code is longer than Arithmetic code.

19. Can you give the arithmetic code for CACG?

20. Please give an algorithm to decode arithmetic code.

Chapter 4

NGS read mapping

4.1 Introduction

Read mapping is the process to align NGS reads on a reference genome. The input consists of a reference genome T and a set of reads $\mathcal{R}$ (represented as a fasta or fastq file (see Section 2.2)). The aim is to align each read in $\mathcal{R}$ on the reference genome T, allowing mismatches, indels, and clipping of some short fragments on the two ends of the read. The output is described in SAM/BAM format (see Section 2.3). Figure 4.1 illustrates the alignments of three reads.

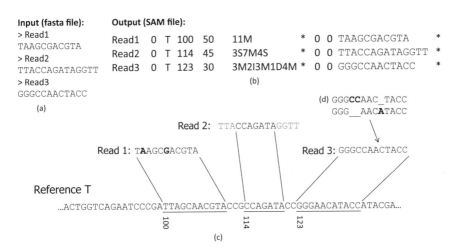

FIGURE 4.1: An example that illustrates the read alignment process. The input consists of three reads represented as a fasta file (see (a)) and a reference genome T. Figure (c) gives the mapping locations of the three reads on T. The first read is aligned at position 100 and the alignment has two mismatches. The second read is aligned at position 114. It is a local alignment with clippings on the left and on the right. The third read is aligned at position 123. It consists of a 2-base insertion and a 1-base deletion. The alignment of the three reads can be described in SAM format (see (b)). (d) visualizes the alignment between read 3 and $T[123..133]$.

Many NGS applications require read mapping, including genome variation calling, transcriptome analysis, transcription factor binding site calling, epigenetic mark calling, metagenomics, etc. Accurate alignment affects the performance of these applications. Furthermore, since the number of reads is huge, it is also important that the mapping step is efficient.

This chapter discusses the computational techniques to address the issues of alignment accuracy and efficiency. It is organized as follows. First, we give an overview of the read mapping problem and give simple brute-force solutions. Then, depending on whether gaps are allowed in the read alignment, different computational methods are given to solve the read mapping problem.

4.2   Overview of the read mapping problem

This section gives an overview of the read mapping problem. Sections 4.2.1 and 4.2.2 define different read mapping computational problems. Section 4.2.3 gives brute-force solutions to them. Section 4.2.4 introduces the mapQ score that evaluates if the hit is by chance or not. Finally, the computational challenges are discussed in Section 4.2.5.

4.2.1   Mapping reads with no quality score

Consider a reference genome $T[1..n]$. For every read $R[1..m]$, the read mapping problem aims to find a region in T that has the highest similarity with the read R. Such a region is called the best hit of R. Roughly speaking, there are three computational problems for aligning reads: the Hamming distance read mapping problem, the edit-distance read mapping problem and the read alignment problem.

Hamming distance read mapping problem: This problem aligns each read on the reference genome allowing mismatches only. Precisely, given a read $R[1..m]$ and a reference genome $T[1..n]$, the mismatch-only read alignment problem aims to find a position i that minimizes $Hamming(R, T[i..i+m-1])$, where $Hamming(R, T[i..i + m - 1])$ is the number of mismatches between R and $T[i..i + m - 1]$. For example, in Figure 4.1, to minimize the Hamming distance, read1 (TAAGCGACGTA) is aligned to position 100 with 2 mismatches.

Edit-distance read mapping problem: Note that our genome is expected to have 1 indel in every 3000 bp [122]. When the read is long, we cannot ignore indels. In this scenario, we define the edit-distance read mapping problem (see Section 3.2.2). Given a read $R[1..m]$ and a reference genome $T[1..n]$, the edit-distance read mapping problem aims to find a region $i..j$ in T that minimizes $edit(R, T[i..j])$, where $edit(R, T[i..j])$ is the minimum number of edit operations (i.e., insert a base, delete a base, and replace a base) that can transform R to $T[i..j]$. For example, in Figure 4.1, to minimize the edit distance, read3 (GGGCCAACTACC) is aligned to position 123 with 2 deletions and 1 insertion.

Read alignment problem: The problem of aligning reads allowing indels can be generalized to the read alignment problem. There are two common alignment schemes: semi-global alignment and local alignment. The semi-global alignment problem aims to align R to a region $T[i..j]$ that maximizes the alignment score between R and $T[i..j]$. An alignment between R and $T[i..j]$ is formed by inserting spaces in arbitrary locations along these two sequences so that they end up having the same length (with no position aligning space with space). Figure 4.1(d) illustrates the alignment between read3 and $T[123..133]$, which contains three spaces. Each aligned position is assigned a score. For positions aligning nucleotides x and y, a substitution score $\delta(x, y)$ is assigned. For positions aligning nucleotides with ℓ consecutive spaces, an affine gap penalty score $g(\ell)$ is assigned (i.e., score = gapopen $+\ell*$ gapextend where gapopen is the penalty for having a gap while gapextend is the penalty for having a gap of length ℓ). The alignment score equals the sum of scores of all aligned positions.

For example, assume $\delta(x, y) = +3$ if $x = y$; and $\delta(x, y) = -2$ otherwise and $g(\ell) = -3 - \ell$, In Figure 4.1(d), to maximize the semi-global alignment score, read3 (GGGCCAACTACC) is aligned to $T[123..133]$, with alignment score 18.

Apart from indels, when the read is long, it is more susceptible to structural variations (see Chapter 7) and misassemblies in the reference genome. We cannot assume that the read is fully aligned to a consecutive region in T. In such case, we can use local alignment. The local alignment problem aims to maximize the alignment score between R' and T' among all substrings R' of R and all substrings T' of T. For example, in Figure 4.1, to maximize the local alignment score, the middle portion of read2 (i.e., CCAGATA) is aligned to $T[114..120]$, with alignment score 21.

4.2.2 Mapping reads with a quality score

The previous subsection aligns reads considering their DNA bases only. Recall that every base in a read is associated with a PHRED quality score. When the quality score of a base is low, the base is likely to be a sequencing error. This principle has been shown to improve the alignment accuracy [279, 171] and motivates us to perform a quality score-aware alignment.

To avoid the effect of low quality bases, the simplest solution is to trim reads by their quality scores before we perform the read mapping. For instance, we can trim the bases at the two ends of the Illumina reads if their quality scores are not good. For the example read R in Figure 4.2, this approach will trim the first base of R since its quality score is only 10.

Read trimming will remove useful information. Instead, a number of methods use the quality score to improve alignment. RMAP [279] proposed to simply give zero penalty for a base mismatch when the quality score is smaller than some fixed threshold. For the example read R in Figure 4.2, if we give zero penalty to bases with quality scores ≤ 10, the best alignment of R is at position 6 of T.

10 40 40 40 40 10 40 30 40 40

Read R: A C G A C T T G C A

Reference T: CCACG<u>TCGACA</u>TGCATCCTCTC<u>ACGCC</u>TTTCATA

FIGURE 4.2: This example illustrates the alignment of a read R on the reference T. Each base of the read R is associated with a quality score. R has two 2-mismatch hits at positions 6 and 23 on T. The mismatch positions are in bold font. For the left alignment (at position 6), the quality scores of the two mismatch positions are both 10. For the right alignment (at position 23), the quality scores of the two mismatch positions are 40 and 30.

Bowtie 1 [155] proposed to align a read to a region with a number of mismatches smaller than some threshold while the total sum of quality scores of the mismatch positions is minimized. For example, the read in Figure 4.2 has two 2-mismatch hits. For the left alignment (at position 6), the sum of quality scores of the two mismatches is $10 + 10 = 20$. For the right alignment (at position 23), the sum of quality scores of the two mismatches is $40 + 30 = 70$. The left alignment is expected to be better.

Another idea is to give less penalty for a mismatch if the position has a low quality score (since such mismatch is likely to be error-prone). For instance, Novoalign, BWA-SW [168], and BatAlign [178] use the base qualities to calculate the error probabilities and to determine the base penalty. Suppose $Q[1..m]$ is the array of m quality scores for the read $R[1..m]$. Recall that the error probability of the ith base equals $P[i] = 10^{-\frac{Q[i]}{10}}$. The base penalty $\delta(R[i], x)$ is defined as follows.

$$\delta(R[i], x) = \begin{cases} -10\log_{10}(1 - P[i]) & \text{if } R[i] = x \\ -10\log_{10}\left(\frac{P[i]}{3}\right) & \text{otherwise} \end{cases}$$

For example, consider the read R in Figure 4.2. The penalty for the left alignment (at position 6) is $-70\log_{10}\left(1 - 10^{-4}\right) - 10\log_{10}\left(1 - 10^{-3}\right) - 20\log_{10}\left(\frac{10^{-1}}{3}\right) = 29.55$. The penalty for the right alignment (at position 23) is $-60\log_{10}\left(1 - 10^{-4}\right) - 20\log_{10}\left(1 - 10^{-1}\right) - 10\log_{10}\left(\frac{10^{-4}}{3}\right) - 10\log_{10}\left(\frac{10^{-3}}{3}\right) = 80.46$. The left alignment has a smaller penalty and it is better.

4.2.3 Brute-force solution

Sections 4.2.1 and 4.2.2 defined several computational problems: the Hamming distance read mapping problem, the edit-distance read mapping problem, and the read alignment problem. Below, we briefly describe the brute-force solutions to these problems.

Hamming distance read mapping problem: For a length-m read, for a fixed position i in the genome $T[1..n]$, $Hamming(R, T[i..i + m − 1])$ can be computed in $O(m)$ time. Hence, we can find the position i that minimizes $Hamming(R, T[i..i + m − 1])$ using $O(mn)$ time. Suppose the maximum allowed Hamming distance is p. Amir et al. [7] gave an algorithm that solves the p-Hamming distance problem in $O(\sqrt{p \log p} n)$ time.

Edit-distance read mapping problem: The edit distance $edit(R[1..m], T[i..j])$ can be computed in $O(m(j − i + 1))$ time by dynamic programming using the algorithm in Figure 3.3. By modifying this algorithm, we can find the interval $i..j$ that minimizes the edit distance using $O(mn)$ time (see Exercise 2). Suppose we aim to find all intervals $i..j$ such that $edit(R[1..m], T[i..j]) \leq p$. This problem can be solved in $O(pn)$ time using the algorithm by Landau and Vishkin [153].

Semi-global alignment problem: Consider a read $R[1..m]$ and a reference genome $T[1..n]$. The semi-global alignment problem aims to find $T[j'..j]$ that maximizes the alignment score between R and $T[j'..j]$. This problem can be solved using dynamic programming. For $1 \leq i \leq n$ and $1 \leq j \leq m$, we define $V(i, j)$ as the maximum global alignment score between $R[1..i]$ and $T[j'..j]$ among $1 \leq j' \leq j$. The optimal semi-global alignment score between R and T is $\max_{j=0}^{n} V(m, j)$. The following lemma states the recursive formula for $V(i, j)$.

Lemma 4.1 *Base case: When either $i = 0$ or $j = 0$, we have*

$$
\begin{aligned}
V(0, j) &= 0 && \text{for } 0 \leq j \leq m \\
V(i, 0) &= \sum_{i'=1}^{i} \delta(R[i'], -) && \text{for } 1 \leq i \leq n.
\end{aligned}
$$

Recursive case: When both $i > 0$ and $j > 0$, we have

$$
V(i, j) = \max \begin{cases}
V(i − 1, j − 1) + \delta(R[i], T[j]) & \texttt{match/mismatch} \\
V(i − 1, j) + \delta(R[i], -) & \texttt{delete} \\
V(i, j − 1) + \delta(-, T[j]) & \texttt{insert.}
\end{cases}
$$

Proof By definition, the base cases are correct.

Below, we derive the equation for the recursive case where both $i > 0$ and $j > 0$. Among all suffixes of $R[1..i]$ and all suffixes of $T[1..j]$, suppose the alignment A between $R[i'..i]$ and $T[j'..j]$ has the highest alignment score. The rightmost column of the alignment A is either (1) substitution, (2) deletion or (3) insertion.

- For (1), after removing the rightmost substitution, the remaining alignment is between $R[i'..i−1]$ and $T[j'..j−1]$ of score $V(i−1, j−1)$. Hence, the alignment score of A is $V(i − 1, j − 1) + \delta(R[i], T[j])$.

- For (2), after removing the rightmost deletion, the remaining alignment is between $R[i'..i-1]$ and $T[j'..j]$ of score $V(i-1,j)$. Hence, the alignment score of A is $V(i-1,j) + \delta(R[i],-)$.

- For (3), after removing the rightmost insertion, the remaining alignment is between $R[i'..i]$ and $T[j'..j-1]$ of score $V(i,j-1)$. Hence, the alignment score of A is $V(i,j-1) + \delta(-,T[j])$.

$V(i,j)$ is the maximum among the three cases. The lemma follows. ∎

By dynamic programming, we can compute $V(i,j)$ for $0 \le i \le n$ and $0 \le j \le m$ by applying Lemma 4.1. Each entry can be computed in $O(1)$ time. Hence, the semi-global alignment score can be computed in $O(nm)$. We can generalize the gap penalty to the affine gap penalty (see Exercise 3). The running time is of the same order.

Local alignment problem: The local alignment problem can be solved using the Smith-Waterman alignment algorithm. Consider a read $R[1..m]$ and a reference genome $T[1..n]$. The Smith-Waterman algorithm defines a recursive formula V such that, for $0 \le i \le n$ and $0 \le j \le m$, $V(i,j)$ equals the maximum score of the global alignment between A and B over all suffixes A of $S[1..i]$ and all suffixes B of $T[1..j]$. Note that an empty string is assumed to be a suffix of both $S[1..i]$ and $T[1..j]$. The optimal local alignment score equals $\max_{i'=0}^{m} \max_{j'=0}^{n} V(i',j')$. The following lemma states the recursive formula for $V(i,j)$.

Lemma 4.2 *Base case: When either $i = 0$ or $j = 0$, we have*

$$V(i,0) = 0 \quad \text{for } 0 \le i \le n$$
$$V(0,j) = 0 \quad \text{for } 0 \le j \le m.$$

Recursive case: When both $i > 0$ and $j > 0$, we have

$$V(i,j) = \max \begin{cases} 0 & \text{align empty strings} \\ V(i-1,j-1) + \delta(R[i],T[j]) & \text{match/mismatch} \\ V(i-1,j) + \delta(R[i],-) & \text{delete} \\ V(i,j-1) + \delta(-,T[j]) & \text{insert.} \end{cases}$$

Proof This lemma can be proved using a similar approach as Lemma 4.1. ∎

Similar to semi-global alignment, local alignment can be solved in $O(nm)$ time and we can also generalize the gap penalty to the affine gap penalty.

4.2.4 Mapping quality

After a read is aligned, some read aligners return a score called the mapping quality, or mapQ, that indicates the likelihood that the read is originating from the mapped location (see column 5 in Figure 2.5(b)).

MAQ [172] is the first method that assigned a mapQ score to each alignment. It defines the mapQ of an alignment as $Q_s = -10 \log_{10} Pr\{$read is wrongly mapped$\}$. In a case where a read is mapped to multiple locations in the reference genome, the aligners will assign a mapQ score of 0 to each such location.

The formal definition of mapQ is as follows. Consider a read $R[1..m]$ with quality score $Q[1..m]$. Suppose R is aligned to a position u in a reference genome $T[1..n]$ without a gap. Assume the sequencing errors are independent at different positions in R. Let $Pr(R|T, u)$ be the probability that R is coming from position u of T. $Pr(R|T, u)$ can be approximated by the product of the error probabilities of the mismatch bases in the alignment between R and $T[u..u + m - 1]$, where the error probability of the base $R[i]$ is $10^{-\frac{Q[i]}{10}}$. For example, the read R in Figure 4.2 has two 2-mismatch alignments. For the left alignment (which is at position 6), $Pr(R|T, 6) \approx 10^{-\frac{10}{10}} 10^{-\frac{10}{10}} = 0.01$. For the right alignment (which is at position 23), $Pr(R|T, 23) \approx 10^{-\frac{40}{10}} 10^{-\frac{30}{10}} = 10^{-7}$. (Note: For a paired read, the error probability is approximated by the product of the error probabilities of the mismatch bases in both reads.)

Assuming a uniform prior distribution $Pr(u|T)$ and by the Bayesian formula, we have

$$Pr(u|T, R) = \frac{Pr(R|T, u)}{\sum_{v=1}^{n-m+1} Pr(R|T, v)}.$$

MAQ [172] defines the mapQ of the alignment to be $-10 \log_{10}(1 - Pr(u|T, R))$. The computation of $\sum_{v=1}^{n-m+1} Pr(R|T, v)$ is time consuming. To speed up, observe that $Pr(R|T, v) \approx 0$ when R and $T[v..v+m-1]$ has many mismatches. Hence, $\sum_{v=1}^{n-m+1} Pr(R|T, v)$ can be approximated by the sum of $Pr(R|T, v)$ for positions v where R and $T[v..v+m-1]$ has a small number of mismatches (say, at most $1 + Hamming(R, T[u..u + m - 1])$) mismatches).

Although the above method can approximate $\sum_{v=1}^{n-m+1} Pr(R|T, v)$ well, such approximation is still time consuming to be computed. Nowadays, almost all aligners approximate $\sum_{v=1}^{n-m+1} Pr(R|T, v)$ by $Pr(R|T, u) + Pr(R|T, u')$ where u and u' are the positions of the best hit and the second best hit, respectively. For the example in Figure 4.2, suppose the left alignment is the best alignment while the right alignment is the second-best alignment. Then, $Pr(6|T, R) = \frac{Pr(6|T,R)}{Pr(6|T,R)+Pr(23|T,R)} \approx 0.99999$. The mapQ score is $-10 \log_{10} (1 - Pr(6|T, R)) \approx 50$.

As a matter of fact, although many methods try to approximate $-10 \log_{10}(1 - Pr(u|T, R))$ using the above strategy. Some method like Bowtie 2 [154] does not use the above formula.

4.2.5 Challenges

Section 4.2.3 presents brute-force methods to align a read on the reference genome. However, the running time is slow.

One solution is to parallelize the dynamic programming. Nowadays, many processors have special vector execution units. Those units enable us to perform multiple operations in one single instruction. For example, consider eight 8-bit integers $a_1, \ldots, a_8$ and another eight 8-bit integers $b_1, \ldots, b_8$. By the special vector execution unit, we can obtain $a_1 + b_1, \ldots, a_8 + b_8$ in one instruction. These special vector execution units can speed up the implementation of the Smith-Waterman alignment algorithm. A number of mapping algorithms take advantage of this technique. They include Bowtie 2 [154] and SHRiMP [255].

Even if we parallelize the alignment routine, the computation is still slow since the dynamic programming algorithm runs in quadratic time. Another choice is to map reads using general aligners like BLAST [5] and BLAT [134]. These methods are also not suitable since they are too slow to process millions of reads generated by next-generation sequencers.

To resolve the efficiency issue, a number of methods have been developed. Different methods provide different trade-offs between speed and mapping accuracy. These methods can be classified in three ways. Firstly, the aligners can be classified depending on whether they can handle indels or not. Most of the old aligners can only handle mismatches. These aligners are called mismatch-only aligners. These aligners are generally faster. A recent study showed that small indels are abundant (representing about 18% of human polymorphisms [209]). Together with the fact that reads are getting longer, the chance of getting a gap in a read is higher. Hence, the new aligners generally can handle gaps. Secondly, the aligners can be classified based on whether they index the reference genome or the reads. Lastly, read aligners can be classified based on the indexing methods. Most aligners can be classified into two types: hashing-based methods and suffix trie-based methods.

Here, we classify the read mappers depending on whether they handle indels or not. We discuss 3 different solutions:

- Aligning short reads allowing a small number of mismatches (see Section 4.3)

- Aligning short reads allowing a small number of mismatches and indels (see Section 4.4)

- Align reads in general (see Section 4.5)

4.3   Align reads allowing a small number of mismatches

This section describes methods that align reads allowing a small number of mismatches. We fixed an upper limit p for the number of mismatches. For every read $R[1..m]$, the problem aims to find all positions i in the reference genome $T[1..n]$ such that $Hamming(R, T[i..i+m-1]) \leq p$. All these positions are called p-mismatch hits of R in T.

(a) Set of reads

1. AAT
2. CCT
3. GCT

(b) Reference

10
...CTTGACTCGGGAT...

(c) Hash table

AAA → 1	CAA	GAA	TAA
AAC → 1	CAC	GAC	TAC
AAG → 1	CAG	GAG	TAG
AAT → 1	CAT → 1,2	GAT → 1,3	TAT → 1
ACA	CCA → 2	GCA → 3	TCA
ACC	CCC → 2	GCC → 3	TCC
ACG	CCG → 2	GCG → 3	TCG
ACT → 1,2,3	CCT → 2,3	GCT → 2,3	TCT → 2,3
AGA	CGA	GGA	TGA
AGC	CGC	GGC	TGC
AGG	CGG	GGG	TGG
AGT → 1	CGT → 2	GGT → 3	TGT
ATA	CTA	GTA	TTA
ATC	CTC	GTC	TTC
ATG	CTG	GTG	TTG
ATT → 1	CTT → 2	GTT → 3	TTT

FIGURE 4.3: (a) is a set of length-3 reads, (b) is an example reference genome, and (c) shows the hash table for all 1-mismatch seeds of the reads.

This problem can be solved by brute force (see Section 4.2.3). However, its running time is slow. Many indexing techniques have been proposed to solve the p-mismatch mapping problem efficiently. Below, we discuss these techniques, which include:

- Mismatch seed hashing approach

- Spaced seed hashing approach

- Reference hashing

- Suffix trie-based approach

4.3.1 Mismatch seed hashing approach

Hashing is a computational technique that speeds up pattern matching (see Section 3.4.1). Here, we describe the mismatch seed approach, which applies hashing to solve the p-mismatch mapping problem more efficiently.

Given a set $\mathcal{R}$ of length-m reads, we index them by hashing as follows. A hash table H with 4^m entries is built for all m-mers (m-mer is defined as a length-m DNA sequence). For every read $R \in \mathcal{R}$, let $\mathcal{S}(R,p)$ be the set of m-mers that have at most p mismatches from R. The read R is included in $H[S]$ for every m-mer $S \in \mathcal{S}(R,p)$. For example, consider the first read $R_1 = $ AAT in Figure 4.3(a). $\mathcal{S}(R_1,1) = \mathcal{S}(\text{AAT},1) = $ {AAA, AAC, AAG, AAT, ACT, AGT, ATT, CAT, GAT, TAT}. We include R_1 in $H[S]$ for all m-mers $S \in \mathcal{S}(R_1,1)$. The remaining two reads can be processed similarly and we obtain the hash table in Figure 4.3(c).

By construction, for each position i in T, $T[i..i+m-1]$ is a p-mismatch hit

---

Algorithm mismatch($T, \mathcal{R}, p$)

Require: A set of reads $\mathcal{R}$, the mismatch threshold p and a reference genome T

Ensure: A list of p-mismatch hits of the reads in $\mathcal{R}$

1: Initialize a size-4^m hash table H for all possible m-mers;
2: **for** each read R in $\mathcal{R}$ **do**
3:    Include R into $H[S]$ for every m-mer $S \in \mathcal{S}(R, p)$;
4: **end for**
5: **for** $i = 1$ to n **do**
6:    All reads $R \in H[T[i..i + m - 1]]$ are reported to occur at position i;
7: **end for**

FIGURE 4.4: The algorithm for finding all p-mismatch hits for a set of reads $\mathcal{R}$ in T using mismatch seed hashing approach.

of a read R if $R \in H[T[i..i+m-1]]$. For example, in Figure 4.3(b), the length-3 DNA sequences at positions 10 and 11 are GAC and ACT, respectively. Since $H[\text{GAC}]$ is empty, there is no read whose 1-mismatch hit is GAC. As $H[\text{ACT}]$ contains all three reads, position 11 is a 1-mismatch hit of all three reads.

The algorihtm for mismatch seed hashing approach is as follows. It has two phases. The first phase constructs the hash table H such that every read $R \in \mathcal{R}$ is included in $H[S]$ for all $S \in \mathcal{S}(R, p)$. After that, the second phase scans the genome $T[1..n]$. For every $i = 1, 2, \ldots, n$, all reads $R \in H[T[i..i + m - 1]]$ are reported to occur at position i since they have at most p mismatches with $T[i..i + m - 1]$. Figure 4.4 details the algorithm.

Now, we analyze the time and space complexity. For time complexity, the first phase of the algorithm needs to include each read R into $H[S]$ for $S \in \mathcal{S}(R, p)$. Since each read R occurs $|\mathcal{S}(R, p)| (= \sum_{i=0}^{p} \binom{m}{i} 3^i = O((3m)^p))$ times in the hash table, $O(|\mathcal{S}(R, p)|m)$ time is required to hash R into the hash table. Then, the second phase of the algorithm scans the reference genome T position by position, which is of length n. The total running time is $O(|\mathcal{R}|(3m)^p m + nm)$. For space complexity, since the hash table has 4^m entries and each read is included in the hash table $|\mathcal{S}(R, p)| = O((3m)^p)$ times, the total space is $O(|\mathcal{R}|(3m)^p + 4^m)$.

4.3.2 Read hashing with a spaced seed

The mismatch seed hashing approach enumerates all p-mismatch patterns of every read, which is of a size exponential in p. Hence, it is time consuming to construct the hash table. To avoid this short-coming, a spaced seed hashing approach is proposed. This approach is used by a number of methods including CASAVA (Illumina's aligner), MAQ [172], SeqMap [121], ZOOM [179] and RMAP [279].

A spaced seed of a read $R[1..m]$ is formed by replacing some bases by wildcard bases, denoted as $*$ (i.e., $*$ matches any base). To define the wildcard positions in a spaced seed, we define the shape $P[1..m]$. The shape $P[1..m]$ is a length-m binary sequence where position i is a wildcard if and only if $P[i] = 0$. The weight of P is defined to be the number of ones in P. Denote $seed_P(R)$ as a spaced seed formed by replacing $R[i]$ by $*$ if $P[i] = 0$, for $i = 1, \ldots, m$. For example, consider a length-6 read ACGTGT and a length-6 shape 110011 (of weight 4); then $seed_{110011}(\text{ACGTGT}) = \text{AC} * *\text{GT}$ is the spaced seed of ACGTGT with respect to 110011. $seed_P(R)$ is said to occur at position i of the reference T if $seed_P(R) = seed_P(T[i..i + m - 1])$. For example, $seed_{110011}(\text{ACGTGT})$ occurs twice (see the two underlines) in the following DNA sequence. Both occurrences are 1-mismatch hits of ACGTGT.

TC<u>ACGGAG</u>TTC<u>ACCTGT</u>ACT

Below, we aim to design a universal set of shapes $\mathcal{P}$ of weight $2p$ which guarantees to cover all possible combinations of p mismatch hits of a read. Precisely, we aim to design $\mathcal{P}$ such that every p-mismatch pattern R' of R satisfies $seed_P(R) = seed_P(R')$ for some $P \in \mathcal{P}$. We partition $1..m$ into $2p$ segments $\{1..\kappa, \kappa+1..2\kappa, 2\kappa+1..3\kappa, \ldots, (2p-1)\kappa..m\}$, where $\kappa = \lfloor \frac{m}{2p} \rfloor$. Let $\mathcal{P}$ be the set of all length-m shapes such that p segments are assigned to be ones while the other p segments are assigned to be zeros. Note that $\mathcal{P}$ contains $\binom{2p}{p}$ shapes.

For example, for $m = 8$ and $p = 2$, we can partition $1..8$ into 4 segments $1..2, 3..4, 5..6, 7..8$. Then, $\mathcal{P}$ is a set of all length-8 binary sequences where 2 of the segments are zeros and the other 2 segments are ones. Precisely, $\mathcal{P} = \{11110000, 11001100, 11000011, 00111100, 00110011, 00001111\}$. We have the following lemma.

Lemma 4.3 *Consider a read $R[1..m]$ and a length-m substring $T[i..i+m-1]$. If $Hamming(R, T[i..i + m - 1]) \le p$, then $seed_P(R) = seed_P(T[i..i + m - 1])$ for some $P \in \mathcal{P}$.*

Proof Let $\kappa = m/(2p)$. Note that $Hamming(R[1..m], T[i..i + m - 1]) = \sum_{j=1}^{2p} Hamming(R[(j-1)\kappa + 1..j\kappa], T[i + (j-1)\kappa..i + j\kappa - 1])$. Conversely, assume we cannot find p parts in the read R that exactly match the p respective parts in $T[i..i + m - 1]$. This means that there are at least $p+1$ parts in the read R that are different from the $p+1$ respective parts in $T[i..i+m-1]$. Then, $Hamming(T[i..i+m-1], R[1..m]) \ge p+1$. We arrive at a contradiction. ∎

The above lemma enables us to efficiently compute a set of candidate p-mismatch hits of a set of length-m reads $\mathcal{R}$ on the genome T. The algorithm is detailed in Figure 4.6. It has two phases. The first phase builds hash tables H_P for all $P \in \mathcal{P}$. For each $P \in \mathcal{P}$, each read $R \in \mathcal{R}$ is hashed to the entry $H_P[seed_P(R)]$.

The second phase scans the genome T to identify candidate hits for all reads in $\mathcal{R}$; then each candidate hit is verified if it is a p-mismatch hit. More

(a)   Reads R_1: ACGTTTAC

(b)

Hash table 1: XXXX****	Hash table 2: XX**XX**	Hash table 3: XX****XX	Hash table 4: **XXXX**	Hash table 5: **XX**XX	Hash table 6: ****XXXX
AAAA****	AA**AA**	AA****AA	**AAAA**	**AA**AA	****AAAA
...	...	...	...	...	...
ACGT**** → R_1	AC**TT** → R_1	AC****AC → R_1	**GTTT** → R_1	**GT**AC → R_1	****TTAC → R_1
...	...	...	...	...	...
TTTT****	TT**TT**	TT****TT	**TTTT**	**TT**TT	****TTTT

(c) **Reference**   ...CGTA$\overset{i}{\text{AC}}$GGTTA$\overset{}{\text{AC}}$TT$\overset{j}{\text{TT}}$GTTGACCGGGAT...

FIGURE 4.5: (a) Consider a read R_1 = ACGTTTAC. We aim to find its 2-mismatch hits in the reference T. (b) We build 6 hash tables corresponding to the set of 6 shapes $\mathcal{P}$ = $\{11110000, 11001100, 11000011, 00111100, 00110011, 00001111\}$. The figure indicates the hash entries that contain the read R_1. (c) is the reference genome. The spaced seed AC$*$$*TT*$$*$ occurs at position i while the spaced seed $*$$*GT*$$*$AC occurs at position j. By verifying the Hamming distance, we report that R_1 has a 2-mismatch hit at position i.

precisely, the algorithm scans the genome T position by position. For every position i, a list of reads R is extracted from $H_P[seed_P(T[i..i + m - 1])]$ for all shapes $P \in \mathcal{P}$. For each such read R, the algorithm checks if R has at most p mismatches from $T[i..i + m - 1]$. If yes, we report $T[i..i + m - 1]$ as a p-mismatch hit of R. Lemma 4.3 guarantees that this algorithm reports all p-mismatch hits of all reads in $\mathcal{R}$.

For example, for a length-8 read R_1 = ACGTTTAC, if we want to find all 2-mismatch hits, Figure 4.5 illustrates the use of the six shapes to find all 2-mismatch hits of R_1.

This technique is used by CASAVA, MAQ [172] and SeqMap [121]. They apply the technique to find all 2-mismatch hits of a length-28 read. Precisely, they partition 1..28 into 4 segments, each segment is of length 7. Then, it obtains a set of 6 shapes $\mathcal{P}$ = $\{1^{14}0^{14}, 1^{7}0^{7}1^{7}0^{7}, 1^{7}0^{14}1^{7}, 0^{7}1^{14}0^{7}, 0^{7}1^{7}0^{7}1^{7}, 0^{14}1^{14}\}$. Using the algorithm in Figure 4.6, we can find all hits of each read with at most 2 mismatches.

Below, we analyze the space and time complexity of the algorithm in Figure 4.6. For space complexity, we need to build $\binom{2p}{p} = O((2p)^p)$ hash tables. Each hash table is of size $O(4^{\frac{m}{2}} + |\mathcal{R}|) = O(2^m + |\mathcal{R}|)$. So, the total space is $O((2p)^p (2^m + |\mathcal{R}|))$.

For time complexity, we analyze the running time of the two phases. The first phase builds the hash tables. The algorithm needs to insert $|\mathcal{R}|$ reads into $\binom{2p}{p}$ hash tables. It takes $O(\binom{2p}{p}|\mathcal{R}|m) = O((2p)^p|\mathcal{R}|m)$ time. The second phase scans the genome $T[1..n]$ position by position. For each position i, it verifies if $R \in H_P[seed_P[T[i..i+m-1]]]$ is a p-mismatch hit of $T[i..i+m-1]$ for every $P \in \mathcal{P}$. There are n positions in T and we need to verify $\binom{2p}{p} = O((2p)^p)$ hash entries for each position. Each hash entry contains $O(1)$ hits assuming

Algorithm mismatch$(T, \mathcal{R}, p)$

Require: A set of length-m reads $\mathcal{R}$, a reference genome T, a mismatch threshold p

Ensure: A list of hits of the reads in $\mathcal{R}$ with at most p mismatches

1: Let $\mathcal{P}$ be a set of length-m shapes that can cover all possible p-mismatch patterns;
2: **for** each read R in $\mathcal{R}$ **do**
3:     **for** each $P \in \mathcal{P}$ **do**
4:         Include R in $H_P[seed_P(R)]$;
5:     **end for**
6: **end for**
7: **for** $i = 1$ to $|T|$ **do**
8:     **for** each $P \in \mathcal{P}$ **do**
9:         **for** each read R in $H_P[seed_P(T[i..i + m - 1])]$ **do**
10:             **if** $Hamming(R, T[i..i + m - 1]) \leq p$ **then**
11:                 return position i as a hit of R;
12:             **end if**
13:         **end for**
14:     **end for**
15: **end for**

FIGURE 4.6: The algorithm finds all p-mismatch hits of R in T, for every length-m reads $R \in \mathcal{R}$, using the spaced seed approach.

there is little collision. It takes $O(m)$ time to verify one hit. Hence, the second phase runs in $O((2p)^p nm)$ expected time. In total, the time complexity is $O((2p)^p(|\mathcal{R}| + n)m)$.

The above approach is not the only way to define a universal set of shapes covering all possible p-mismatch patterns of length m. The following lemma gives another approach. RMAP [279] proposed to define $\mathcal{P}' = \{P_1, \ldots, P_{p+1}\}$, where P_j is a binary pattern consisting of exactly one segment $[(j-1)\kappa+1..j\kappa]$ equals ones, with $\kappa = \lfloor \frac{m}{p+1} \rfloor$ for $j = 1, \ldots, p+1$. For example, to find all 2-mismatch hits of a length-28 read, $\mathcal{P}' = \{1^9 0^{19}, 0^9 1^9 0^{10}, 0^{18} 1^{10}\}$.

Lemma 4.4 *Consider a read $R[1..m]$ and a length-m substring $T[i..i + m - 1]$ of the reference genome T. If $Hamming(R, T[i..i + m - 1]) \leq p$, then $seed_P(R) = seed_P(T[i..i + m - 1])$ for some $P \in \mathcal{P}'$.*

Proof Let $\kappa = m/(p+1)$ and $R_j = R[(j-1)\kappa+1..j\kappa]$, $T_j = T[i+(j-1)\kappa..i+j\kappa-1]$. Then, $Hamming(T[i..i+m-1], R[1..m]) = \sum_{j=1}^{p+1} Hamming(T_j, R_j)$. Conversely, assume $R_j \neq T_j$ for all $j = 1, \ldots, p + 1$. Then, we have $Hamming(T[i..i + m - 1], R[1..m]) = \sum_{j=1}^{p+1} Hamming(T_j, R_j) \geq p + 1$. We arrived at a contradiction. ∎

Since $\mathcal{P}'$ covers all hits with p mismatches, after replacing $\mathcal{P}$ by $\mathcal{P}'$ in the algorithm in Figure 4.6, the algorithm can still report all p-mismatch hits of all reads in $\mathcal{R}$. However, the time and space complexity are different since $\mathcal{P}$ and $\mathcal{P}'$ use different shape sets (see Exercise 8).

From the above discussion, we know that there are many ways to generate a universal set of shapes. A universal set of shapes should (1) cover all p-mismatch hits and (2) minimize the number of shapes. Given the weight of the shapes, ZOOM [179] proposed an algorithm to construct a universal set of shapes of minimum size. For example, ZOOM showed that the following set of four shapes is the smallest weight-13 shape set that covers all two-mismatch hits for any length-33 read.

$$1111111111111000000000000000000000$$
$$0000000111111111111110000000000000$$
$$0000000000000000000001111111111111$$
$$1111111000000111111100000000000000$$

4.3.3 Reference hashing approach

The previous two subsections build hash tables for reads. Another choice is to build hash tables for the reference genome. The process is exactly reversed. There are two phases. The first phase uses hashing to index every position i of the reference genome T; then, the second phase scans reads in $\mathcal{R}$ one by one to identify their p-mismatch hits in T.

Similar to the read hashing approach, we can use a mismatch seed or spaced seed to index the reference genome $T[1..n]$. Below, we describe the

---

Algorithm mismatch$(T, \mathcal{R}, p)$

Require: A set of length-m reads $\mathcal{R}$, a reference genome T and a mismatch threshold p

Ensure: A list of hits of the reads in $\mathcal{R}$ with at most p mismatches

1: Let $\mathcal{P}$ be a set of length-m shapes that can cover all possible p-mismatch hits;
2: **for** $i = 1$ to $|T|$ **do**
3:   **for** each $P \in \mathcal{P}$ **do**
4:     Include i in $H_P[seed_P(T[i..i + m - 1])]$;
5:   **end for**
6: **end for**
7: **for** each read R in $\mathcal{R}$ **do**
8:   **for** each $P \in \mathcal{P}$ **do**
9:     **for** each position i in $H_P[seed_P(R)]$ **do**
10:       Report position i as a hit of R if $Hamming(T[i..i+m-1], R) \leq p$;
11:     **end for**
12:   **end for**
13: **end for**

FIGURE 4.7: The algorithm for finding p-mismatch hits for a set of reads $\mathcal{R}$ in T using the reference hashing approach.

spaced seed reference hashing approach. Let $\mathcal{P}$ be a universal set of shapes that cover all p-mismatch hits. The algorithm has two phases. The first phase builds hash tables. For every $P \in \mathcal{P}$, $T[i..i + m - 1]$ is hashed to the entry $H_P[seed_P(T[i..i + m - 1])]$ for $i = 1, \ldots, n$. The second phase scans every read in $\mathcal{R}$. For each read $R \in \mathcal{R}$, for each $P \in \mathcal{P}$, for each $T[i..i + m - 1]$ in $H_P[seed_P(R)]$, if $Hamming(R, T[i..i + m - 1]) \leq p$, we report $T[i..i + m + 1]$ as a hit of R. Figure 4.7 details the method.

The advantage of the reference hashing approach is that the hash tables can be precomputed since the reference genome is known (i.e., Steps 1−6 in Figure 4.7 can be precomputed).

However, when the reference genome is long, the hash tables for the reference genome may be too big to be stored in memory. Interestingly, we don't need to store all $|\mathcal{P}|$ hash tables explicitly. For instance, to report 1-mismatch hits using the algorithm in Figure 4.7, the algorithm uses two shapes, $P_1 = 1^{m/2}0^{m/2}$ and $P_2 = 0^{m/2}1^{m/2}$. We observe that the two hash tables H_{P_1} and H_{P_2} are almost the same except that the positions are shifted by $m/2$. To reduce the redundancy, we use one shape $P_0 = 1^{m/2}$ and store one hash table H_{P_0}. Observe that $H_{P_1}[Q] = H_{P_0}[Q[1..m/2]]$ and $H_{P_2}[Q] = H_{P_0}[Q[m/2 + 1..m]] - m/2$. For $p > 1$, we can use a similar strategy to reduce the redundancy. For example, to support 2-mismatch hit reporting, we need to build 3 hash tables instead of 6. However, it is still too big. This limited the usefulness of the reference hashing approach.

A number of methods use the reference hashing approach, including SOAP [175], BFAST [103] and MOSAIK [161].

4.3.4 Suffix trie-based approaches

Section 4.3.3 uses hashing to index the reference genome. However, a hash table is not space-efficient and it is not suitable to index a long reference genome. This section discusses suffix trie-based methods.

Consider a reference genome $T[1..n]$. Its suffix trie τ_T is a rooted tree where every edge is labeled by a nucleotide in $\{A, C, G, T\}$ and every suffix of T is represented by a path from the root to a leaf (see Section 3.5.1). For every node u in τ_T, $plabel(u)$ (or $plabel_T(u)$ for clarity) is denoted as the path label of u, that is, the concatenation of the edge labels on the path from the root of τ_T to the node u. The depth of u, $depth(u)$, is the length of $plabel(u)$. Denote $Leaf(u)$ (or $Leaf_T(u)$ for clarity) as the set of leaves under u. For example, Figure 4.8(a) gives the suffix trie for $T = \texttt{ACACAG\$}$. We have $plabel(v_1) = \texttt{A}$, $plabel(v_7) = \texttt{ACA}$, $depth(v_1) = 1$, $depth(v_7) = 3$, $Leaf(v_1) = \{1, 3, 5\}$ and $Leaf(v_7) = \{1, 3\}$.

Storing suffix trie τ_T explicitly is space-inefficient. It requires $O(n^2)$ words, where n is the length of T. Fortunately, the suffix trie τ_T can be simulated by the FM-index of the reverse sequence of T (see Section 3.5.4). The FM-index is a compressed index that enables us to store τ_T using $O(n)$ bits. For instance, the FM-index of the reference human genome can be stored in 2 gigabytes, which is small enough to be loaded into the memory of a typical PC. The FM-index not only enables us to represent the suffix trie space-efficiently, but also provides efficient operations to traverse the suffix trie. For any node u in the suffix trie τ_T, we can compute its child node $child(u, \sigma)$ for every $\sigma \in \{A, C, G, T\}$ in $O(1)$ time. $Leaf(u)$ can be computed in $O(|Leaf(u)| \log^2 n)$ time (see Exercise 11).

With the suffix trie τ_T, the exact occurrences of a read $R[1..m]$ can be found efficiently. Basically, we just traverse down the suffix trie and locate the node u with $R = plabel(u)$; then, the occurrences of R is the set of leaf labels $Leaf(u)$. Hence, the exact occurrences of R can be found in $O(m + |Leaf(u)| \log^2 n)$ time.

We can also apply the suffix trie τ_T to find all occurrences of R allowing at most p mismatches. This problem is equivalent to finding all depth-m nodes u in τ_T such that $Hamming(R, plabel(u)) \leq p$. For example, suppose we want to find 1-mismatch hits of $\texttt{CCC}$ in $T = \texttt{ACACAG\$}$. We first find nodes of depth 3 in the suffix trie τ_T. There are 3 such nodes (see Figure 4.8(a–c)). By comparing $\texttt{CCC}$ with the path labels of these three nodes, we discover that the 1-mismatch hit of $\texttt{CCC}$ occurs at position 2 of T.

Below, we describe the basic recursive solution for finding p-mismatch hits of a read R using the suffix trie τ_T. First, we define the recursive equation. For every node u in τ_T, denote $HSet(u)$ (or $HSet_{T,R,p}(u)$ for clarity) as the set of descendants v of u such that $depth(v) = m$ and $Hamming(R, plabel(v)) \leq p$.

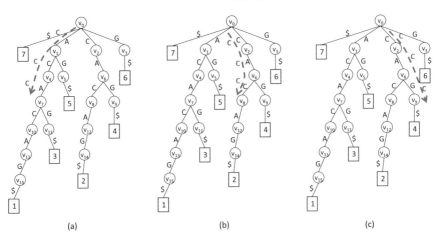

(a) (b) (c)

FIGURE 4.8: Consider a genome $T = $ ACACAG$. There are three nodes of depth 3. They occur on the paths corresponding to the first, second and fourth suffixes. (a), (b) and (c) illustrate the alignment between CCC and the path labels of these three nodes (see the dotted lines).

Then, the p-mismatch hits of R occur at positions in $\bigcup_{v \in S} Leaf(v)$, where $S = HSet_{T,R,p}(root(\tau_T))$. The following lemma gives the recursive equation for computing $HSet_{T,R,p}(u)$ for every u in τ_T.

Lemma 4.5 *Consider a read R, the suffix trie τ_T for a reference genome T and a mismatch threshold p. For any node u in τ_T, $HSet_{T,R,p}(u)$ can be computed in two cases. For the base case ($Hamming(R[1..depth(u)], plabel(u)) > p$ or $depth(u) \geq m$), we have:*

$$HSet_{T,R,p}(u) = \begin{cases} \emptyset & \text{if } depth(u) > m \text{ or } Hamming(R, plabel(u)) > p \\ \{u\} & \text{otherwise.} \end{cases}$$

For the recursive case ($Hamming(R[1..depth(u)], plabel(u)) \leq p$ and $depth(u) < m$), we have:

$$HSet_{T,R,p}(u) = \bigcup_{v=child(u,\sigma)} HSet_{T,R,p}(v).$$

Proof For the base case ($Hamming(R[1..depth(u)], plabel(u)) > p$ or $depth(u) \geq m$), by definition, $HSet_{T,R,p}(u) = \emptyset$ if $depth(u) > m$ or $Hamming(R, plabel(u)) > p$; and $HSet_{T,R,p}(u) = \{u\}$ otherwise.

For the recursive case ($Hamming(R[1..depth(u)], plabel(u)) \leq p$ and $depth(u) < m$), for every node w in $HSet_{T,R,p}(u)$, w is in $HSet_{T,R,p}(v)$ for some child of u. Hence, we have $HSet_{T,R,p}(u) = \bigcup_{v=child(u,\sigma)} HSet_{T,R,p}(v)$.

∎

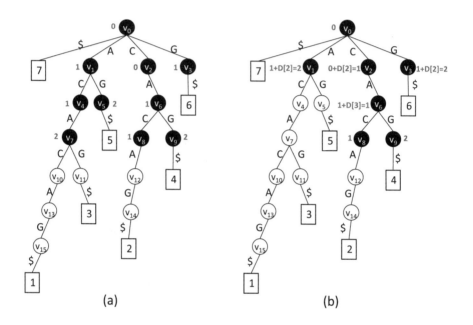

FIGURE 4.9: Consider a genome $T = \texttt{ACACAG\$}$ and a pattern $R = \texttt{CCC}$. This figure illustrates the algorithm to find all 2-mismatch hits of R in T. (a) and (b) show the suffix trie τ_T of T. In (a), the black nodes are the nodes visited by $HSet_{T,R,2}(root(\tau_T))$. For each black node u, the gray number is $Hamming(R[1..depth(u)], plabel(u))$. In (b), the black nodes are the nodes visited by $HSet_{T,R,2}(root(\tau_T))$ after the if-condition of Step 3 is replaced by $Hamming(R[1..depth(u)], plabel(u)) + D[depth(u) + 1] > p$. For each black node u, the gray number is $Hamming(R[1..depth(u)], plabel(u)) + D[depth(u) + 1]$.

Algorithm $HSet_{T,R,p}(u)$

Require: T is the genome, $R[1..m]$ is a read, p is the mismatch threshold, u is any node in τ_T

Ensure: A set of nodes v such that v is descendants of u, $depth(v) = m$ and $Hamming(R, plabel(v)) \leq p$.

1: **if** $depth(u) > m$ **then**
2:     Return $\emptyset$;
3: **else if** $Hamming(R[1..depth(u)], plabel(u)) > p$ **then**
4:     Return $\emptyset$;
5: **else if** $depth(u) = m$ **then**
6:     Return $\{u\}$;
7: **else**
8:     $Ans = \emptyset$;
9:     **for** every child v of u **do**
10:       Set $Ans = Ans \cup HSet_{T,R,p}(v)$;
11:     **end for**
12:     Return Ans;
13: **end if**

FIGURE 4.10: By calling $HSet_{T,R,p}(root(\tau_T))$, we can find all nodes v in τ_T such that $Hamming(R, plabel(v)) \leq p$.

Figure 4.10 gives the recursive algorithm to compute $HSet_{T,R,p}(root(\tau_T))$ based on Lemma 4.5. The algorithm recursively traverses down τ_T and computes $HSet_{T,R,p}(u)$. For example, when $R = $ CCC, $T = $ ACACAG\$ and $p = 1$, Figure 4.9(a) shows the set of nodes in τ_T visited by the algorithm $HSet_{T,R,p}()$.

We can further speed up the basic algorithm in Figure 4.10. The following subsections detail four techniques.

4.3.4.1 Estimating the lower bound of the number of mismatches

The algorithm $HSet_{T,R,p}()$ in Figure 4.10 aligns a read $R[1..m]$ to every suffix of T by traversing down the suffix trie τ_T. The algorithm runs in $O(n^2)$ time if it visits all nodes in τ_T. To reduce running time, the algorithm stops traversing down τ_T when it visits a node u such that $Hamming(R[1..depth(u)], plabel(u)) > p$ (see Step 3 in Figure 4.10). This rule reduces the running time of the algorithm.

Although the above rule prunes some subtrees in τ_T, Li and Durbin [167] suggested that we can prune more nodes if we estimate the lower bound of the number of mismatches contributed by $R[i..m]$. Their solution has two phases. First, it estimates $D[i]$, for $i = 1, \ldots, m$, where $D[i]$ is the lower bound of the number of mismatches between $R[i..m]$ and any length-$(m-i+1)$ substring in

Algorithm CalculateD(R)

1: $z = 0$; $j = m$;
2: **for** $i = m$ to 1 **do**
3:   **if** $R[i..j]$ is not a substring of T **then**
4:     $z = z + 1$; $j = i - 1$;
5:   **end if**
6:   $D[i] = z$;
7: **end for**

FIGURE 4.11: The algorithm CalculateD estimates the lower bound of the number of mismatches between $R[i..m]$ and any length-$(m - i + 1)$ substring in T.

T. Then, the second phase runs the recursive algorithm $HSet_{T,R,p}(root(\tau_T))$ with a better pruning strategy that utilizes the array $D[]$.

We first discuss the first phase, that is, estimating $D[i]$, for $i = 1, \ldots, m$. Suppose $R[i..m]$ equals $S_z S_{z-1} \ldots S_1$ where all segments $S_1, \ldots, S_z$ do not appear in T. Then, we can conclude that $D[i] \geq z$. This observation suggests that the lower bound $D[i]$ can be estimated by iteratively partitioning $R[i..m]$ into non-overlapping segments that do not appear in T.

To illustrate the idea, consider a genome $T = $ ACACAG\$. Suppose the query string is CGTA. We scan the query string from right to left. Observe that $S_1 = $ TA does not appear in T; then $S_2 = $ CG does not appear in T. Since CGTA $= S_2 S_1$, we conclude that the mismatch lower bound for CGTA is 2. Based on this approach, Figure 4.11 gives the algorithm CalculateD to compute $D[i]$. $CalculateD(R)$ can compute $D[1], \ldots D[m]$ in $O(m)$ time (see Exercise 12).

Now, we discuss the second phase. Given $D[]$, the following lemma gives a better rule (called the look-ahead rule) to prune subtrees of τ_T.

Lemma 4.6 *Consider any node u in τ_T and a read $R[1..m]$. If $depth(u) > m$ or $Hamming(R[1..depth(u)], plabel(u)) + D[depth(u) + 1] > p$, $HSet_{T,R,p}(u) = \emptyset$.*

Proof Suppose $HSet_{T,R,p}(u) \neq \emptyset$. Then, there exists a depth-m descendant v of u such that $Hamming(R, plabel(v)) \leq p$. By definition, $plabel(v)$ is a substring of T, say $T[j..j + m - 1]$. Let $\kappa = depth(u)$. We have:

$$Hamming(R, plabel(v))$$
$$= Hamming(R[1..\kappa], plabel(u)) + Hamming(R[\kappa + 1..m], T[j + \kappa..j + m - 1])$$
$$\geq Hamming(R[1..\kappa], plabel(u)) + D[\kappa + 1] > p.$$

We arrived at a contradiction. The lemma follows. ∎

The above lemma can improve the efficiency of the algorithm $HSet_{T,R,p}()$ in Figure 4.10. The original algorithm stops traversing down τ_T when it visits a node u such that $Hamming(R[1..depth(u)], plabel(u)) + D[depth(u)+1] > p$ (see the if-condition of Step 3 in Figure 4.10). The above lemma implies that this if-condition can be replaced by $Hamming(R[1..depth(u)], plabel(u)) + D[depth(u) + 1] > p$. This if-condition can prune more nodes while the algorithm can still report all p-mismatch hits of R. Figure 4.9(b) illustrates that the updated algorithm prunes more nodes in τ_T while it still can compute all correct p-mismatch hits.

4.3.4.2 Divide and conquer with the enhanced pigeon-hole principle

This section applies the divide-and-conquer strategy to find all p-mismatch hits of $R[1..m]$ in T. This strategy is used by BatMis [294]. Bowtie 1 [155] also applies this strategy when $p = 2$ (it is called double-indexing).

The divide-and-conquer strategy partitions $R[1..m]$ into two equal halves $R_l = R[1..m/2]$ and $R_r = R[m/2 + 1..m]$. Then, two phases are executed. The first phase identifies the $\lfloor p/2 \rfloor$-mismatch hits of both R_l and R_r. After that, the second phase extends these hits and verifies if they are p-mismatch hits of R. By the pigeon-hole principle, the above idea guarantees to report all p-mismatch hits of R.

Figure 4.12 illustrates an example for $p = 2$. In the example, phase 1 finds three 1-mismatch hits for R_l and three 1-mismatch hits for R_r (see Figure 4.12(a,c)). Phase 2 extends these two sets of hits. After extending the hits for R_l, two 2-mismatch hits of R are identified at positions i' and i'' (see Figure 4.12(b)). Similarly, after extending the hits for R_r, two 2-mismatch hits of R are identified at positions i and i' (see Figure 4.12(d)). Combining the results, the set of 2-mismatch hits of R is $\{i, i', i''\}$.

Although the above solution can work, it performs redundant computation. For the example in Figure 4.12, the hit at position i' is found twice, which generates unnecessary computation.

Here, we show that redundant computation can be avoided by using an enhanced pigeon-hole principle. The detail is stated in Lemma 4.7. This lemma ensures that all p-mismatch hits of R can be found while no redundant candidate hit will be generated.

Lemma 4.7 *Consider a read R and a substring R' in T of equal length such that $Hamming(R, R') \leq p$, where $p \geq 1$. We have two cases.*

- *Case 1: p is even. We have either $Hamming(R_l, R'_l) \leq p/2$ or $Hamming(R_r, R'_r) \leq p/2 - 1$.*

- *Case 2: p is odd. We have either $Hamming(R_l, R'_l) \leq (p-1)/2$ or $Hamming(R_r, R'_r) \leq (p-1)/2$.*

Proof For case 1, when p is even, we have either (1) $Hamming(R_l, R'_l) \leq p/2$ or (2) $Hamming(R_l, R'_l) > p/2$. Condition (2) means that $Hamming(R_l, R'_l) \geq p/2 + 1$, which implies $Hamming(R_r, R'_r) \leq p - $

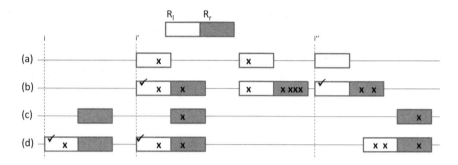

FIGURE 4.12: An example that demonstrates the simple divide-and-conquer method to find 2-mismatch hits of R. (a) We find all the 1-mismatch hits of R_l. (b) After extension of the hits of R_l, 2 hits at positions i' and i'' are found (marked by ticks). (c) We find all the 1-mismatch hits of R_r. (d) After extension of the hits of R_r, 2 hits at positions i and i' are found (marked by ticks). Note that the hit at position i' is discovered twice, which generates redundant computation.

$Hamming(R_l, R_l') \leq p - p/2 - 1 = p/2 - 1$. Hence, Case 1 follows. Case 2 can be proved similarly. ∎

For the example in Figure 4.12, using Lemma 4.7, phase 1 still generates three 1-mismatch hits of R_l. However, it generates one 0-mismatch hit of R_r instead of three 1-mismatch hits of R_r. This modification reduces the number of candidate hits. Phase 2 extends the candidate hits of both R_l and R_r. For R_l, after extending the three hits of R_l, we identify positions i' and i'' as the 2-mismatch hits of R. For R_r, after extending the only hit of R_r, we identify position i as the 2-mismatch hit of R. Observe that we did not find position i' twice. Hence, we avoid the redundant computation.

To further speed up, we can combine the above idea with the idea of suffix trie traversal in the algorithm *HSet* (see Figure 4.10). Instead of finding all positions i in T such that $Hamming(R, T[i..i+m-1]) \leq p$, we find all nodes u in τ_T such that $Hamming(R, plabel_T(u)) \leq p$. Then, every position i in $Leaf_T(u)$ is a p-mismatch hit of R in T.

Denote DivideAndConquer(R, p, T) as the algorithm that reports all nodes in τ_T such that $Hamming(R, plabel_T(u)) \leq p$. Let τ_T and $\tau_{\overline{T}}$ be the suffix tries of T and $\overline{T}$, respectively. ($\overline{T}$ is the reverse of T.) We need functions $child_T(u, \sigma)$ and $child_{\overline{T}}(u', \sigma)$ for traversing τ_T and $\tau_{\overline{T}}$, respectively. Furthermore, we need a function that converts any node u in τ_T into the corresponding node u' in $\tau_{\overline{T}}$ such that $\overline{plabel_T(u)} = plabel_{\overline{T}}(u')$, and vice versa. Precisely, we denote $u' = reverse_T(u)$ and $u = reverse_{\overline{T}}(u')$. Using the data structure in Lemma 3.9, $child_T(u, \sigma)$ and $reverse_T(u)$ can be computed in $O(|\mathcal{A}|)$ time and $O(1)$ time respectively.

Figure 4.13 presents the detail of the algorithm DivideAndConquer(R, p, T).

Algorithm DivideAndConquer(R, p, T)

Require: A read R, a mismatch threshold p and two suffix tries τ_T and $\tau_{\overline{T}}$

Ensure: A set of nodes v in τ_T such that $Hamming(R, plabel_T(v)) \le p$.

1: **if** $p = 0$ or $|R|$ is small **then**
2:     Report $HSet_{T,R,p}(root(\tau_T))$;
3: **end if**
4: $R_l = R[1..m/2]$ and $R_r = R[m/2 + 1..m]$;
5: **if** p is even **then**
6:     $H_l = \text{DivideAndConquer}(R_l, p/2, T)$;
7:     $H_r = \text{DivideAndConquer}(R_r, p/2 - 1, T)$;
8: **else**
9:     $H_l = \text{DivideAndConquer}(R_l, (p-1)/2, T)$;
10:     $H_r = \text{DivideAndConquer}(R_r, (p-1)/2, T)$;
11: **end if**
12: $Ans = \emptyset$;
13: **for** every node $u \in H_l$ **do**
14:     $Ans = Ans \cup HSet_{T,R,p}(u)$;
15: **end for**
16: **for** every node $u \in H_r$ **do**
17:     $u' = reverse_T(u)$;
18:     $A' = HSet_{\overline{T},\overline{R},p}(u')$;
19:     $Ans = Ans \cup \{reverse_{\overline{T}}(v') \mid v' \in A'\}$;
20: **end for**
21: Report Ans;

FIGURE 4.13: The algorithm DivideAndConquer(R, p, T) reports all p-mismatch hits of the read R in T. (For the algorithm $HSet_{T,R,p}(root(\tau_T))$, please refer to Figure 4.10.)

DivideAndConquer(R, p, T) is a recursive procedure. When R is short or $p = 0$, we just run $HSet_{T,R,p}(root(\tau_T))$ (see Steps 1–3) to report all nodes u such that $Hamming(R, plabel_T(u)) \leq p$. Otherwise, the read R is partitioned into two equal halves $R_l = R[1..m/2]$ and $R_r = [m/2 + 1..m]$. Two phases are executed. The first phase (Steps 5–11) finds the candidate hits of R_l and R_r recursively. By Lemma 4.7, the computation of the first phase depends on whether p is even or odd. When p is even, it computes H_l =DivideAndConquer$(R_l, p/2, T)$ and H_r =DivideAndConquer$(R_r, p/2 - 1, T)$. When k is odd, it computes H_l =DivideAndConquer$(R_l, (p-1)/2, T)$ and H_r =DivideAndConquer$(R_r, (p-1)/2, T)$.

The second phase extends all nodes in H_l and H_r to obtain the p-mismatch hits of R. For each node $u \in H_l$, we need to find all nodes v in τ_T such that $Hamming(R, plabel_T(v)) \leq p$ and $plabel_T(u)$ is a prefix of $plabel_T(v)$. In other words, we need to find all descendant nodes v of u in τ_T such that $Hamming(R, plabel_T(v)) \leq p$. This can be done by running $HSet_{T,R,p}(u)$ (see Steps 13–15).

For each node $u \in H_r$, we need to find $A = \{v \text{ in } \tau_T \mid Hamming(R, plabel_T(v)) \leq p \text{ and } plabel_T(u) \text{ is a suffix of } plabel_T(v)\}$. A cannot be found directly from τ_T. The following lemma suggests that such $A' = \{reverse_T(v) \mid v \in A\}$ can be found by traversing $\tau_{\overline{T}}$.

Lemma 4.8 *For every node $u \in H_r$, let $u' = reverse_T(u)$. Denote $A = \{v \text{ in } \tau_T \mid Hamming(R, plabel_T(v)) \leq p \text{ and } plabel_T(u) \text{ is a prefix of } plabel_T(v)\}$. Denote $A' = \{v' \text{ in } \tau_{\overline{T}} \mid Hamming(\overline{R}, plabel_{\overline{T}}(v')) \leq p \text{ and } plabel_{\overline{T}}(u') \text{ is a suffix of } plabel_{\overline{T}}(v')\}$. We have $A = \{reverse_{\overline{T}}(v') \mid v' \in A'\}$ and $A' = \{reverse_T(v) \mid v \in A\}$.*

Proof For any node $v \in A$, let $v' = reverse_T(v)$. Observe that $Hamming(R, plabel_T(v)) = Hamming(\overline{R}, plabel_{\overline{T}}(v'))$. The lemma follows. ∎

By Lemma 4.8, for every $u \in H_r$, A can be found in three steps (see Steps 16–20). The first step computes $u' = reverse_T(u)$. The second step finds $A' = \{v' \text{ in } \tau_{\overline{T}} \mid Hamming(\overline{R}, plabel_{\overline{T}}(v')) \leq p \text{ and } plabel_{\overline{T}}(u') \text{ is a suffix of } plabel_{\overline{T}}(v')\}$. This is equivalent to finding all descendant nodes v' of u' in $\tau_{\overline{T}}$ such that $Hamming(\overline{R}, plabel_{\overline{T}}(v')) \leq p$. Hence, $A' = HSet_{\overline{T}, \overline{R}, p}(u')$. The third step set $v = reverse_{\overline{T}}(v')$ for every $v' \in A'$.

As a side note, although the above algorithm reports all p-mismatch hits of R in T, the hits are reported in random order. Tennakoon et al. [294] showed that the algorithm can be modified such that it can efficiently report the hits in the order of increasing number of mismatches (see Exercise 16).

4.3.4.3 Aligning a set of reads together

Most suffix trie-based methods align reads on the reference T one by one. Note that reads may share prefixes. By processing them together, the running time can be improved. This idea is used by MASAI [278].

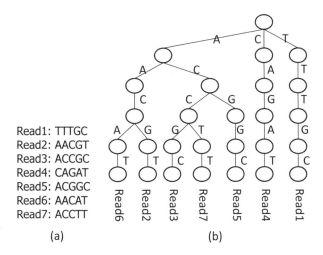

Read1: TTTGC
Read2: AACGT
Read3: ACCGC
Read4: CAGAT
Read5: ACGGC
Read6: AACAT
Read7: ACCTT

(a)   (b)

FIGURE 4.14: This example illustrates the construction of a radix tree from 7 reads. (a) gives 7 reads. (b) gives the radix tree of all reads.

A set of reads $\mathcal{R}$ can be represented by a radix trie $\tau_\mathcal{R}$. A radix trie is a tree where every edge is labeled by a nucleotide in $\{a, c, g, t\}$ and every path from root to leaf represents a read in $\mathcal{R}$. Figure 4.14 illustrates an example.

Given a radix trie $\tau_\mathcal{R}$ for all reads in $\mathcal{R}$ and the suffix trie τ_T of T, we can align the radix trie to the suffix trie to find all p-mismatch occurrences of $\mathcal{R}$. The algorithm for finding all p-mismatch occurrences of $\mathcal{R}$ is shown in Figure 4.15.

For any node u in τ_T and any node v in $\tau_\mathcal{R}$, denote $PSet_{T,\mathcal{R},p}(u, v)$ as a set of pairs (u', v') where $depth(u') = m$, v' is a leaf in $\tau_\mathcal{R}$ and $Hamming(plabel(u'), plabel(v')) \le p$. The following lemma gives the recursive formula for $PSet_{T,\mathcal{R},p}(u, v)$.

Lemma 4.9 *Consider a radix trie $\tau_\mathcal{R}$ for a set of reads $\mathcal{R}$, a suffix trie τ_T for a genome T and a mismatch threshold p. For any node u in τ_T and any node v in $\tau_\mathcal{R}$, $PSet_{T,\mathcal{R},p}(u, v)$ can be computed in two cases. For the base case ($Hamming(plabel(u), plabel(v)) > p$ or v is a leaf), we have:*

$$PSet_{T,\mathcal{R},p}(u, v) = \begin{cases} \emptyset & Hamming(plabel(u), plabel(v)) > p \\ \{(u, v)\} & otherwise. \end{cases}$$

For the recursive case ($Hamming(plabel(u), plabel(v)) \le p$ and v is not a leaf), we have:

$$PSet_{T,\mathcal{R},p}(u, v) = \bigcup_{\sigma,\sigma' \in \{\mathtt{A,C,G,T}\}} PSet_{T,\mathcal{R},p}(child(u, \sigma), child(v, \sigma)).$$

Proof For the base case, we have $Hamming(plabel(u), plabel(v)) > p$ or v is a

---

Algorithm $PSet_{T,\mathcal{R},p}(u,v)$

Require: u is node in the suffix trie τ_T of T, v is a node in the radix treee $\tau_{\mathcal{R}}$ and p is the mismatch threshold

Ensure: Among all leaves u' under u and all leaves v' under v, report all (u',v') such that $Hamming(plabel_T(u'), plabel_{\mathcal{R}}(v')) \leq p$

1: **if** $Hamming(plabel_T(u), plabel_{\mathcal{R}}(v)) > p$ **then**
2:     Return $\emptyset$;
3: **else if** v is a leaf **then**
4:     Return $\{(u,v)\}$;
5: **else**
6:     $Ans = \emptyset$;
7:     **for** every (u',v') with $u' = child(u,\sigma)$ and $v' = child(u,\eta)$ **do**
8:         $Ans = Ans \cup PSet_{T,\mathcal{R},p}(u',v')$;
9:     **end for**
10:     Return Ans;
11: **end if**

FIGURE 4.15: By calling $PSet_{T,\mathcal{R},p}(root(\tau_T), root(\tau_{\mathcal{R}}))$, we find all nodes (u',v') such that v' is a leaf and $Hamming(plabel_T(u'), plabel_{\mathcal{R}}(v')) \leq p$.

leaf. If $Hamming(plabel(u), plabel(v)) > p$, $PSet_{T,\mathcal{R},p}(u,v) = \emptyset$ by definition. Otherwise, we have v is a leaf and $Hamming(plabel(u), plabel(v)) \leq p$. Then, $PSet_{T,\mathcal{R},p}(u,v) = \{(u,v)\}$.

For the recursive case, we have $Hamming(plabel(u), plabel(v)) \leq p$ and v is not a leaf. It is easy to verify that

$$PSet_{T,\mathcal{R},p}(u,v) = \bigcup_{\sigma,\sigma' \in \{\text{A,C,G,T}\}} PSet_{T,\mathcal{R},p}(child(u,\sigma), child(v,\sigma')).$$

∎

By implementing the recursive equation in the above lemma, Figure 4.15 gives the recursive algorithm to compute $PSet_{T,R,p}(root(\tau_T), root(\tau_{\mathcal{R}}))$.

4.3.4.4   Speed up utilizing the quality score

Previous methods align reads by their DNA sequences only. Moreover, every base of a read is associated with PHRED quality scores. We expect that mismatches occur mostly in bases with low quality scores. One question is whether the quality scores can improve read alignment. This problem is denoted as a quality score-aware alignment problem. Bowtie 1 [155] defines this problem as follows. The input consists of a reference $T[1..n]$, a read $R[1..m]$, a mismatch threshold p and a quality score threshold E. Bowtie 1 aims to report a position i in T such that $Hamming(R, T[i..i+m-1]) \leq p$ and $QS(R, T[i..i+m-1]) \leq E$ where $QS(R, R') = \sum_{R[i] \neq R'[i]} Q[i]$.

For example, consider an input $T = \texttt{ACACAG\$}$, $p = 1$, $E = 30$, $R = \texttt{ACGG}$ and the corresponding quality score is $(40, 20, 25, 40)$. Then, Bowtie 1 will report position 3 since $Hamming(R, T[3..6]) = 1$ and $QS(R, T[3..6]) = 25$.

Utilizing the suffix trie τ_T, Bowtie 1 proposed two techniques to resolve the problem: (1) double indexing and (2) a quality score-aware mapping algorithm. Double indexing is described in Section 4.3.4.2. Below, we detail the quality-aware alignment algorithm.

For any node w in τ_T, denote $QSmap_{T,R,p,E}(w)$ as a function that returns a descendant node u of w such that $depth(u) = m$, $Hamming(R, plabel(u)) \leq p$ and $QS(R, plabel(u)) \leq E$. If node u does not exist, $QSmap_{T,R,p,E}(w)$ returns ϵ. (Note that $QSmap_{T,R,p,E}(w)$ may not have a unique solution.) By definition, Bowtie 1 aims to compute $QSmap_{T,R,p,E}(root(\tau_T))$. (In Bowtie 1, $p = 2$ and $E = 70$ by default.) The following lemma states the recursive formula.

Lemma 4.10 *Denote P_u as the path from u to v in τ_T, where v is the deepest descendant node of u such that the path label of P_u equals $R[depth(u) + 1..depth(v)]$. Let $Child(P_u)$ be all nodes w such that $w \notin P_u$ and the parent of w is in P_u. For the base case $(depth(v) = m)$, we have*

$$QSmap_{T,R,p,E}(u) = \begin{cases} v & Hamming(R, R') \leq p \text{ and } QS(R, plabel(v)) \leq E \\ \epsilon & otherwise. \end{cases}$$

For the recursive case $(depth(v) < m)$, we have $QSmap_{T,R,p,E}(u) \in \{QSmap_{T,R,p,E}(w) \mid w \in Child(P_u)\}$.

Proof When $depth(v) = m$, by definition, $QSmap_{T,R,p,E}(v) = v$ if $Hamming(R, R') \leq p$ and $QS(R, plabel(v)) \leq E$.

When $depth(v) < m$, let $P_u = (u = u_1, u_2, \ldots, v = u_t)$ be the path from u to v in τ_T. Let $P_{u_j} = (u_j, u_{j+1}, \ldots, u_t)$.

We claim that $QSmap_{T,R,p,E}(u_j) \in \{QSmap_{T,R,p,E}(w) \mid w \in Child(P_{u_j})\}$ for any $j = 1, \ldots, t$.

We prove the claim by induction. When $j = t$, P_{u_t} contains exactly one node u_t. $QSmap_{T,R,p,E}(u_t) \in \{QSmap_{T,R,p,E}(w) \mid w \in Child(u_t)\}$.

For $j < t$, suppose $QSmap_{T,R,p,E}(u_{j+1}) \in \{QSmap_{T,R,p,E}(w) \mid w \in Child(P_{u_{j+1}})\}$. Then, $QSmap_{T,R,p,E}(u_j) \in \{QSmap_{T,R,p,E}(w) \mid w \in Child(u_j)\} = \{QSmap_{T,R,p,E}(w) \mid w \in Child(u_j) - \{u_{j+1}\}\} \cup \{QSmap_{T,R,p,E}(u_{j+1})\}$. Since $QSmap_{T,R,p,E}(u_{j+1}) \in \{QSmap_{T,R,p,E}(w) \mid w \in Child(P_{u_{j+1}})\}$, we have $QSmap_{T,R,p,E}(u_j) \in \{QSmap_{T,R,p,E}(w) \mid w \in Child(P_{u_t})\}$. The claim is true. ∎

By the above lemma, a recursive algorithm can be developed that computes a depth-m node v in τ_T such that $Hamming(R, plabel(v)) \leq p$ and $QS(R, plabel(v)) \leq E$. Bowtie 1 suggested that we can speed up the algorithm using a quality score. Since we want to reduce the sum of quality scores in mismatch positions, the greedy approach is to try mismatches in the lowest quality score position first. Hence, we can try to run $QSmap_{T,R,p,E}(w)$

Algorithm $QSmap_{T,R,p,E}(u)$

Require: T is the genome, $R[1..m]$ is a read, p is the mismatch threshold, E is the quality score threshold and u is any node in τ_T.

Ensure: A depth-m descendant v of u such that $Hamming(R, plabel(v)) \le p$ and $QS(R, plabel(v)) \le E$.

1: **if** $depth(u) > m$ or $Hamming(R[1..depth(u)], plabel(u)) > p$ or $QS(R, plabel(u)) > E$ **then**
2:     Return ϵ;
3: **end if**
4: Identify the lowest descendant v of u in τ_T such that the path label of $P = u, \ldots, v$ equals $R[depth(u) + 1..depth(v)]$;
5: Return v if $depth(v) = m$;
6: **for** every $w \in Child(P)$ in increasing order of $Q[depth(w)]$ **do**
7:     Return $QSmap_{T,R,p,E}(w)$ if $QSmap_{T,R,p,E}(w) \ne \epsilon$;
8: **end for**
9: Return ϵ;

FIGURE 4.16: By calling $QSmap_{T,R,p,E}(root(\tau_T))$, we report a node v in τ_T such that $Hamming(R, plabel(v)) \le p$ and $QS(R, plabel(v)) \le E$.

for $w \in Child(P_u)$ in increasing order of $Q[depth(w)]$. Figure 4.16 states the detail of the algorithm $QSmap_{T,R,p,E}(u)$.

Figure 4.17 illustrates the method. In the example, the genome is $T =$ ACACAG$. The read is $R =$ ACGG, whose corresponding quality scores are $(40, 20, 25, 40)$. We aim to find a hit of R on T allowing at most $p = 1$ mismatch and the quality score threshold is $E = 30$. We start by calling $QSmap_{T,R,1,E}(v_0)$ (see Figure 4.17(b)). $QSmap_{T,R,1,E}(v_0)$ determines $P_{v_0} = (v_0, v_1, v_4)$ and computes $Child(v_0, v_1, v_4) = \{v_2, v_3, v_5, v_7\}$. Note that $Q[depth(v_2)] = 40, Q[depth(v_3)] = 40, Q[depth(v_5)] = 20, Q[depth(v_7)] = 25$. $QSmap_{T,R,1}(v_0)$ selects the node with the lowest quality score, which is v_5; then, it calls $QSmap_{T,R,1,E}(v_5)$ (see Figure 4.17(c)). $QSmap_{T,R,1,E}(v_5)$ tries to find a depth-m descendant of v_5 whose path label is a 1-mismatch hit of R. However, it is impossible to find such a node and $QSmap_{T,R,1,E}(v_5)$ return ϵ. After that, among $\{v_2, v_3, v_5, v_7\}$, $QSmap_{T,R,1,E}(v_0)$ selects the node with the second lowest quality score, which is v_7; then, it calls $QSmap_{T,R,1,E}(v_7)$ (see Figure 4.17(d)). $QSmap_{T,R,1,E}(v_7)$ finds $P_{v_7} = (v_7, v_{11})$. Since $depth(v_{11}) = m = 4$, $QSmap_{T,R,1,E}(v_7)$ return v_{11}.

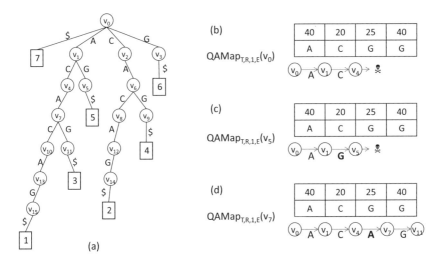

FIGURE 4.17: (a) is the suffix trie of $T = $ ACACAG\$. Let R be the read ACGG, whose list of quality scores is $(40, 20, 25, 40)$. (b) illustrates the alignment obtained when we call $QSmap_{T,R,1,E}(v_0)$ where $E = 30$. $QSmap_{T,R,1,E}$ first recursively calls $QSmap_{T,R,1,E}(v_5)$, which returns ϵ. Then, $QSmap_{T,R,1,E}$ recursively calls $QSmap_{T,R,1,E}(v_7)$, which returns v_{11}.

4.4 Aligning reads allowing a small number of mismatches and indels

The previous section describes the mismatch-only read alignment. However, it fails to capture the wide spectrum of indel variants which are expected to occur once in every 3000 bp [122]. To capture indels, this section discusses alignment methods that allow both mismatches and indels. Given a reference $T[1..n]$, this section discusses methods for aligning a read $R[1..m]$ to some region $T[i..j]$ that minimizes the edit distance or maximizes the semi-global alignment score or local alignment score (the alignment score is defined in Section 4.2.1).

Below, we describe a few techniques to solve this problem. They include the q-mer approach and suffix-trie approach.

4.4.1 q-mer approach

The hashing methods find p-mismatch hits of a read R on the reference genome. When there are insertion or deletion errors (like 454, Ion-torrent, PacBio reads), we need to compute the edit distance between the read and the reference genome instead. However, hashing is not designed for computing edit distance. Here, we describe the q-mer approach. q-mer is a technique that speeds up the edit-distance computation. It is based on the following observation.

Lemma 4.11 *Given a length-m sequence R and another sequence Y, if $edit(R, Y) \le p$, then R and Y share at least t common q-mers (length-q substrings) where $t = m + 1 - (p + 1)q$.*

Proof Suppose R and Y have at most p differences (i.e., indels or substitutions). R has $(m + 1 - q)$ q-mers. Note that a q-mer in R overlaps with some difference if and only if R and Y do not share that q-mer. Each difference overlaps with at most q q-mers. Since there are at most p differences, at most pq q-mers overlap with the differences. Thus, R and Y share at least $(m + 1 - q - pq)$ q-mers. The lemma follows. ∎

   To illustrate the correctness of the lemma, consider two sequences $R = $ GACGCAT and $Y = $ GACGCAG with $edit(R, Y) = 1 = p$. Lemma 4.11 implies that R and Y share at least $t = m + 1 - (p + 1)q = 7 + 1 - (1 + 1)3 = 2$ 3-mers. This is true since the number of 3-mers shared by R and Y is $4 > t$. For another example, consider two sequences $R = $ AAATGGATT and $Y = $ AAACAACGG with $edit(R, Y) = 2 = p$. We have $t = m + 1 - (p + 1)q = 9 + 1 - (2 + 1)3 = 1$. Since the number of 3-mers shared by R and Y is $1 = t$, Lemma 4.11 is tight.
   Note that the reverse statement of the lemma is not correct. In other words, R and Y share at least $t = m + 1 - (p + 1)q$ q-mers does not imply $edit(R, Y) \le p$. For example, consider $R = $ ACAGCAT and $Y = $ GCGGCAT. They share $t = 2$ 3-mers. By substituting $m = 7, q = 3, t = 2$ into $t = m + 1 - (p + 1)q$, we have $p = 1$. However, $edit(R, Y) = 2 > p$.
   Using Lemma 4.11, we can find all positions i such that $edit(R, T[i..j]) \le p$ for some j as follows. We scan the reference genome T to find all positions i such that R and $T[i..i + m + p - 1]$ share at least t q-mers; then, each position i is reported as a hit of R if $\min_{i \le j \le i+m+p-1} edit(R, T[i..j]) \le p$.
   Figure 4.18 details the algorithm. The algorithm assumes the suffix array of T is constructed (see Section 3.5.2). Hence, we have the list of occurrences of every q-mer. Steps 1−7 of the algorithm try to compute $Count_i$, which is the number of q-mers shared between R and $T[i..i + m + p - 1]$, for every position i in the genome T. By Lemma 4.11, we know that $edit(R, T[i..i + |R| - 1]) \le p$ only if $Count_i > t = |R| + 1 - (p + 1)q$. Hence, Steps 8−11 will only compute edit distance for position i when $Count_i > t$.
   A number of methods use the q-mer approach or its variants, including: SHRiMP [255], RazerS [311], and SeqAlto [208]. SeqAlto uses long q-mers, where $17 \le q \le 32$, to improve the efficiency.
   SHRiMP [255] and RazerS [311] generalize q-mer to spaced q-mer. Spaced q-mer is a noncontiguous q-mer. Its idea is similar to the spaced seed in Section 4.3.2. When the number of mismatches in a read increases, we may fail to have q-mer hits. Spaced q-mer increases the chance of having hits. More precisely, the minimum number of spaced q-mers shared by two length-m strings

---

Algorithm p-difference(R, T)

Require: A read $R[1..m]$ and a reference genome $T[1..n]$

Ensure: A list of hits of the read R with at most p differences

1: **for** every position i in T **do**
2:     Set $Count_i = 0$;
3: **end for**
4: **for** each q-mer Q in R **do**
5:     Find the hitlist, that is, the list of positions in T where Q occurs;
6:     For every position i in the hitlist, increment the counters $Count_i, Count_{i-1}, \ldots, Count_{i-m+p-q}$ by 1;
7: **end for**
8: Set $t = m + 1 - (p + 1)q$;
9: **for** position i with $Count_i > t$ **do**
10:     If $\min_{i \le j \le i+m+p-1} edit(R, T[i..j]) \le p$, report position i is a hit of R;
11: **end for**

FIGURE 4.18: The algorithm for finding p-difference hits for a read R in the reference genome T.

of Hamming distance p is actually bigger than that for a contiguous q-mer [33]. This number can be computed by dynamic programming (see Exercise 18). Exercise 17 gives an example showing that a spaced q-mer approach is more sensitive than the q-mer approach.

RazerS [311] relaxes the constraint used in Lemma 4.11. Lemma 4.11 showed that when R and Y share more than $t = m + 1 - (p + 1)q$ q-mers, R and Y have at most p differences. What happens if we filter reads using a threshold θ bigger than $t = m + 1 - (p+1)q$? In this case, some true hits may be missed. Weese et al. [311] generalizes Lemma 4.11 and describes a dynamic programming method to compute the probability that a randomly chosen true match satisfies the threshold θ. Then, instead of fixing the threshold to be t, RazerS allows users to control the threshold. If the user chooses a threshold θ bigger than t, more reads are filtered, which improves the running time; however, we will also miss some true hits. Hence, by controlling the threshold θ, RazerS can provide a seamless trade-off between sensitivity and running time.

4.4.2 Computing alignment using a suffix trie

Section 4.3.4 presents the suffix trie approach to solve the Hamming distance read mapping problem. The suffix trie approach can also be used to solve the edit-distance read mapping problem and the read alignment problem. This section discusses these methods.

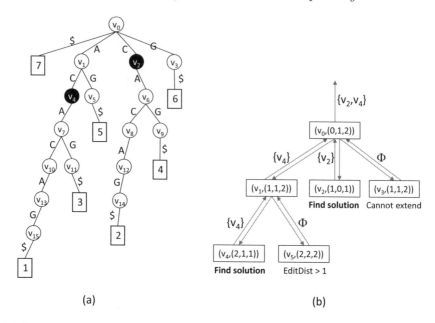

(a)   (b)

FIGURE 4.19: (a) is the suffix trie of $T = $ ACACAG\$. The two nodes v_2 and v_4 are the most ancestral nodes such that $edit(R, plabel(v_2)), edit(R, plabel(v_4)) \leq 1$. (b) illustrates the recursive computation of $EditDist_{T,\text{CC},1}(v_0, (0, 1, 2))$. We start from problem instance $(v_0, (0, 1, 2))$. It calls three sub-instances. For one sub-instance $(v_1, (1, 1, 2))$, it further calls two sub-instances. The label of the upward edge is the set of nodes returned by the problem instance.

4.4.2.1 Computing the edit distance using a suffix trie

Consider a reference genome $T[1..n]$ and an edit-distance threshold p. For any read $R[1..m]$, the edit-distance problem aims to find all positions i in T such that $edit(R, T[i..j]) \leq p$ for some $j \geq i$.

Similar to Section 4.3.4, the edit-distance problem can be solved using a suffix trie as follows. First, we find the set of nodes $\mathcal{Z} = \{u \in \tau_T \mid edit(R, plabel(u)) \leq p\}$. Second, we report $\bigcup_{u \in \mathcal{Z}} Leaf_T(u)$. For example, consider $T = $ ACACAG\$, $R = $ CC and $p = 1$. Figure 4.19(a) shows the corresponding suffix trie τ_T. It is easy to check that $\mathcal{Z} = \{v_2, v_4, v_6\}$. Then, the answer is $Leaf(v_2) \cup Leaf(v_4) \cup Leaf(v_6) = \{1, 2, 4\}$.

However, $\mathcal{Z}$ has redundancy. Since v_6 is a descendant of v_2, $Leaf(v_6) \subseteq Leaf(v_2)$. The redundancy can be avoided by reporting only the most ancestral nodes u such that $edit(R, plabel(u)) \leq p$. ($u$ is a most ancestral node if $edit(R, plabel(w)) > p$ for all ancestors w of u.) For the above example, the set of all of the most ancestral nodes is $\{v_2, v_4\}$, which does not have any redundancy.

This section describes an algorithm $EditDist_{T,R,p}()$ to compute all of the most ancestral nodes u such that $edit(R, plabel(u)) \leq p$. First, we have a definition. Denote $d_v[i] = edit(R[1..i], plabel(v))$. The following lemma gives the recursive formula for $d_v[i]$.

Lemma 4.12 *Denote $\delta(x, y) = 0$ if $x = y$; and $\delta(x, y) = 1$ otherwise.*

- *Base case ($v = root(\tau_T)$): $d_{root(\tau_T)}[i] = i$.*

- *Recursive case ($v \neq root(\tau_T)$): Let u be the parent of v in τ_T and σ be the edge label for the edge (u, v) (i.e., $v = child(u, \sigma)$). We have $d_v[0] = d_u[0] + 1$ and, for $i > 0$,*

$$d_v[i] = \min \begin{cases} d_u[i-1] + \delta(R[i], \sigma) & match/mismatch \\ d_u[i] + 1 & insert \\ d_v[i-1] + 1 & delete. \end{cases} \quad (4.1)$$

Proof Exercise 20. ∎

Now, we describe the recursive algorithm $EditDist_{T,R,p}(u, d_u[0..m])$. The recursive algorithm required two parameters u and $d_u[0..m]$. To compute the set of the most ancestral nodes w such that $edit(R, plabel(w)) \leq p$, we call $EditDist_{T,R,p}(root(\tau_T), d_{root(\tau_T)}[0..m])$. (By Lemma 4.12, $d_{root(\tau_T)}[0..m] = (d_{root(\tau_T)}[0], d_{root(\tau_T)}[1], \ldots, d_{root(\tau_T)}[m]) = (0, 1, \ldots, m)$.) We have the following lemma.

Lemma 4.13 $EditDist_{T,R,p}(u, d_u[0..m])$ *reports the following:*

- *$\{u\}$ if $d_u[m] \leq p$.*

- *$\emptyset$ if $d_u[i] > p$ for $i = 0, \ldots, m$.*

- *$\bigcup_{v=child(u,\sigma)} \{EditDist_{T,R,p}(v, d_v[0..m])\}$ otherwise.*

(Note: $d_v[0..m]$ can be computed from $d_u[0..m]$ using Lemma 4.12.)

Proof Exercise 21. ∎

$EditDist_{T,R,p}(u, d_u[0..m])$ runs in three steps. First, it checks if $d_u[m] \leq p$. If yes, it returns $\{u\}$. Second, it checks if $d_u[i] > p$ for $i = 0, \ldots, m$. If yes, it returns $\emptyset$. Otherwise, the last step returns $\bigcup_{v=child(u,\sigma)} \{EditDist_{T,R,p}(v, d_v[0..m])\}$. (The correctness follows from Lemma 4.13.) The detail of the algorithm is stated in Figure 4.20. For the above example ($T = $ ACACAG$, $R = $ CC and $p = 1$), $\{u \in \tau_T \mid edit(R, plabel(u)) \leq 1$ and u is a most ancestral node$\}$ can be found by executing $EditDist_{T,R,p}(v_0, d_{v_0}[0..m])$, where $d_{v_0}[0..m] = (0, 1, 2)$. Figure 4.19(b) shows the recursion tree for $EditDist_{T,R,p}(v_0, d_{v_0}[0..m])$.

The algorithm can be further sped up by estimating the lower bound of the edit distance similar to that in Section 4.3.4.1 (see Exercise 19).

Algorithm EditDist$_{T,R,p}(u, d_u[0..m])$

Require: u is a node in τ_T and $d_u[i]$ is the edit distance between $R[1..i]$ and $plabel(u)$ for $i = 0, 1, \ldots, m$.

Ensure: All most ancestral nodes u in τ_T such that $edit(R, plabel(u)) \leq p$.

1: **if** $d_u[m] \leq p$ **then**
2:     Return $\{u\}$;
3: **else if** $d_u[i] > p$ for $i = 0, \ldots, m$ **then**
4:     Return $\emptyset$;
5: **else**
6:     $Ans = \emptyset$;
7:     **for** every v with $v = child(u, \sigma)$ **do**
8:         $d_v[0] = d_u[0] + 1$;
9:         **for** $i = 1$ to m **do**
10:            $d_v[i] = \min\{d_u[i-1] + \delta(R[i], \sigma), d_u[i] + 1, d_v[i-1] + 1\}$
11:        **end for**
12:        $Ans = Ans \cup EditDist_{T,R,p}(v, d_v)$;
13:    **end for**
14:    Return Ans;
15: **end if**

FIGURE 4.20: This algorithm returns all of the most ancestral nodes u in τ_T such that $edit(R, plabel(u)) \leq p$ by calling $EditDist_{T,R,p}(root(\tau_T), d_{root(\tau_T)}[0..m])$ where $d_{root(\tau_T)}[0..m] = (0, 1, \ldots, m)$.

4.4.2.2 Local alignment using a suffix trie

Given a reference genome T and a read R, the local alignment problem aims to compute the local alignment score between R and T. By definition, the local alignment score between R and T equals the maximum global alignment score $score(R[i..j], T[i'..j'])$ among all substrings $R[i..j]$ of R and all substrings $T[i'..j']$ of T.

The computation of the local alignment score can be accelerated by a suffix trie. Recall that any substring of T can be represented by $plabel(u)$ for some u in τ_T. For any $i \in [1..m]$ and every node v in τ_T, we define $V_v(i)$ as the maximum alignment score between any suffix of $R[1..i]$ and $plabel(v)$. It is easy to verify that the local alignment score between R and T equals $\max_{i \in [1..m], v \in \tau_T} V_v(i)$.

The following lemma gives the recursive formula for $V_v(i)$.

Lemma 4.14 *For any nucleotide* $x, y \in \{a, c, g, t\}$*, let* $\delta(x, y)$ *be the substitution score,* $\delta(x, -)$ *be the deletion score, and* $\delta(-, x)$ *be the insertion score. (For example, we can set* $\delta(x, y) = +2$ *if* $x = y$ *and* $\delta(x, y) = -1$ *if* $x \neq y$*.)*

- *Base case (v is the root):* $V_{root(\tau_T)}(i) = 0$

- *Recursive case (v = child(u, \sigma)):* $V_v(0) = V_u(0) + \delta(-, \sigma)$
$$V_v(i) = \max \begin{cases} V_v(i-1) + \delta(R[i], \sigma) & match/mismatch \\ V_v(i-1) + \delta(R[i], -) & delete \\ V_u(i) + \delta(-, \sigma) & insert \end{cases}$$

Proof Exercise 20. ∎

By applying dynamic programming on the recursive equation in Lemma 4.14, $V_v(i)$ can be computed for all $i \in [1..m]$ and all nodes v in τ_T; then, we report (i, v) that maximizes $V_v(i)$. Note that τ_T contains $O(n^2)$ nodes. The above solutions will take $O(n^2 m)$ time, which is slower than the Smith-Waterman algorithm.

Observe that it is unnecessary to compute $V_v(i)$ for all pairs (i, v). First, we need a definition. A pair (i, v) is called meaningful if any prefix of the alignment for $(R[1..i], plabel(v))$ has a positive score; otherwise, it is called meaningless. (See Figure 4.21 for examples.)

We have the following lemma, which is useful to avoid unnecessary computation.

Lemma 4.15 *If* (i, v) *is a meaningless alignment, there exists another node* v' *such that* $plabel(v')$ *is a suffix of* $plabel(v)$ *and* $V_{v'}(i) \geq V_v(i)$*.*

Proof Suppose (i, v) corresponds to a meaningless alignment between $R[j..i]$ and $plabel(v)$, with score $V_v(i)$. The meaningless alignment can be partitioned into two halves: $(R[j..i'], P_1)$ and $(R[i' + 1..i], P_2)$ where (1) $plabel(v)$ is the concatenation of P_1 and P_2 and (2) the alignment score for $(R[j..i'], P_1)$ is

ACTGG-GCA   AGTCGCA
ACTCGTG-A   ACAGG-A
(a)   (b)

FIGURE 4.21: (a) and (b) show two alignments. Suppose the scores for match, mismatch, and indels are $+2, -1, -1$, respectively. (Assume linear gap penalty.) For (a), it is a meaningful alignment of score 9 since any prefix alignment has a positive score. For (b), it is a meaningless alignment of score 4 since the alignment of the first 4 bases has the score $+2 - 1 - 1 - 1 = -1$. Note that its length-3 suffix alignment is a meaningful alignment, with score 3.

non-positive. This implies that the alignment score s of $(R[i' + 1..i], P_2)$ must be at least $V_v(i)$. Let v' be the node such that $plabel(v') = P_2$. Then, $V_{v'}(i) \geq s \geq V_v(i)$. ∎

The above lemma implies that it is unnecessary to extend an alignment for (i, v) if it is meaningless. We update the definition of $V_v(i)$ as follows. We define $V_v(i)$ as the maximum meaningful alignment score between any suffix of $R[1..i]$ and $plabel(v)$. After the modification, we have the following lemma.

Lemma 4.16 *Let $\delta(x, y)$ be the substitution score for $x, y \in \{a, c, g, t, -\}$. Let $\Delta(z) = z$ if $z > 0$; otherwise, $\Delta(z) = -\infty$.*

- *Base case ($i = 0$ or $v = root(\tau_T)$): $V_{root(\tau_T)}(i) = 0$ and $V_v(0) = -\infty$ if v is not the root of τ_T*

- *Recursive case ($v = child(u, \sigma)$):*

$$V_v(i) = \Delta \left(\max \left\{ \begin{array}{l} V_u(i-1) + \delta(R[i], \sigma) \\ V_v(i-1) + \delta(R[i], -) \\ V_u(i) + \delta(-, \sigma) \end{array} \right\} \right) \tag{4.2}$$

Proof (Sketch) By definition, the base case follows. Let A be the meaningful alignment between $R[1..i]$ and $plabel(v)$ of score $V_v(i)$. Let $v = child(u, \sigma)$. The rightmost column of the alignment A is either (1) substitution, (2) deletion or (3) insertion. For (1), we have $V_v(i) = V_u(i-1) + \delta(R[i], \sigma)$. For (2), we have $V_v(i) = V_v(i-1) + \delta(R[i], -)$. For (3), we have $V_v(i) = V_u(i) + \delta(-, \sigma)$. To ensure that the alignment is meaningful, we require $V_v(i) > 0$; otherwise, we set $V_v(i) = -\infty$. The lemma follows. ∎

By the above lemma, we can compute $V_v(i)$ recursively by Equation (4.2). To compute $(V_v(0), \ldots, V_v(m))$, we traverse all nodes u in τ_T in top-down order. For each node u, we compute $(V_v(0), \ldots, V_v(m))$; then, we report $\max_{i=0,\ldots,m, v \in \tau_T} V_v(i)$. To speed up, we can implement a pruning strategy based on the following lemma.

Algorithm LocalAlign(V_u, u)

Require: $R[1..m]$ is the read, τ_T is the suffix trie of the reference genome T, u is a node in τ_T and $V_u = (V_u(0), \ldots, V_u(m))$

Ensure: Compute the optimal local alignment score between R and T (assuming linear gap penalty)

  1: $s = \max\{-\infty, \max_{i=0,\ldots,m} V_u(i)\}$;

  2: **if** (u is a leaf) or ($s = -\infty$) **then**

  3:     Report s;

  4: **end if**

  5: **for** every v with $v = child(u, \sigma)$ **do**

  6:     **for** $i = 0$ to m **do**

  7:       Compute $V_v(i)$ by Equations (4.2);

  8:     **end for**

  9:     $s = \max\{s, \text{LocalAlign}(V_v, v)\}$;

10: **end for**

11: Report s;

FIGURE 4.22: The dynamic programming algorithm to compute the optimal local alignment between R and T. To execute the program, run $LocalAlign((0, \ldots, 0), root(\tau_T))$.

Lemma 4.17 *Suppose* $(V_u(0), \ldots, V_u(m)) = (-\infty, \ldots, -\infty)$. *Then,* $V_v(i) = -\infty$ *for any* $i = 0, \ldots, m$ *and any descendant* v *of* u.

Proof See Exercise 23. ∎

By Lemma 4.17, when $(V_u(0), \ldots, V_u(m)) = (-\infty, \ldots, -\infty)$, we don't need to compute $V_v(i)$ for any descendant v of u. Together with Lemma 4.16, we have the algorithm in Figure 4.22. Figure 4.23 illustrates the execution of algorithm.

Lam et al. [151] showed that this algorithm runs in $O(n^{0.628}m)$ time on average, which is faster than the standard Smith-Waterman algorithm. However, its worst-case running time is still $O(n^2 m)$. The worst-case running time can be improved by replacing the suffix trie with the directed acyclic word graph (DAWG) [67].

Last but not least, this method can be generalized to handle the affine gap penalty (Exercise 24).

4.5 Aligning reads in general

Sections 4.3 and 4.4 discuss techniques that align reads allowing a small number of mismatches and indels. The running time of these methods are

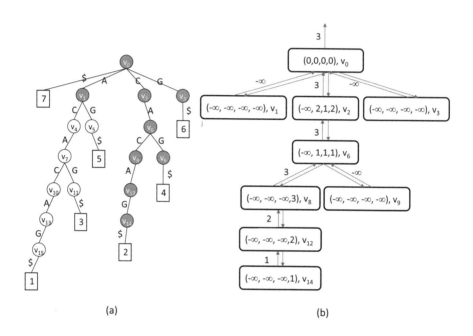

(a)

(b)

FIGURE 4.23: (a) is the suffix trie τ_T of $T = \texttt{ACACAG\$}$. Suppose the scores for match, mismatch, and indels are $+2, -1, -1$, respectively. We aim to compute the optimal local alignment score between $Q = \texttt{CTC}$ and T by executing $LocalAlign(V_{v_0}, v_0)$, where $V_{v_0} = (V_{v_0}(0), V_{v_0}(1), V_{v_0}(2), V_{v_0}(3)) = (0, 0, 0, 0)$. The gray nodes are nodes visited by the algorithm. (b) illustrates the recursive computation of $LocalAlign((0, 0, 0, 0), v_0)$. We start from problem instance $((0, 0, 0, 0), v_0)$. The algorithm recursively calls the sub-instances. The label of the upward edge is the value returned by the problem instance. The optimal local alignment score is 3.

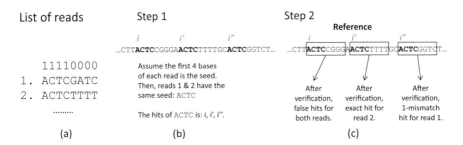

FIGURE 4.24: Illustration of the seed-and-extension approach. (a) is a list of reads. (b) illustrates the seeding step. We first extract the seed (which is the first 4 bases in this example) of each read. Then, some efficient method is applied to identify the list of hits of the seeds. This list of hits is treated as the candidate hit list of the reads. (c) illustrates the extension step. Each candidate hit is aligned with the full read to verify if it is a correct hit.

exponential in the number of errors. As the reads get long, these methods cannot work.

This section discusses the general strategy for aligning reads. It uses methods in the previous two sections as the building blocks and solve the read alignment problem in general. Roughly speaking, the methods can be classified into two classes: (1) the seed-and-extension approach and (2) the filtering approach. Below, we will discuss these two approaches in detail.

4.5.1 Seed-and-extension approach

Observe that errors are not uniformly distributed in an alignment. Some regions of an alignment will have more mismatches/indels while some regions will have less. This suggests the following seed-and-extend strategy to align reads (which is illustrated in Figure 4.24):

- Step 1 (Seeding step): This step extracts some substrings of each read as seeds. Then, hits of the seeds allowing a small number of errors are computed. These hits are the candidate hits of the read. This step is accelerated by the efficient algorithms in Sections 4.3−4.4.

- Step 2 (Extension step): This step validates the goodness of the candidate hits. Since there are not many candidate hits, accurate but slow methods (see Section 4.2.3) can be used to verify them.

Note that the seed-and-extend strategy is a trade-off between efficiency and sensitivity. To reduce the running time, we can adjust Step 1 to identify fewer candidate hits with some stringent criteria; however, this will reduce the sensitivity. On the other hand, we can adjust Step 1 to use loose criteria so that the optimal hits will not be missed; however, this will increase the running time.

Many short read aligners use the seed-and-extend strategy. They include BWA-SW [168], Bowtie 2 [154], BatAlign [178], LAST[?], Cushaw2 [182], BWA-MEM , etc. Below, we demonstrate the use of the seed-and-extend strategy in different aligners.

4.5.1.1 BWA-SW

Given a reference genome $T[1..n]$ and a read $R[1..m]$, Figure 4.22 gives the algorithm LocalAlign, which performs local alignment using a suffix trie to find substrings X of R and Y of T such that $score(X,Y)$ is maximized, where $score(X,Y)$ is the global alignment score between X and Y. Although it is faster than the Smith-Waterman algorithm, it is still too slow to align short reads.

Li and Durbin [168] proposed BWA-SW to further speed up this process. BWA-SW uses the seed-and-extend strategy and it consists of two steps: a seeding step and an extension step. The seeding step finds a set of good candidate hits (X,Y) such that $score(X,Y) > 0$, X is a substring of R and Y has at most γ occurrences in T (by default, $\gamma = 3$). Then, the extension step of BWA-SW uses the Smith-Waterman algorithm to extend the good candidate hits and reports the hit with the highest alignment score.

The extension step is straightforward. Below, we detail the seeding steps. Note that every substring Y of T equals $plabel(u)$ for some node u in the suffix trie τ_T. Hence, finding good candidate hits is equivalent to finding

$$\begin{aligned} \mathcal{H} &= \{(i,u) \mid |Leaf(u)| \le \gamma, max_{i' \le i} score(R[i'..i], plabel(u)) > 0\} \\ &= \{(i,u) \mid |Leaf(u)| \le \gamma, V_u(i) > 0\}, \end{aligned}$$

where $V_u(i)$ is the best alignment score between some suffix of $R[1..i]$ and $plabel(u)$. $\mathcal{H}$ can be found efficiently by running a modified version of the algorithm in Figure 4.22 (see Exercise 25).

However, when the read R is long, $\mathcal{H}$ may contain many node pairs (i,u). Li and Durbin proposed two heuristics to further reduce the size of $\mathcal{H}$. The first heuristic is the Z-best strategy. For every position i in R, let $\{(i,u_1),\ldots,(i,u_\alpha)\}$ be the set of pairs in $\mathcal{H}$. The Z-best strategy only keeps Z pairs which have the highest scores $V_{u_j}(i)$. Using simulation and real data, [168] showed that the Z-best strategy does not filter high-homology hits.

Another heuristic is the overlap filter. We retain pairs in $\mathcal{H}$ that are largely non-overlapping. Precisely, consider two pairs $(i_1,u_1),(i_2,u_2) \in \mathcal{H}$, where the pair (i_s,u_s) corresponds to the alignment between $R[i'_s..i_s]$ and $plabel(u_s)$, for $s = 1,2$. The two pairs (i_1,u_1) and (i_2,u_2) are said to be non-overlapping if the overlap length between $R[i'_1..i_1]$ and $R[i'_2..i_2]$ is less than $\frac{1}{2}\min\{i_1 - i'_1 + 1, i_2 - i'_2 + 1\}$.

To illustrate the seeding step of BWA-SW, we use the example in Figure 4.23. Suppose we set $\gamma = 1$. Then, $\mathcal{H} = \{(3,u_8),(3,u_{12}),(3,u_{14})\}$. Note that $(3, u_8)$ corresponds to the alignment between $R[1..3]$ and $T[2..4]$ with score $V_{u_8}(3) = 3$, $(3, u_{12})$ corresponds to the alignment between $R[1..3]$ and $T[2..5]$ with score $V_{u_{12}}(3) = 2$, and $(3, u_{14})$ corresponds to the alignment between $R[1..3]$ and $T[2..6]$ with score $V_{u_{14}}(3) = 1$.

Suppose we apply the Z-best strategy with $Z = 2$ to $\mathcal{H}$. Then, among the pairs in $\mathcal{H} = \{(3, u_8), (3, u_{12}), (3, u_{14})\}$, we filter $(3, u_{14})$ since $V_{u_8}(3), V_{u_{12}}(3) > V_{u_{14}}(3)$. For the overlap filter, as $(3, u_8)$ and $(3, u_{12})$ are overlapping, we filter $(3, u_{12})$ which has a lower score. After applying the Z-best strategy and overlap filter, $\mathcal{H}$ only contains one pair $\{(3, u_8)\}$.

To further speed up the algorithm, the original paper of BWA-SW suggested that we can replace the read R by the directed acyclic word graph (DAWG) of R. Please refer to [168] for details.

4.5.1.2 Bowtie 2

Bowtie 2 [154] is a gapped aligner that aligns a read $R[1..m]$ onto the reference genome $T[1..n]$ allowing gaps. It extends Bowtie 1 (an ungapped aligner) by the seed-and-extend strategy.

The basic assumption is that, if $T[i..j]$ is a correct hit of the read R, then $T[i..j]$ and R share some length-22 substrings which have very few occurrences in the reference genome T.

The algorithm of Bowtie 2 consists of two steps: a seeding step and an extension step. The seeding step of Bowtie 2 extracts length-22 seeds from the read R for every $\lfloor 1 + 0.75\sqrt{m} \rfloor$-bp interval. For example, when $m = 100$, we extract length-22 seeds from R for every 8 bp. Each seed is aligned to the suffix trie τ_T allowing 0 mismatches (without gap) by Bowtie 1. Then, the candidate hits of all seeds are obtained. The number of candidate hits may be huge. The following paragraphs describe the candidate hit filters used by Bowtie 2. The extension step of Bowtie 2 performs a gapped extension for these candidate hits. The gapped extension is based on the Smith-Waterman algorithm, which is sped up by SIMD (single-instruction multiple-data) accelerated dynamic programming.

It is too time consuming to extend the candidate hits of all length-22 seeds. To speed up, Bowtie 2 proposes a few heuristics to select a subset of candidate hits for gapped extension. Bowtie 2 assumes that seeds with a smaller number of occurrences are likely to be correct. It prioritizes the seeds as follows. Each candidate hit at position i in the list is assigned a priority $\frac{1}{s^2}$ if the seed that appears in position i has s occurrences in T. Bowtie 2 iteratively selects random candidate hits based on the priority weight. To avoid executing an excessive number of dynamic programming iterations, Bowtie 2 has a stopping strategy. Bowtie 2 maintains the alignment score of the alignment of every gapped extension it performed. An alignment is said to be failed if the alignment score is worse than the current best or the second-best alignment found so far. Bowtie 2 will stop if it fails some fixed number of attempts (by default, 15). Finally, among all alignments computed, Bowtie 2

Algorithm Bowtie2(R, τ_T)

Require: τ_T is the suffix trie of T and R is the length-m read

Ensure: A position i with good local alignment score with R

1: Obtain a set of length-22 seeds $\{seed_1, \ldots, seed_k\}$ from the read R for every $\lfloor 1 + 0.75\sqrt{m} \rfloor$ bp interval;

2: From τ_T, we obtain a list L_j of exact hits of seed $seed_j$, for $j = 1, \ldots, k$;

3: For $j = 1, \ldots, k$, for each loci i in L_j, its weight is $\frac{1}{|L_j|^2}$;

4: $count = 0$; $S = \emptyset$;

5: **repeat**

6:   Randomly select a position i from $\bigcup_{j=1}^{k} L_j$ according to the weight;

7:   Compute the alignment score $score_i$ between R and the sequence in position i;

8:   If $score_i$ is smaller than the second best alignment score in S, then $count + +$;

9:   Include $(i, score_i)$ into the set S;

10: **until** $count > 15$ or all positions have been selected

11: Report the position i which has the best alignment score in S;

FIGURE 4.25: Bowtie 2 algorithm.

Read R: ACGTACGTACGTACGTACGTACGT

Reference T: ...CTGCGTCT**ACGGACGACGGACGACGGACGG**CCGATCGACCGCGGGACGTA...
                        i

FIGURE 4.26: In this example, the correct hit of the read R is at position i. However, none of the exact occurrences of any length-4 seed of R is a correct hit.

reports the best alignment. The detail of the Bowtie 2 algorithm is stated in Figure 4.25.

4.5.1.3 BatAlign

Many seed-and-extend methods compute the hit list of short seeds in the seeding step. The hit list is defined to be the exact hits or 1-mismatch hits of the short seeds in the reference genome T. When the read has a high number of mismatches and/or indels, it is likely that the hit list of the short seeds does not contain the correct hit of the queried read. (See Figure 4.26 for an example.)

To address the problem of missing correct hits from low-edit-distance short seeds, BatAlign [178] uses high-edit-distance long seeds instead. Given a read R, it partitions R into a set of length-75 seeds. Then, the seeding step finds hits of each long seed in the reference genome T. Precisely, the seeding step

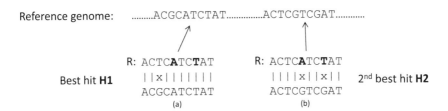

FIGURE 4.27: In this example, the read R has two low-quality bases at positions 5 and 8. When R is aligned on the genome, the best hit is $H1$ which has one mismatch. The second-best hit is $H2$ which has two mismatches. (a) and (b) show the alignments of R with the best hit and the second-best hit, respectively. The two low-quality score bases of R are in bold font. Although (b) has more mismatches than (a), the mismatches are at the low-quality score bases. If we use the quality score-aware alignment in Section 4.2.2, (b) is a better hit. This example shows that the hit having the minimum number of errors may not be correct.

enumerates putative candidate hits of each long seed in increasing order of alignment cost (i.e., increasing number of mismatches and gaps). To be sensitive, BatAlign allows up to 5 mismatches and at most one gap. We observe that the hit having the minimum number of errors (mismatch and/or gap) may not be correct (see Figure 4.27 for an example). BatAlign tries to obtain hits with the minimum number of errors and the second minimum number of errors. Simulation study showed that it will not generate many candidate hits while it improves sensitivity. Using the algorithm in Exercise 16, these hits can be found efficiently. After that, for every hit found in the seeding step, the extension step extends the hit to verify the full alignment of the read.

4.5.1.4 Cushaw2

Cushaw2 [182] is a seed-and-extend method for aligning long reads. It proposed to define the seeds of a read by maximal exact matches (MEMs). Given a read $R[1..m]$ and a reference genome $T[1..n]$, MEMs of R are defined to be maximal subsequences of R that also occur in T. Precisely, each MEM is defined to be a segment $R[i..j]$ that occurs in the reference genome T while $R[i-1..j]$ and $R[i..j+1]$ do not occur in T. Figure 4.28 gives an example to illustrate the definition of MEM. In this example, we have 4 MEMs: $R[1..6], R[4..16], R[8..21]$ and $R[20..24]$.

Based on MEMs, we detail the algorithm of Cushaw2, which consists of the seeding step and the extension step. The seeding step of Cushaw2 identifies

Read R: CTGACAGATCAGAGAGGAATCCGA

Reference T: ...CTGCGTCA**ACAGATCAGAGAG**C**TCCGA**TC**TCCGA**CCG**CTGACA**T**ATCAGAGAGGAATC**GGA|CGTA...

FIGURE 4.28: In this example, the read R has a 2-mismatch alignment in the rectangle region of the reference genome T. There are 4 MEMs between R and T, which are represented in T as the bold font.

all MEMs of the read R as seeds. By the suffix trie, MEMs can be found in $O(n + m)$ time (see Exercise 27). For each seed $R[i..j]$, the extension step of Cushaw2 extends every candidate hit $T[i'..j']$ of $R[i..j]$ by Smith-Waterman algorithm; then, the significant hits are reported.

We use the example in Figure 4.28 to illustrate the algorithm of Cushaw2. The seeding step of Cushaw2 identifies 4 MEMs as seeds. For $R[1..6]$, $R[4..16]$ and $R[8..21]$, each has one hit in T. For $R[20..24]$, it has two hits in T. All 5 hits of the MEM seeds are shown in bold font in Figure 4.28. In the extension step, Cushaw2 extends every candidate hit of each MEM by Smith-Waterman algorithm. In particular, for the hits of $R[1..6]$ and $R[8..21]$, after the seed extension, we discover the 2-mismatch hit of R, which is shown in the rectangular box.

Cushaw2 is a very fast method. However, Cushaw2 may miss correct seeds since the maximal feature of the MEM definition expands the seed aggressively. The following example illustrates this issue. Consider $R =$ CCCCGTTTT and a reference genome $T = \ldots$ CCCCATTTT $\ldots$ CCCCG $\ldots$ GTTTT $\ldots$. The substring CCCCATTTT of T is a 1-mismatch hit of R, which is the correct hit. Note that MEMs of R are CCCCG and GTTTT. Since the 1-mismatch hit of CCCCGTTTT does not contain both MEMs, Cushaw2 fails to find this 1-mismatch hit.

4.5.1.5 BWA-MEM

BWA-MEM is a general aligner for aligning reads of length 100 bp or longer. It uses the seed-and-extend approach. Its seeding step is specially designed so that BWA-MEM will not report many seeds and sensitivity will not be compromised.

In the seeding step, BWA-MEM identifies seeds by three steps: (1) seeding, (2) reseeding and (3) chaining.

As in Cushaw2, seeding identifies all MEMs of R. However, as stated above, MEM seeds may miss correct seeds. To avoid missing seeds, BWA-MEM performs re-seeding when the seed is long (more than 28 bp by default). Precisely, suppose a MEM of length ℓ occurs k times in T. The re-seeding step finds the longest exact matches that cover the middle base of the MEM that occur at least $k + 1$ times in T. This idea ensures the seeds identified from seeding and reseeding steps can cover almost all true hits of R.

Given all seeds obtained from seeding and reseeding, we can perform seed extension for all of them. However, there may be many seeds and many of them are redundant. To reduce the number of redundant extensions, BWA-MEM performs the chaining step to further filter the seeds. The chaining step aims to find groups of seeds (called chains) that are co-linear and close to each other. Chains can be found by greedy clustering of all seeds. From the set of chains, a chain is filtered out if (1) it is largely contained in a longer chain and (2) it is shorter than the long chain (by default, both 50% and 38 bp shorter than the long chain).

In the extension step, every selected seed is ranked by the length of the chain it belongs to and then by the seed length. Then, seeds are extended one by one according to the ranking. If a seed is contained in an alignment found before, we skip that seed. To save time, BWA-MEM extends each seed using a banded affine-gap-penalty dynamic programming.

BWA-MEM can be applied to align Illumina reads of length > 70 bp. It also can be used to align long reads generated by PacBio reads and Oxford Nanopore (ONT) reads.

4.5.1.6 LAST

LAST[139] is a software for aligning two genomes. Now, it is also used to align long reads R on the reference genome T. LAST also uses seed-and-extend approach. Its seeding step uses adaptive seed (variable-length seed) to identify hits of R; then, every hit is extended to compute the alignment between R and T.

For a fixed frequency threshold f, adaptive seed is a seed which is lengthened until its number of hits in the reference genome is at most f. Precisely, given a read $R[1..m]$, for every position i, LAST finds the smallest index ℓ such that $R[i..i + \ell - 1]$ has at most f hits in the reference genome T.

As long reads have a high error rate, adaptive seed cannot perform well. To skip the mismatch errors in long reads, LAST actually uses adaptive spaced seed. Recall that a shape $P[1..p]$ is a binary bit vector of length p. For any DNA sequence $Q[1..q]$, denote $Aseed_P(Q)$ as an adaptive spaced seed formed by replacing $Q[i]$ by '$*$' if $P[i \bmod p] = 0$ for every $i = 1, \ldots, q$. For example, suppose the shape $P = 110$, $Aseed_{110}(\texttt{ACGTCAGA}) = \texttt{AC} * \texttt{TC} * \texttt{GA}$. Given a shape P, a read $R[1..m]$, a text T, $R[i..i+\ell-1]$ has a spaced hit in $T[j..j+\ell-1]$ if $Aseed_P(R[i..i + \ell - 1]) = Aseed_P(T[j..j + \ell - 1])$.

In other words, the seeding step of LAST is as follows. For a fixed shape P, for every position i in R, LAST finds the smallest length ℓ such that $R[i..i + \ell - 1]$ has at most f spaced hits in the reference genome T.

Spaced hits can be found efficiently using spaced suffix array. Given a text $T[1..n]$, a spaced suffix array $SSA[1..n]$ is a permutation of $\{1, \ldots, n\}$ such that $seed_P(T[SSA[1]..n]) \leq seed_P(T[SSA[2]..n]) \leq \ldots \leq seed_P(T[SSA[n]..n])$. With spaced suffix array $SSA[1..n]$, for every position i in R, we can find the smallest length ℓ such that $R[i..i + \ell - 1]$ has at most f spaced hits in the reference genome T using $O(\ell \log n)$ time (see Exercise 29).

Apart from BWA-MEM and LAST, BLASR [39] is another aligner that can align long reads.

4.5.2   Filter-based approach

A filter-based approach is used by a number of methods like SeqAlto [208], GEM [189], MASAI [278], etc.

A filter-based approach aims to align a read $R[1..m]$ on the reference genome $T[1..n]$ allowing at most k errors. The basic idea is as follows. The read R is partitioned into s equal-length segments $R_1, \ldots, R_s$. By the pigeon-hole principle, for any p-error occurrence of R, some R_i has at most $\lfloor \frac{p}{s} \rfloor$ errors.

By this observation, p-error hits of R can be found in two steps. First, for each segment R_j, we identify all hits of R_j with at most $\lfloor \frac{p}{s} \rfloor$ errors and we denote this hit list as L_j. Second, for every hit in $L_1, \ldots, L_s$, we extend it and verify if it is a p-error hit.

For example, Figure 4.29(a) gives an example to illustrate the idea. To find 3-error hits of a length-11 read R, we partition R into 4 segments. Then, we extract all exact hits of the four segments. Each exact hit is verified as a 3-error hit. By the pigeon-hole principle, this method will not miss any correct hit.

Although the above approach can find all p-error hits, its running time may be slow when some segment has a long list of candidate hits. To speed up, a few heuristics are proposed.

The first heuristic is proposed by MASAI [278]. It is time consuming to find $\lfloor \frac{p}{s} \rfloor$-error hits. MASAI suggested finding $\lfloor \frac{p}{s} \rfloor$-mismatch hits of each seed instead. Although this approach will miss hits, empirical study showed that such change will only reduce the sensitivity by 1%.

Another heuristic is to skip segments with a high number of hits. This will reduce the number of seed extensions, where each seed extension requires quadratic time. Since some segments are skipped, this approach may miss answers.

GEM[189] proposed another variant, which is called the region-based adaptive filtering approach. Instead of fixing the length of each segment, GEM restricts the number of hits of each segment. It fixed each segment can have at most t hits in the reference genome, where t is some user-defined threshold. This idea significantly reduces the number of candidate hits, which makes GEM a very fast aligner. Figure 4.30 describes the algorithm GEMseg(R,t) that is used by GEM to generate the segments. Let τ_T be the suffix trie of the genome T. The algorithm searches the suffix trie τ_T and finds the shortest prefix $R[1..i]$ of R whose number of occurrences in T is less than t. Such prefix $R[1..i]$ is reported. For the remaining portion $R[i+1..m]$, the algorithm iteratively finds the next shortest prefix of $R[i+1..m]$ until the whole read is processed. GEMseg(R,t) can compute all segments in $O(|R|)$ time. Figure 4.29(b) gives an example to illustrate the segmentation method.

Although GEM is fast, GEM cannot guarantee that the number of seg-

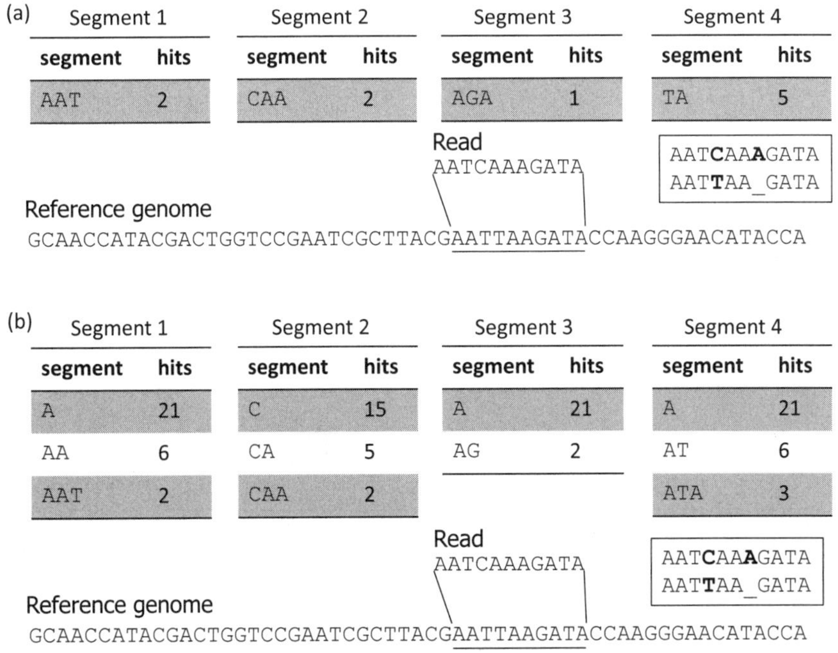

(a)

Segment 1		Segment 2		Segment 3		Segment 4	
segment	**hits**	**segment**	**hits**	**segment**	**hits**	**segment**	**hits**
AAT	2	CAA	2	AGA	1	TA	5

Read
AATCAAAGATA

AAT**CAAA**GATA
AAT**TAA**_GATA

Reference genome
GCAACCATACGACTGGTCCGAATCGCTTACG<u>AATTAAGATA</u>CCAAGGGAACATACCA

(b)

Segment 1		Segment 2		Segment 3		Segment 4	
segment	**hits**	**segment**	**hits**	**segment**	**hits**	**segment**	**hits**
A	21	C	15	A	21	A	21
AA	6	CA	5	AG	2	AT	6
AAT	2	CAA	2			ATA	3

Read
AATCAAAGATA

AAT**CAAA**GATA
AAT**TAA**_GATA

Reference genome
GCAACCATACGACTGGTCCGAATCGCTTACG<u>AATTAAGATA</u>CCAAGGGAACATACCA

FIGURE 4.29: Consider a read R = AATCAAAGATA. We aim to find its 3-error hit in the reference genome. (a) The read R is partitioned into 4 segments, each of size 3 (the last segment is of length 2). Then, the hits of each segment in the reference genome are obtained. The best hit is a 2-error hit and it is underlined. (b) The read R is segmented using the GEM approach, which aims to find segments with at most $t = 4$ hits. Note that the GEM approach generates fewer hits.

Algorithm GEMseg(R, t)

Require: A suffix trie τ_T of the genome $T[1..n]$ a read $R[1..m]$ and a threshold t

Ensure: A set of segments of R such that the number of hits per segment is at most t

1: Set $j = 1$ and $u = $ root of τ_T;
2: **for** $i = 1$ to m **do**
3:     Set $u = child(u, R[i])$; /* As an invariant, $plabel(u) = R[j..i]$. */
4:     **if** ($|Leaf(u)| < t$) **then**
5:         Report $R[j..i]$ as a segment;
6:         Set $j = i + 1$ and $u = $ root of τ_T;
7:     **end if**
8: **end for**

FIGURE 4.30: The algorithm GEMseg(R, t) partitions $R[1..m]$ into a set of segments such that the number of hits per segment in the genome T is less than t.

ments is at least $(p + 1)$, where p is the number of allowed errors (says, 3). Moreover, GEM showed that this approach works well in practice.

4.6   Paired-end alignment

Consider a set of DNA fragments, whose insert size (i.e., fragment length) is expected to be within ($span_{\min}..span_{\max}$). Paired-end sequencing lets us read the two ends of every DNA fragment. This read pair is called a paired-end read. When a paired-end read (R_1, R_2) is mapped to the reference genome, the two reads R_1 and R_2 are expected to map nearby and their distance is expected to be within ($span_{\min}..span_{\max}$). This property can be used as a criterion to improve the alignment accuracy.

Different aligners utilize the insert size information differently. Many aligners use the following strategy. The two reads R_1 and R_2 are aligned independently. Let H_1 and H_2 be the sets of top hits (i.e., hits with a high mapQ score or alignment score) of R_1 and R_2, respectively, reported by an aligner. Then, a pair $(h_1, h_2) \in H_1 \times H_2$ is reported if they map on the same chromosome, have a correct orientation, and a correct insert size; otherwise, read rescue is performed. Precisely, for $i = 1, 2$, for each hit $h_i \in H_i$ of R_i, the Smith-Waterman alignment algorithm is applied to find the best alignment location of its mate within ($span_{\min}..span_{\max}$) of h_i. This location is denoted as $mate(h_i)$. If the alignment score of the mate is above a user-defined threshold, we report $(h_i, mate(h_i))$. Otherwise, if we fail to find any pairing within the correct insert size, we report (h_1, h_2) where h_1 and h_2 are the best hits (with the highest mapQ score or alignment score) of R_1 and R_2, respectively.

This strategy increases the chance of a pairing with the correct insert size. However, it is biased to give a pairing with the correct insert size.

BatAlign and BWA-MEM try to perform unbiased pairing. Similar to the standard approach, we align R_1 and R_2 independently and obtain the set of hits H_1 and H_2, respectively. Then, let $C_1 = \{(h_1, mate(h_1)) \mid h_1 \in H_1\}$ and $C_2 = \{(mate(h_2), h_2) \mid h_2 \in H_2\}$. Third, among all hit pairs (h_1, h_2) in $(H_1 \times H_2) \cup C_1 \cup C_2$, we report (h_1, h_2) with the highest sum of alignment score.

4.7 Further reading

This chapter describes various techniques used to map NGS reads on a reference genome. A number of topics are not covered in this chapter. We do not cover methods that handle color bases in SOLiD reads.

Another topic is parallel processing. Since the number of reads is very high, it is important to speed up the mapping using parallel processing. Most existing methods like Bowtie and BWA can be accelerated by multi-threading. Some methods have been developed using specialized parallel hardware like GPU and FPGA. SARUMAN [22], SOAP3 [181] and CUSHAW [183] are GPU-based methods. SARUMAN [22] uses a NVIDIA graphics card to accelerate the time-consuming alignment step. SOAP3 [181] and CUSHAW [183] achieve performance improvements by parallelizing the BWT-approach on GPUs.

This chapter did not give a comprehensive comparison of different read aligners and did not study the performance of different methods. For the classification of different methods, please read [171] and [81]. Li and Homer [171] classified the methods according to the indexing techniques. Fonseca et al. [81] provided a comprehensive overview of the characteristics of different methods.

For performance comparisons, please refer to [254], [102], [264] and [98]. Ruffalo et al. [254] compare the accuracy of Bowtie, BWA, Novoalign, SHRiMP, mrFAST, mrsFAST and SOAP2. Holtgrewe et al. [102] presented a benchmarking method, Rabema, for comparing sensitivity of different methods. Rabema was applied to compare the performance of SOAP2, Bowtie, BWA and Shrimp2. Schbath et al. [264] compared the sensitivity of 9 methods (BWA, Novoalign, Bowtie, SOAP2, BFAST, SSAHA2, MPscan, GASSST, and PerM) with a controlled benchmark. Hatem et al. [98] also benchmarked the accuracy of 9 methods: Bowtie, Bowtie2, BWA, SOAP2, MAQ, RMAP, GSNAP, Novoalign and mrsFAST(mrFAST). Overall, the comparison showed that different aligners have different strengths and weaknesses. There is no single method that outperforms the other methods in all metrics.

4.8   Exercises

1. Consider a length-100 read where all bases are Q20. Can you compute the probability that there are at most 2 sequencing errors?

2. Consider a read $R[1..m]$ and a reference genome $T[1..n]$. Please give an $O(mn)$-time algorithm to compute $\min_{1 \leq i \leq j \leq n} edit(R[1..m], T[i..j])$.

3. Consider a read $R[1..m]$ and a reference genome $T[1..n]$. Let $\delta(x, y)$ be the similarity score for bases x and y. For a gap of size c, let the gap penalty be $-(q + cr)$. Please give an $O(mn)$-time algorithm to compute the local alignment score.

4. Consider a reference genome $T[1..n]$ and a read $R[1..m]$ where $n > m$. We want to find the positions i such that $Hamming(R, T[i..i+m-1]) \leq k$. Please propose an $O(kn)$-time algorithm. (Hint: We can build the suffix tree for $T\#R\$$ and the corresponding longest common prefix data structure using $O(n + m)$ time.)

5. We aim to generate spaced seeds for length-12 reads so that we can find all their 3-mismatch hits. Can you give the set of shapes using (a) MAQ approach and (b) RMAP approach?

6. Consider a length-20 read R. We want to find all 2-mismatch hits of R. We try two approaches:

   - Mismatch seed: A length-10 1-mismatch seed (i.e. all 1-mismatch patterns of $R[1..10]$
   - Gapped seed: Six length-20 weight-10 shapes $1^{10}0^{10}$, $1^50^51^50^5$, $1^50^{10}1^5$, $0^51^{10}0^5$, $0^51^50^51^5$, $0^{10}1^{10}$.

   For each approach, can we recover all 2-mismatch hits of R? How many hash entries do we need to scan? Among the two approaches, which one is better?

7. The spaced seed approach in the previous question requires 6 spaced seeds. Suppose we still want to use length-20 weight-10 spaced seeds only. Is it possible to use fewer spaced seeds while we still finding all the 2-mismatch hits? Please explain your answer.

8. Consider a reference genome T of length n. We aim to align K reads of length m on T allowing p mismatches. Using the RMAP approach, what is the time and space complexity?

9. Suppose we want to find 1-mismatch hits of a set of reads using MAQ and RMAP. What is the difference between the two algorithms?

10. Given a reference genome T, we aim to report all 2-mismatch hits of a length-m read using the reference hashing approach. If we use the spaced seeds of the MAQ approach, we need to use 6 shapes (i.e. we need to build 6 hash tables). Can we reduce the number of shapes? Please detail the set of shapes (or hash tables) we need to build.

11. Given a reference genome T, we aim to report all p-mismatch hits of a set of reads $\mathcal{R}$ using the reference hashing approach. Suppose we use the mismatch seed approach. Can you detail the algorithm?

12. Figure 4.11 describes an algorithm to estimate the lower bound $D[i]$ of the number of differences between $R[i..m]$ and the reference genome T. Given the FM-index for T (which includes the $BW[]$, $C(.)$ and $Occ(.,.)$ data structures), can you describe an implementation of CalculateD(R) which runs in $O(m)$ time?

13. Figure 4.13 describes the algorithm DivideAndConquer(R, p) to find all p-mismatch hits of R. For each recursive call of DivideAndConquer, the algorithm will call the function *reverse* twice. Can you restructure the algorithm so that *reverse* is called once for each recursive call of DivideAndConquer?

14. Below is an algorithm for computing hits of a read R in the reference genome T with at most 4 mismatches.

    1: Partition R into $R_l = R[1..m/2]$ and $R_r = R[m/2 + 1..m]$;
    2: Let H_l be the set of hits of R_l in the reference genome with at most 1 mismatches.
    3: Let H_r be the set of hits of R_r in the reference genome with at most 3 mismatches.
    4: **for** every hit h in H_l **do**
    5:    Report h if $Hamming(R, T[h..h + m - 1]) \leq 4$;
    6: **end for**
    7: **for** every hit h in H_4 **do**
    8:    Report $h - m/2$ if $Hamming(R, T[h - m/2..h + m/2 - 1]) \leq 4$;
    9: **end for**

    - Can the above algorithm report all hits of R in the reference genome T with at most 4 mismatches? Please explain your answer.

    - Will the above algorithm report the same hit twice? Please explain your answer. If the algorithm reports the same hit twice, can you suggest a solution to avoid this?

15. Given a read $R[1..m]$, a reference genome T and a mismatch threshold p, please modify the algorithm $HSet_{T,R,p}(u)$ in Figure 4.10 so that

the algorithm reports all descendant nodes v of u such that either (1) $Hamming(plabel(v), R) = p$ and $depth(v) < m$ or (2) $depth(v) = m$ and $Hamming(plabel(v), R) < p$.

16. The algorithm in Figure 4.13 does not report the occurrences of R in the order of an increasing number of mismatches. Can you update the algorithm so that it reports the occurrences in the order of an increasing number of mismatches? (Hint: You can use the algorithm in Exercise 15 as a subroutine.)

17. Consider X and Y of length 11 with at most 3 mismatches. For a q-mer of shape Q, let $t_Q(m, k)$ be the minimum number of q-mers of shape Q shared by two length-m strings X and Y, where $Hamming(X, Y) \leq k$. What are the values of $t_{111}(11, 3)$ and $t_{1101}(11, 3)$?

18. Consider a shape Q of length w and weight q. Denote $t_Q(m, k)$ as the minimum number of spaced q-mers of shape Q shared by length-m strings X and Y, where $Hamming(X, Y) \leq k$. Can you give a dynamic programming algorithm to compute $t_Q(m, k)$? What is its running time?

19. The algorithm in Figure 4.20 computes the minimum edit distance between the read R and the reference T. Can you speed up this algorithm by estimating the lower bound of the edit distance?

20. Please prove Lemmas 4.12 and 4.14.

21. Please prove Lemma 4.13.

22. Section 4.3.4.1 mentioned a trick to estimate the lower bound of the edit distance. Can you describe how to use this trick to speedup the algorithm $EditDist_{T,R,p}(u, d_u[0..m])$ in Figure 4.20?

23. Please prove Lemma 4.17.

24. Please generalize the algorithm LocalAlign in Figure 4.22 to handle the affine gap penalty. (We assume q is the penalty for openning a gap and r is the penalty per space.)

25. Consider a read $R[1..m]$ and a reference genome $T[1..n]$. Let τ_T be the suffix trie of T. For any node u in τ_T, let $V_u(i)$ is the maximum meaningful alignment score between any suffix of $R[1..i]$ and $plabel(u)$. Please give an algorithm that computes $\{(i, u) \mid |Leaf(u)| \leq \gamma, |Leaf(p(u))| > \gamma, V_u(i) > 0\}$ where $p(u)$ is the parent of u in τ_T.

26. Consider a read $R[1..m]$ and a reference genome $T[1..n]$. Is the following statement correct? Please explain your answer.

The number of MEMs of R with respect to T is at most m.

27. Consider a read $R[1..m]$ and a reference genome $T[1..n]$. Suppose you are given the suffix trie τ_T of T. You are also provided a function suffix link $sl()$. (For any node u in τ_T, $sl(u)$ is the node v such that $plabel(u) = \sigma \cdot plabel(v)$ for some $\sigma \in \{A, C, G, T\}$.) Assume $sl(u)$ can be computed in constant time for any node u. Can you propose an $O(m)$-time algorithm to compute all MEMs of R in T?

28. Consider a text $T[1..17] = $ ACTGCATGGCATGGACT. Suppose the shape is $P = 110$. What is the corresponding spaced suffix array? Can you list the spaced hits of TGTA?

29. Given a text $T[1..n]$ and a shape $P[1..p]$, For a query $Q[1..q]$, can you give an $O(q \log n)$-time algorithm that find all the spaced hits of Q?

30. Consider the following reference genome T.

    ACTGGTCAGAATCCCGATTAGCAACGTACCGCCAGATACCGACCAAGGCATATACGA

    We aim to find all 3-error occurrences of the read $R = $ ACCAAGCCT.

    (a) Can you use the standard filter-based approach to find all 2-error occurrences of R? You can assume every segment is of length 2 (except the last segment is of length 3).

    (b) Can you use a region-based adaptive filtering approach to find all 2-error occurrences of R? You can assume every segment has at most 4 hits in the reference genome.

    For both methods, please also state the number of hits that will be generated.

Chapter 5

Genome assembly

5.1 Introduction

Reconstructing genomes through DNA sequencing is an important problem in genomics. This problem looks simple if the genome can be read base-by-base from 5' to 3'. Unfortunately, existing biotechnologies cannot read through the whole chromosome since it is too long. Instead, we reconstruct the genome indirectly. First, the genome is broken into DNA fragments using the whole genome shotgun approach (see Section 5.2); then, the sequencing machine decodes DNA sequences from these fragments. These DNA sequences are called reads. Due to random sampling, the extracted reads cover the entire genome uniformly. By stitching the reads together, we can computationally reconstruct the genome. This process is known as de novo genome assembly.

Sanger sequencing was the earliest available sequencing technique. Two assemblers were developed for assembling Sanger sequencing reads. They are the OLC assembler Celera [213] and the de Bruijn graph assembler Euler [231, 37]. Our human reference genome is assembled using these two approaches. However, since Sanger sequencing is low throughput and expensive, only a few genomes are assembled with Sanger sequencing.

The rise of second-generation sequencing has changed the game. We can sequence hundreds of millions of reads efficiently and cost-effectively. However, the second-generation sequencing reads are short. Some tailor-made genome assemblers are developed for reconstructing genomes from short reads. Their success led to a number of successful de novo genome assembly projects, including the reconstruction of James Watson's genome [315] and the panda genome [174]. Although this approach is cost effective, this approach usually gives fragmented genomes since the reads are short and the repeat regions are long.

Recently, third-generation sequencing is available, which can sequence long reads (of length ∼10,000 bp). Although long reads can resolve the ordering of repeat regions, these long reads have a high error rate (15%−18%). A number of computational methods have been developed to correct errors in third-generation sequencing reads.

The chapter discusses different methods for de novo genome assembly and it is organized as follows. Section 5.2 first discusses how the reads are sequenced from the genome. Then, Section 5.3 describes genome assembly methods for assembling short reads generated by second-generation sequencing. The techniques for assembling long reads generated by third-generation sequencing is covered in Section 5.4. Finally, the chapter ends with criteria for evaluating the goodness of a genome assembly and some further reading.

5.2 Whole genome shotgun sequencing

To assemble a genome, the first step is to sequence a set of reads from the sample genome. There are two wet-lab protocols for this purpose: (1) whole genome sequencing and (2) mate-pair sequencing. Below, we briefly describe them.

5.2.1 Whole genome sequencing

Whole genome sequencing involves three steps as shown in Figure 5.1. First, the sample genome is randomly broken into DNA fragments by sonication or enzymatic cutting; then, the size selection step extracts DNA fragments of a certain fixed length (called the insert size). Finally, single-end or paired-end sequencing of the DNA fragments is performed. For single-end sequencing, the sequencer reads one end of each DNA fragment. For paired-end sequencing, the sequencer reads both ends of the DNA fragment. (Note: Third-generation sequencers read the whole DNA fragment. We consider it as a single-end read.) Figure 5.2 gives an example. For single-end sequencing, we obtain one read from the 5' end of the forward template of the DNA fragment, that is, ACTCAGCACCTTACGGCGTGCATCA. For paired-end sequencing, we obtain the 5' reads from both forward and reverse templates of the DNA fragment (in inward orientation, i.e., reading the template from 5' to 3'), that is,

- ACTCAGCACCTTACGGCGTGCATCA

- AGTTTGTACTGCCGTTCAGAACGTA.

Note that the second read AGTTTGTACTGCCGTTCAGAACGTA is the reverse complement of TACGTTCTGAACGGCAGTACAAACT since this read is obtained from the reverse template.

For genome assembly, the insert size (the length of the DNA fragment) is an important parameter. Different sequencing technologies have different limitations on insert size. For instance, Illumina Hi-seq can only perform paired-end sequencing for DNA fragments of insert size < 1000 bp. For third-generation sequencing, the insert size limit can be as large as 10,000 bp.

Genomes of the sample

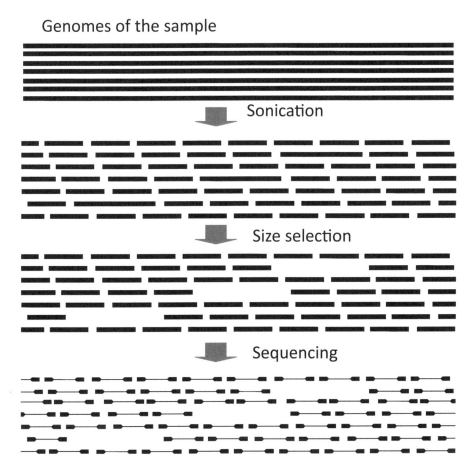

FIGURE 5.1: Illustration of the shotgun protocol for whole genome sequencing. It involves 3 steps: (1) the sonication step breaks the genome, (2) the size selection step selects fragments of certain size, (3) the sequencing step reads the end(s) of every fragment.

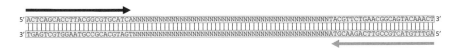

FIGURE 5.2: Consider a DNA fragment extracted from the sample genome. The paired-end read is formed by extracting two DNA substrings (also called "tags") from the two ends of the DNA fragment.

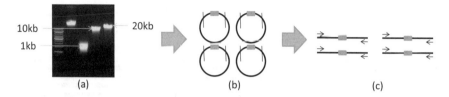

FIGURE 5.3: Illustration of the mate-pair sequencing protocol. (a) Extraction of all fragments of certain size, say 10,000 bp, by cutting gel. (b) Each fragment is circularized with an adaptor (represented by the gray color block) in the middle; then, we cut the two flanking regions around the adaptor. (c) Finally, paired-end reads are extracted by paired-end sequencing.

5.2.2   Mate-pair sequencing

Second-generation sequencers can extract paired-end reads from the two ends of some short DNA fragments (of insert size < 1000 bp). To extract paired-end reads from long DNA fragments, we can use the mate-pair sequencing protocol. Figure 5.3 describes the mate-pair sequencing protocol. First, long DNA fragments of some fixed insert size (say 10,000 bp) are selected by cutting the gel. Then, the long fragments are circularized with an adaptor. The circularized DNAs are fragmented and only fragments containing the adaptors are retained (say, by biotin pull-down). Finally, by paired-end sequencing, paired-end reads are sequenced from the DNA fragments with adaptors.

Note that the orientation of the paired-end read reported by mate-pair sequencing is different from that reported by paired-end sequencing. Mate-pair sequencing reports two reads at the two ends of each DNA fragment in outward orientation instead of inward orientation (Exercise 1). For the example DNA fragment in Figure 5.2, the mate-pair sequencing will report:

- TGATGCACGCCGTAAGGTGCTGAGT

- TACGTTCTGAACGGCAGTACAAACT

Although the mate-pair sequencing protocol can extract paired-end reads with long insert size, it requires more input DNA to prepare the sequencing library and it is prone to ligation errors.

5.3   De novo genome assembly for short reads

Second-generation sequencing enables us to obtain a set of single-end or paired-end short reads for the whole genome. The de novo assembly problem aims to overlap the reads in the correct order and to reconstruct the genome.

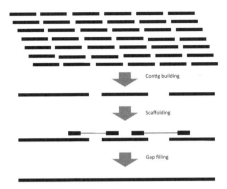

FIGURE 5.4: The four steps in genome assembly. 1. Read correcting prepro-cessing: Correct the sequencing errors in the reads. 2. Contig building: Stitch the reads to form contigs. 3. Scaffolding: Use paired-end reads to link the con-tigs to form scaffolds. 4. Gap filling: For adjacent contigs in the scaffolds, try to fill in the gaps.

The genome assembly problem is computationally difficult. Even when there is no sequencing error, this problem is equivalent to the superstring problem, which is known to be NP-hard. (The superstring problem is as follows. Given a set $\mathcal{S}$ of strings, we aim to find the superstring, that is, the shortest string P such that every string $S \in \mathcal{S}$ is a substring of P. For example, if $\mathcal{S} = \{\texttt{ACATGC}, \texttt{ATGCGTGT}, \texttt{GTGTACGT}\}$, then the superstring is $\texttt{ACATGCGTGTACGT}$.)

Many de novo assemblers are proposed to assemble short reads. The general solution involves four steps as shown in Figure 5.4. The first step corrects the sequencing errors in the reads. Given the corrected reads, the second step stitches the reads by overlapping them. Ideally, we want to stitch all reads to form the complete genome. Due to repetitions, ambiguity exists and we fail to stitch the reads to recover the complete genome. Existing methods output a set of contigs, where each contig represents some subregion of the sample genome. Next, using paired-end reads, we try to recover the ordering of the contigs to form scaffolds. (Each scaffold is a chain of contigs. It is also called a supercontig or metacontig.) Finally, reads are realigned on the scaffolds to fill some gaps between adjacent contigs.

This section is organized as follows. Section 5.3.1 discusses the read er-ror correction method. Then, we present the contig building step. Roughly speaking, there are two approaches to building contigs: (1) the base-by-base extension approach and (2) the de Bruijn graph approach. They will be cov-ered in Sections 5.3.2 and 5.3.3, respectively. Finally, the scaffolding step and gap-filling step are described in Sections 5.3.4 and 5.3.5, respectively.

5.3.1   Read error correction

Errors in the sequencing reads may mislead the de novo assemblers. It is better if we can correct them prior to genome assembly.

Assume that the genome is sampled at a high coverage. A simple method is to identify highly similar reads and pile them up. If a read contains a base that is different from the consensus, such a base is likely to be a sequencing error.

For example, consider the following set of reads $\mathcal{R} = \{R_1 = \text{AAGTGAA}, R_2 = \text{AGTGCAG}, R_3 = \text{ACTTCAC}, R_4 = \text{TGAAGTG}\}$. Figure 5.5(a) shows the multiple sequence alignment of $R_1, R_2, \overline{R}_3, R_4$ ($\overline{R}_3$ is the reverse complement of R_3). The base C in the 5th position of R_2 is inconsistent with the bases A in the other 3 reads. The base C is likely to be a sequencing error. To correct the error, we can convert the base C in the 5th position of the read R_2 to the base A.

The above solution is time consuming since it requires the comparison of all reads. Below, we present the k-mer frequency-based error correction method. This idea is used by a number of methods, including ALLPATHS [34], Euler [231, 37], and ABySS [275].

The k-mer frequency-based error correction method was proposed by Pevzner et al. [231]. Its observation is as follows. Fix k to be some integer bigger than $\log_4 N$ where N is the genome size. Provided the genome is sampled at a high coverage, any k-mer (length-k substring) that occurs in the genome is likely to occur in many input reads. Suppose a particular k-mer occurs once (or very sparingly) in the input reads; this k-mer is unlikely to occur in the target genome and is likely to be a result of a sequencing error. On the other hand, a high-frequency k-mer is likely to be error-free.

More precisely, let $\mathcal{R}$ be the set of reads $\{R_1, \ldots, R_n\}$. For every read $R_i \in \mathcal{R}$, let $\overline{R}_i$ be its reverse complement. Let $\overline{\mathcal{R}}$ be $\{\overline{R}_i \mid R_i \in \mathcal{R}\}$. Let $S_{\mathcal{R}}$ be the set of k-mers occurring in $R_1, \ldots, R_n$ and $\overline{R}_1, \ldots, \overline{R}_n$. For each k-mer t, denote $freq_{\mathcal{R}}(t)$ (or $freq(t)$ when the context is clear) as the number of occurrences of t in $\mathcal{R} \cup \overline{\mathcal{R}}$.

Figure 5.5(b) lists all 4-mers in $S_{\mathcal{R}}$. It also shows the frequencies $freq_{\mathcal{R}}(t)$ for every 4-mers $t \in S_{\mathcal{R}}$. Observe that all 4-mers overlapping the error in R_2 occur exactly once. This suggests that a low-frequency k-mer is likely to be error-prone.

By the above observation, we define a k-mer t to be solid if $freq_{\mathcal{R}}(t) \geq M$ (M is a user-defined threshold). The solid k-mer is likely to be error-free.

However, it is not easy to select the correct threshold M. For example, if $M = 2$, the set of solid 4-mers is $\mathcal{T} = \{\text{AAGT}, \text{ACTT}, \text{AGTG}, \text{CACT}, \text{CTTC}, \text{GAAG}, \text{GTGA}, \text{TCAC}, \text{TGAA}, \text{TGCA}, \text{TTCA}\}$. $\mathcal{T}$ includes all correct 4-mers, except TGCA. If $M = 3$, the set of solid 4-mers is $\mathcal{T} = \{\text{AAGT}, \text{ACTT}, \text{AGTC}, \text{CACT}, \text{TGAA}, \text{TTCA}\}$. Although all k-mers in $\mathcal{T}$ are correct, many error-free 4-mers are also filtered.

The above example presented the issue in selecting the correct threshold M. If M is too big, we will throw away error-free k-mers. If M is too small,

$R_1 =$ AAGTGAA
$R_2 =$ AGTG**C**AG
$\overline{R}_3 =$ GTGAAGT
$R_4 =$ TGAAGTG

(a)

4-mer	AAGT	ACTT	AGTG	CACT	CTTC	**CTGC**	GAAG	**GCAC**	GCAG	GTGA	**GTGC**	TCAC	TGAA	**TGCA**	TTCA
freq(t)	3	3	3	3	2	1	2	1	1	2	1	2	3	2	3

(b)

FIGURE 5.5: (a) shows the alignment of the 4 reads R_1, R_2, $\overline{R}_3$ and R_4, where $R_1 =$ AAGTGAA, $R_2 =$ AGTGCAG, $R_3 =$ ACTTCAC and $R_4 =$ TGAAGTG. In the alignment, the C in position 5 of R_2 is in bold font, which is inconsistent with the A's in the other 3 reads. (b) shows the frequencies of all 4-mers in $\mathcal{R}$.

many solid k-mers are error-prone. We aim to select an M that maximizes the number of error-free k-mers and minimizes the number of error-prone k-mers. M can be estimated as follows. Suppose the coverage is C (i.e., the total length of all reads is C-fold the genome size) and the sequencing error rate is e. Assume the reads are uniformly distributed. For any position p in the genome, let Ψ_p be the set of k-mers covering positions $p..p + k - 1$. The size of Ψ_p is expected to be C. The chance that one k-mer in Ψ_p is error-prone is $(1 - (1 - e)^k)$. Then, the expected number of error-prone k-mers in Ψ_p is $C(1 - (1 - e)^k)$. If a k-mer occurs more than $C(1 - (1 - e)^k)$ times, it is unlikely that such a k-mer is error-prone. Hence, the threshold M can be set as $C(1 - (1 - e)^k)$ (or any bigger value). For example, when $C = 10$, $k = 20$ and $e = 0.02$, $C(1 - (1 - e)^k) = 3.32$. We can set $M = 4$.

Below, Section 5.3.1.1 discusses the spectral alignment problem, which corrects errors in each read given a set of solid k-mers. Since k-mer counting is a bottleneck step, we will discuss methods for k-mer counting in Section 5.3.1.2.

5.3.1.1 Spectral alignment problem (SAP)

Let $\mathcal{R} = \{R_1, \ldots, R_n\}$ be a set of reads sequenced from a sample genome T. Denote $\mathcal{T}$ as the set of k-mers appearing in T. As suggested in the previous section, $\mathcal{T}$ can be approximated by $\{t \in S_{\mathcal{R}} \mid t$ is a k-mer appearing in at least M reads$\}$.

A read R is denoted as a $\mathcal{T}$-string if every k-mer of R is in $\mathcal{T}$. Since $\mathcal{T}$ consists of all k-mers appearing in T, every read from the sample genome T is expected to be a $\mathcal{T}$-string. If $R[1..m]$ is not a $\mathcal{T}$-string, R is expected to have an error. To correct the error of R, the spectral alignment problem [37, 231] is proposed. The spectral alignment problem aims to find a minimum number of corrections (insertion, deletion or mutation) that transform R into a $\mathcal{T}$-string R'. If there is a unique way to correct R and the number of corrections is at most Δ, then R' is said to be a corrected read of R.

For the previous example where $\mathcal{R} = \{R_1 =$ AAGTGAA$, R_2 =$ AGTGCAG$, R_3 =$ ACTTCAC$, R_4 =$ TGAAGTG$\}$. $\mathcal{T}$ can be approximated by

$\{t \in S_{\mathcal{R}} \mid t$ is a k-mer appearing in at least 2 reads$\}$. That is, $\mathcal{T} = \{$AAGT, ACTT, AGTG, CACT, CTTC, GAAG, GTGA, TCAC, TGAA, TGCA, TTCA$\}$. Note that R_1, R_3 and R_4 are $\mathcal{T}$-strings. Hence, they are assumed to be correct. For the read $R_2 =$ AGTGCAG, R_2 is not a $\mathcal{T}$-string since the 4-mers GTGC and GCAG are not in $\mathcal{T}$. To correct R_2, we can introduce one mutation C $\rightarrow$ A at position 5 and obtain $R' =$ AGTGAAG, which is a $\mathcal{T}$-string. If $\Delta = 1$, this correction is accepted since this is the only way to convert R into a $\mathcal{T}$-string by at most one mutation.

The spectral alignment problem can be solved in polynomial time [37]. To simplify the discussion, we assume there is no indel error in the first k bases of R. We need two definitions. First, for any k-mer t and any base $b \in \{$A, C, G, T$\}$, denote $b \cdot t[1..k-1]$ as the k-mer formed by concatenating b with $t[1..k-1]$. (For example, if t is a 4-mer GAAG and $b =$ T, then $b \cdot t[1..k-1] =$ TGAA.) Second, we define $dist(i,t)$ as the minimum edit distance between $R[1..i]$ and any $\mathcal{T}$-string that ends at the k-mer t. The following lemma states the recursive formula for $dist(i,t)$.

Lemma 5.1 *Denote* $\rho(x,y) = 0$ *if* $x = y$; *and* 1 *otherwise.*

- *For* $i = k$ *and* $t \in \mathcal{T}$ *(base case),* $dist(k,t) = Hamming(R[1..k], t)$.

- *For* $i > k$ *and* $t \in \mathcal{T}$ *(recursive case), we have* $dist(i,t) =$

$$\min \begin{cases} \min_{b \in \{A,C,G,T\}} \{dist(i-1, b \cdot t[1..k-1]) + \rho(R[i], t[k])\} & \texttt{match} \\ dist(i-1, t) + 1 & \texttt{delete} \\ \min_{b \in \{A,C,G,T\}} \{dist(i, b \cdot t[1..k-1]) + 1\} & \texttt{insert.} \end{cases}$$

Proof For the base case, it is correct by definition. When $i > k$, $dist(i,t)$ is the minimum of three cases: (1) $R[i]$ is replaced by $t[k]$, (2) $R[i]$ is a deleted base and (3) $t[k]$ is an inserted base.

For case (1), $dist(i,t)$ equals the sum of (i) $\rho(R[i], t[k])$ (the replacement cost) and (ii) the edit distance between $R[1..i-1]$ and any $\mathcal{T}$-string that ends at $b \cdot t[1..k-1]$ for some $b \in \{$A, C, G, T$\}$. Hence, we have $dist(i,t) = \min_{b \in \{A,C,G,T\}} \{dist(i-1, b \cdot t[1..k-1]) + \rho(R[i], t[k])\}$.

For case (2), $dist(i,t)$ equals the sum of (i) 1 (the deletion cost) and (ii) the edit distance between $R[1..i-1]$ and any $\mathcal{T}$-string that ends at t. Hence, we have $dist(i,t) = dist(i-1, t) + 1$.

For case (3), $dist(i,t)$ equals the sum of (i) 1 (the insertion cost) and (ii) the edit distance between $R[1..i]$ and any $\mathcal{T}$-string that ends at $b \cdots t[1..k-1]$ for some $b \in \{$A, C, G, T$\}$. Hence, we have $dist(i,t) = \min_{b \in \{A,C,G,T\}} \{dist(i, b \cdot t[1..k-1]) + 1\}$. ∎

The minimum edit distance between R and any $\mathcal{T}$-string equals $\min_{t \in \mathcal{T}} dist(|R|, t)$.

Given the recursive formula in Lemma 5.1, the first thought is to use dynamic programming to compute $dist(|R|,t)$ for all $t \in \mathcal{T}$; then, we can compute $\min_{t\in\mathcal{T}} dist(|R|,t)$. However, this recursive formula has cyclic dependencies and dynamic programming cannot work.

Below, we illustrate that cyclic dependency may exist. We first define the dependency graph G for the recursive formula $dist(i,t)$. The dependency graph $G = (V,E)$ is a graph with vertex set $V = \{v_s\} \cup \{(i,t) \mid i = k,\ldots,|R|, t \in \mathcal{T}\}$. With respect to Lemma 5.1, for every node (i,t) (recursive case) where $i > k$, Lemma 5.1 implies that $dist(i,t)$ is the minimum among $dist(i-1, b \cdot t[1..k-1]) + \rho(R[i], t[k])$, $dist(i-1,t) + 1$ and $dist(i, b \cdot t[1..k-1]) + 1$ where $b \in \{\text{A}, \text{C}, \text{G}, \text{T}\}$. We create three types of edges in the dependency graph G:

- **match edge**: $((i-1, b \cdot t[1..k-1]), (i,t))$ whose weight is $\rho(t[k], R[i])$

- **delete edge**: $((i-1,t), (i,t))$ whose weight is 1

- **insert edge**: $((i, b \cdot t[1..k-1]), (i,t))$ whose weight is 1

For every node (k,t) (base case), $dist(k,t) = Hamming(R[1..k],t)$ and does not depend on any other values. So, we include an edge $(v_s, (k,t))$ of weight $Hamming(R[1..k],t)$.

Figure 5.6 presents the dependency graph for the read $R_2 = \text{AGTGCAG}$ and $\mathcal{T} = \{\text{AAGT}, \text{AGTG}, \text{GAAG}, \text{GTGA}, \text{TGAA}, \text{TGCA}\}$. For example, the node $(6, \text{TGAA})$ depends on three edges: (1) a match edge $((5, \text{GTGA}), (6, \text{TGAA}))$, (2) a delete edge $((5, \text{TGAA}), (6, \text{TGAA})$, and (3) an insert edge $((6, \text{GTGA}), (6, \text{TGAA}))$. This means that $dist(6, \text{TGAA})$ can be computed by Lemma 5.1 given $dist(5, \text{GTGA}), dist(5, \text{TGAA})$ and $(6, \text{GTGA})$.

This example graph has cycles. Hence, dynamic programming cannot solve this problem. Fortunately, we have the following lemma which is the key to computing $dist(i,t)$.

Lemma 5.2 *$dist(i,t)$ equals the length of the shortest path in G from v_s to (i,t).*

Proof This lemma follows from Lemma 5.1. The detail of the proof is left as an exercise. ∎

By the above lemma, $\min_{t\in\mathcal{T}} dist(|R|,t)$ is just the shortest path in G from v_s to $(|R|,t)$ among all $t \in \mathcal{T}$. Since G does not have a negative cycle (Exercise 4), we can run Dijkstra's algorithm to find the shortest paths from v_s to all nodes in G; then, we can determine $\min_{t\in\mathcal{T}} dist(|R|,t)$. Figure 5.7 gives the detail of the algorithm SpectralEdit for computing the minimum edit distance between R and any $\mathcal{T}$-string.

For the example in Figure 5.6, the bold path is the minimum weight path and its weight is 1. Note that there is no other path with weight 1. Hence, the minimum edit distance between $R_2 = \text{AGTGCAG}$ and any $\mathcal{T}$-string is 1.

For time complexity, i has $|R|$ choices and t has $|\mathcal{T}|$ choices. Hence, the

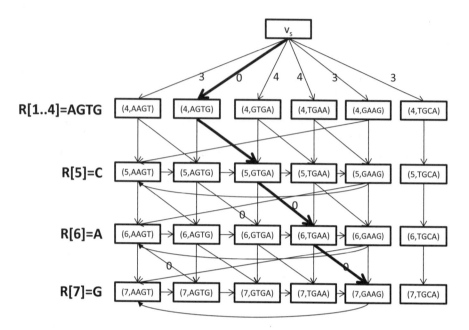

FIGURE 5.6: Consider $R = $ AGTGCAG and $\mathcal{T} = \{$AAGT, AGTG, GAAG, GTGA, TGAA, TGCA$\}$. This figure shows the corresponding dependency graph. The insert edges are all horizontal edges. The delete edges are all vertical edges. The match/mismatch edges are all slanted edges. All edges with no label are of weight 1. The bold path from v_s to $(7, $GAAG$)$ is the shortest path among all nodes $(7, t)$ for $t \in \mathcal{T}$. Its weight is 1.

Algorithm SpectralEdit$(\mathcal{T}, R)$

Require: $\mathcal{T}$ is a set of k-mers, R is a read
Ensure: Find the minimum edit distance between R and any $\mathcal{T}$-string
  1: Create the dependency graph G;
  2: Compute single-source shortest path from the node v_s using Dijkstra's algorithm;
  3: Report $\min_{t \in \mathcal{T}} dist(|R|, t)$;

FIGURE 5.7: The algorithm SpectralEdit$(\mathcal{T}, R)$ computes the minimum edit distance between R and any $\mathcal{T}$-string.

graph has $|\mathcal{T}||R|$ nodes and $O(|\mathcal{T}||R|)$ edges. The running time of the algorithm is dominated by the execution of the Dijkstra algorithm, which takes $O(|\mathcal{T}||R|\log(|\mathcal{T}||R|))$ time. It is possible to improve the running time to $O(|\mathcal{T}||R|)$ (see Exercise 8). Note that $|\mathcal{T}|$ can be as big as 4^k. This algorithm can work when k is small. When k is big, it will be inefficient.

To speed up the algorithm, heuristics is proposed. The algorithm first identifies a long enough substring of R which is a $\mathcal{T}$-string. (In [37], the substring is of length at least $k+|R|/6$.) This region is assumed to be error-free. Then, the algorithm finds the minimum number of edits on the remaining portion of R such that the string can become a $\mathcal{T}$-string.

5.3.1.2 k-mer counting

One conceptually simple yet fundamental problem is k-mer counting. This is the subroutine used by read error correction. It is also used in the assembly step (Section 5.3.3), repeat detection, and genomic data compression (Section 10.4.1). The problem is defined as follows. The input is a set of reads $\mathcal{R}$ and a parameter k. Let Z be the set of all possible k-mers (length-k substrings) that appear in $\mathcal{R}$. Our aim is to compute the frequencies of the k-mers in Z. Below, we discuss four solutions: (1) Simple hashing, (2) JellyFish, (3) BFCounter, and (4) DSK.

Simple hashing: The k-mer counting problem can be solved by implementing an associative array using hashing (see Section 3.4.1). When k is small (say, less than 10), we use perfect hashing. Note that every k-mer z can be encoded as a $2k$-bit binary integer $b(z)$ by substituting A, C, G, and T in z by 00, 01, 10, and 11, respectively. We build a table $Count[0..4^k-1]$ of size 4^k such that each entry $Count[b(z)]$ stores the frequency of the k-mer z in Z. Precisely, we initialize every entry in $Count[0..4^k-1]$ as zero. Then, we iteratively scan every k-mer z in Z and increment $Count[b(z)]$ by one. Finally, all non-zero entries in $Count[]$ represent k-mers occurring in Z and we can report their counts. Figure 5.8(a) gives an example to illustrate this simple counting method.

Suppose $N=|Z|$. The above approach is very efficient. It runs in $O(N+4^k)$ time. Since we need to build a table of size 4^k, the space complexity is $O(4^k)$. When k is big, the above algorithm cannot work since it uses too much space.

JellyFish: It is possible to reduce the hash table size using an open-addressing mechanism (see Section 3.4.1). Let $h()$ be a hash function and $H[0..\frac{N}{\alpha}-1]$ be a hash table that stores an array of k-mers where α is the load factor ($0 < \alpha \le 1$). We also build a table $Count[0..\frac{N}{\alpha}-1]$ where $Count[i]$ stores the count for the k-mer $H[i]$.

For each k-mer $z \in Z$, we hash z into some entry $H[i]$ where $i = h(z)$. If $H[i]$ is not empty and $H[i] \ne z$, we cannot store z in $H[i]$. This situation is called a collision. A collision can be resolved using the open-addressing mechanism. For instance, we can resolve a collision by linear probing. Linear probing tries to increment the index i by 1 when collision occurs until either $H[i] = z$

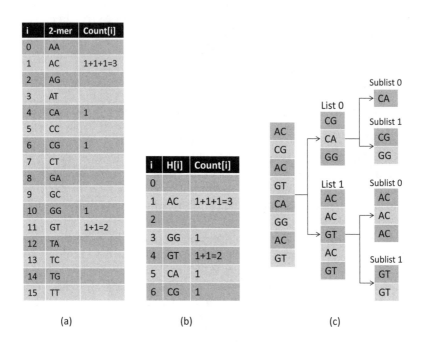

(a)   (b)   (c)

FIGURE 5.8: Consider a set of 4-mers $Z = \{\text{AC}, \text{CG}, \text{AC}, \text{GT}, \text{CA}, \text{GG}, \text{AC}, \text{GT}\}$. (a) illustrates the simple k-mer counting method that uses a count table of size 4^k. (b) illustrates the Jellyfish k-mer counting method that uses a hash table of size 7. The hash function is $h(z) = b(z) \mod 7$. For example, GT is stored in the 4th entry since $h(\text{GT}) = 4$. There is one collision in this example. Since $h(\text{CA}) = 4$, CA collides with GT. By linear probing, CA is stored in the 5-th entry instead. (c) illustrates the DSK k-mer counting method. We assume $h(z) = b(z)$, $n_{list} = 2$ and $n_{sublist} = 2$. DSK partitions Z into $4(= n_{list} * n_{sublist})$ sublists. Then, it runs Jellyfish to count k-mers in each sublist.

Algorithm Jellyfish(Z, α, h)

Require: Z is a set of N's k-mers, α is a load factor that controls the hash table size and $h(.)$ is the hash function

Ensure: The count of every k-mer appearing in Z

1: Set $H[1..\frac{N}{\alpha}]$ be a table where each entry requires $2k$ bits
2: Set $Count[1..\frac{N}{\alpha}]$ be a table where each entry requires 32 bits
3: Initialize T to be an empty table
4: **for** each k-mer z in Z **do**
5:   $i = hashEntry(z, h, \frac{N}{\alpha})$;
6:   **if** $H[i]$ is empty **then**
7:     $H[i] = z$ and $Count[i] = 1$;
8:   **else**
9:     $Count[i] = Count[i] + 1$;
10:   **end if**
11: **end for**
12: Output $(H[i], Count[i])$ for all non-empty entries $H[i]$;

Algorithm hashEntry$(z, h, size)$

1: $i = h(z) \mod size$;
2: **while** $H[i] \neq z$ **do**
3:   $i = i + 1 \mod size$; /* *linear probing* */
4: **end while**
5: Return i;

FIGURE 5.9: The Jellyfish algorithm for k-mer counting algorithm.

or $H[i]$ is an empty entry. The function $hashEntry()$ in Figure 5.9 illustrates the linear probing scheme to resolve a collision. If $hashEntry(z, h, \frac{N}{\alpha})$ returns an empty entry $H[i]$, then z does not exist in the hash table and we set $H[i] = z$ and $Count[i] = 1$. Otherwise, if $hashEntry(z, h, \frac{N}{\alpha})$ returns an entry $H[i] = z$, we increment $Count[i]$ by 1. After all k-mers in Z are processed, we report $(H[i], Count[i])$ for all non-empty entries $H[i]$.

The above algorithm is detailed in Figure 5.9. Figure 5.8(b) gives an example to illustrate this algorithm. It is efficient if there is no collision. In practice, the number of collisions is expected to be low when the load factor $\alpha \leq 0.7$. Then, the running time is expected to be $O(N)$. For space complexity, the tables $H[]$ and $Count[]$ require $\frac{N}{\alpha}(2k + 32)$ bits, assuming the count takes 32 bits.

The above idea is used in Jellyfish [188].

Although Jellyfish uses less space than the naive counting method, Jellyfish's hash table must be of a size that is at least the number of unique k-mers in Z. Jellyfish still requires a lot of memory when the number of unique k-mers in Z is high.

BFCounter: In many applications, we are only interested in k-mers that occur at least q times. If we can avoid storing k-mers that occur less than q times, we can save a lot of memory space. Melsted et al. [199] proposed BF-Counter, which counts k-mers occurring at least q times. It uses the counting Bloom filter to determine if a k-mer occurs at least q times. The definition of the counting Bloom filter is in Section 3.4.3. It is a space-efficient data structure that allows us to insert any k-mers into it and queries if a k-mer occurs at least q times. Note that the counting Bloom filter is a probabilistic data structure. Although it may give a false positive (i.e., it may incorrectly report a k-mer exists or over-estimate the count of the k-mer), it does not give a false negative.

BFCounter maintains a counting Bloom filter B and a hash table H. It has two phases. The first phase starts with an empty counting Bloom filter B and an empty hash table H. It scans the k-mers in Z one by one. For each k-mer $z \in Z$, it checks if z occurs at least $q - 1$ times in the counting Bloom filter B by verifying if $countBloom(z, B) \geq q - 1$. If not, we insert z into B by $insertBloom(z, B)$. Otherwise, z occurs at least q times. We check if z is in the hash table H. If not, we insert z into some empty entry $H[i]$ and set $count[i] = 0$;

The second phase performs the actual counting. It scans the k-mers in Z one by one. For each k-mer $z \in Z$, if z occurs in the hash entry $H[i]$, we increment $count[i]$ by one. The detail pseudocode is shown in Figure 5.10.

The running time of BFCounter is $O(N)$ expected time. For space complexity, the counting Bloom filter requires $O(N \log q)$ bits. The space for $H[]$ and $Count[]$ is $\frac{N'}{\alpha}(2k + 32)$ bits, where N' is the number of k-mers occurring at least q times. Note that $N' \leq \frac{N}{q}$.

DSK: Although BFCounter is space efficient, its space complexity still depends on the number N of k-mers in Z. Suppose the memory is fixed to be M bits and the disk space is fixed to be D bits. Can we still count the occurrences of k-mers efficiently? Rizk et al. [245] gave a positive answer and proposed a method called DSK. The idea of DSK is to partition the set Z of k-mers into different lists so that each list can be stored in the disk using D bits; then, for each list, the k-mers in the list are further partitioned into sublists so that each sublist can be stored in memory using M bits; lastly, the frequencies of the k-mers in each sublist are counted by the algorithm Jellyfish in Figure 5.9.

More precisely, the k-mers in Z are partitioned into n_{list} lists of approximately the same length. Since the disk has D bits and each k-mer can be represented in $2k$ bits, each list can store $\ell_{list} = \frac{D}{2k}$ k-mers. As there are N k-mers in Z, we set $n_{list} = \frac{N}{\ell_{list}} = \frac{2kN}{D}$. This partition is done by a hash function $h()$ that uniformly maps all k-mers to the n_{list} lists. Precisely, for each k-mer $z \in Z$, z is assigned to the i-th list if $h(z) \mod n_{list} = i$.

Then, each list is further partitioned into sublists, each of length $\ell_{sublist}$. Each sublist will be processed in memory using Jellyfish, which requires $\frac{\ell_{sublist}}{0.7}(2k + 32)$ bits. As the memory has M bits, we have $\ell_{sublist} = \frac{0.7M}{(2k+32)}$.

Algorithm BFCounter(Z, q, α, h)

Require: Z is a set of N's k-mers, α is a load factor that controls the hash table size and $h(.)$ is the hash function

Ensure: For every k-mer in Z occurring at least twice, report its count.

1: Set $H[1..\frac{N}{2\alpha}]$ be an empty hash table where each entry requires $2k$ bits;
2: Set $Count[1..\frac{N}{q\alpha}]$ be an empty vector where each entry requires 32 bits;
3: Set $B[0..m-1]$ be an empty vector where each entry requires $\lceil \log q \rceil$ bits and $m = 8N$;
4: **for** each k-mer z in Z **do**
5:    **if** countBloom$(z, B) \geq q - 1$ **then**
6:       $i = hashEntry(z, h, \frac{N}{q\alpha})$;
7:       **if** $H[i]$ is empty **then**
8:          Set $H[i] = z$ and $Count[i] = 0$;
9:       **end if**
10:   **else**
11:       insertBloom(z, B);
12:   **end if**
13: **end for**
14: **for** each k-mer z in Z **do**
15:    $i = hashEntry(z, h, \frac{N}{q\alpha})$;
16:    **if** $H[i] = z$ **then**
17:       $Count[i] = Count[i] + 1$;
18:    **end if**
19: **end for**
20: Output $(H[i], Count[i])$ for all non-empty entries $H[i]$ where $Count[i] \geq q$;

FIGURE 5.10: A space efficient k-mer counting algorithm that only counts k-mers occurring at least q times.

So, the number of sublists equals $n_{sublist} = \frac{\ell_{list}}{\ell_{sublist}} = \frac{D(2k+32)}{0.7(2k)M}$. Similarly, every list is partitioned into sublists by the hash function $h()$. Precisely, for each k-mer s in the ith list, s is assigned to the jth sublist if $(h(s)/n_{list})$ mod $n_{subilst} = j$.

For each sublist of length $\ell_{sublist} = \frac{0.7M}{(2k+32)}$, using M bit memory, we count the number of occurrences of every k-mer in the sublist using Jellyfish$(d_j, 0.7, h)$ in Figure 5.9.

Figure 5.11 presents the DSK algorithm. The algorithm iterates n_{list} times. For the ith iteration, the algorithm runs two phases. The first phase enumerates all k-mers in Z and identifies all k-mers belonging to the ith list; then, it partitions them into sublists and writes them to the disk. The second phase reads each sublist into memory one by one. For each sublist, with the help of the algorithm Jellyfish in Figure 5.9, the frequencies of the k-mers in each sublist are counted.

Figure 5.8(c) gives an example to illustrate the execution of DSK. We assume $n_{list} = 2$, $n_{sublist} = 2$ and $h(z) = b(z)$ for every $z \in Z$. Since $n_{list} = 2$, the algorithm runs two iterations. Here, we describe the 0th iteration. (The 1st iteration is executed similarly.) The first phase of the 0th iteration scans all k-mers in Z and identifies every k-mer $z \in Z$ which belongs to the 0th list. For example, $h(\mathtt{GG}) = 10$, since $h(\mathtt{GG})$ mod $n_{list} = 0$ and $(h(z)/n_{list})$ mod $n_{sublist} = 1$, $\mathtt{GG}$ belongs to the 0th list and the 1st sublist. After that, the 0th list is partitioned into the 0th sublist $\{\mathtt{CA}\}$ and the 1st sublist $\{\mathtt{CG}, \mathtt{GG}\}$. Both sublists are written into the disk. The second phase reads each sublist into the memory and counts the k-mers using the Jellyfish algorithm.

Observe that this algorithm will only write each k-mer in Z once, though it will read each k-mer n_{list} times. Hence, the algorithm will not generate many disk write accesses. Next, we analyze the time complexity. For the ith iteration, the algorithm enumerates all k-mers in Z, which takes $O(N)$ time. Then, it identifies $\frac{D}{2k}$'s k-mers belonging to the ith list and writes them out to the disk, which takes $O(\frac{D}{k})$ time. After that, the algorithm reads in $\frac{D}{2k}$'s k-mers and performs counting, which takes $O(\frac{D}{k})$ expected time. So, each iteration takes $O(N + \frac{D}{k}) = O(N)$ expected time, as $N > \frac{D}{2k}$. Since there are $n_{list} = \frac{2kN}{D}$ iterations, the algorithm runs in $O(\frac{kN^2}{D})$ expected time. When $D = \Theta(N)$, the algorithm runs in $O(kN)$ expected time.

5.3.2   Base-by-base extension approach

After all reads are corrected, we can stitch the corrected reads to form sub-fragments of the sample genome, which are called contigs. There are two approaches to constructing contigs: the base-by-base extension approach and the de Bruijn graph approach. This section covers the base-by-base extension approach.

The base-by-base extension approach reconstructs each contig by extending it base by base. The method starts with a randomly selected read as a

---

Algorithm DSK(Z, M, D, h)

Require: Z is a set of N's k-mers, target memory usage M (bits), target disk space D (bits) and hash function $h(.)$

Ensure: The count of every k-mer appearing in Z

1: $n_{list} = \frac{2kN}{D}$;

2: $n_{sublist} = \frac{D(2k+32)}{0.7(2k)M}$;

3: **for** $i = 0$ to $n_{list} - 1$ **do**

4:      Initialize a set of empty sublists $\{d_0, \dots, d_{n_{sublist}-1}\}$ in disk;

5:      **for** each k-mer z in Z **do**

6:          **if** $h(z) \mod n_{list} = i$ **then**

7:              $j = (h(z)/n_{list}) \mod n_{sublist}$;

8:              Write z to disk in the sublist d_j;

9:          **end if**

10:     **end for**

11:     **for** $j = 0$ to $n_{sublist} - 1$ **do**

12:         Load the jth sublist d_j in memory;

13:         Run Jellyfish$(d_j, 0.7, h)$ (see Figure 5.9) to output the number of occurrences of every k-mer in the sublist d_j;

14:     **end for**

15: **end for**

FIGURE 5.11: The DSK algorithm.

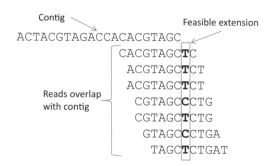

FIGURE 5.12: The top sequence is the template. There are 7 reads aligned at the 3' end of the template. The rectangle shows that bases C and T are the feasible extensions of the template. Since the consensus is T, the base-by-base extension method will extend the contig with the base T.

template. Then, reads are aligned on the two ends (3' end and 5' end) of the template. From the alignments, a consensus base is obtained and the template is extended by such base. Figure 5.12 illustrates the base-by-base extension step. The extension is repeated until there is no consensus. Then, the extension is stopped and we obtain a contig. We perform extension on both the 3' and 5' ends of the template. Figure 5.13 gives the pseudocode of the method.

SSAKE [309], VCAKE [117] and SHARCGS [69] are example methods that use this approach to construct contigs. Although base-by-base extension approach is simple, it tends to give short contigs due to two problems. First, the initial template is an arbitrary read. If the read contains sequencing errors or is in a repeat region, it will affect the extension. The second problem occurs when we extend the template into some repeat regions. The repeat creates branches and the above approach cannot resolve them.

To resolve the first problem, we select a read as a template if it is unlikely to contain a sequencing error and it is unlikely to be in a repeat region. Using the idea in Section 5.3.1, the frequencies of all k-mers of all reads are counted. A read R is selected as a template if the frequencies of all its k-mers are within some user-defined thresholds $\theta_{\min}$ and $\theta_{\max}$. If the number of occurrences of some k-mer is smaller than $\theta_{\min}$, R is likely to a contain sequencing error. If the number of occurrences of some k-mer is bigger than some $\theta_{\max}$, R is likely to be repeated. The two thresholds can be determined by studying the histogram of the frequencies of k-mers of the input sequencing reads.

For the second problem, one solution is to use the connectivity information of paired-end reads to resolve ambiguity. This approach is used by PE-assembler [9]. Figure 5.14 illustrates the idea. Suppose we can extend the template using two different reads (black and gray reads). We cannot decide which one is correct (see Figure 5.14(a)). Since each read has a mate, we may be able to make a decision. There are two cases: For the first case, if the mate

---

Algorithm SimpleAssembler($\mathcal{R}$)

Require: $\mathcal{R}$ is a set of reads
Ensure: A set of contigs
 1: **while** some read R are not used for reconstructing contigs **do**
 2:   Set the template $T = R$;
 3:   **repeat**
 4:     Identify a set of reads that align to the 3' end (or 5' end) of the template T;
 5:     Identify all feasible extensions of the template;
 6:     If there is a consensus base b, set $T = Tb$ if the extension is at 3'end and set $T = bT$ if the extension is at 5' end;
 7:   **until** there is no consensus base;
 8:   Report the template as a contig;
 9:   Mark all reads aligned to the contig as used;
10: **end while**

FIGURE 5.13: A simple base-by-base extension assembler.

of the black read can align on the template, we will trust the black read (Figure 5.14(b)). For the second case, suppose there exists some reads R that align on the template and the mate of R can align with the mate of the black read (see Figure 5.14(c)); Then, we can also trust the black read. In other words, the connectivity information of paired-end reads can help to filter those false positive alignments.

5.3.3 De Bruijn graph approach

The previous section covers the base-by-base extension approach to reconstruct the contigs. Another approach is based on de Bruijn graph. The de Bruijn graph approach is introduced in Idury and Waterman [113]. Today, it is the mainstream approach for assembling short reads. Existing methods include Velvet [329], SOAPdenovo [176], Euler-SR [38], IDBA [227], ABySS [275], ALLPATHS [34] and Edena [100].

We first define the de Bruijn graph. Consider a set of reads $\mathcal{R}$ and a parameter k; a de Bruijn graph is a graph $H_k = (V, E)$ where V is the set of all k-mers of $\mathcal{R}$ and two k-mers u and v form an edge $(u, v) \in E$ if u and v are the length-k prefix and suffix of some length-$(k + 1)$ substring of $\mathcal{R}$, respectively. Suppose the total length of all reads in $\mathcal{R}$ is N, the de Bruijn graph can be constructed in $O(N)$ time.

For example, consider the set of strings $\mathcal{R} = \{\texttt{ACGC}, \texttt{CATC}, \texttt{GCA}\}$. Suppose we want to build a de Bruijn graph for 2-mers. Then, the vertex set is $\{\texttt{AC}, \texttt{AT}, \texttt{CA}, \texttt{CG}, \texttt{GC}, \texttt{TC}\}$. Figure 5.15(a) gives the de Bruijn graph.

The sample genome can be predicted by identifying a Eulerian path (a Eulerian path is a path that uses every edge of H_k exactly once) of H_k. Note

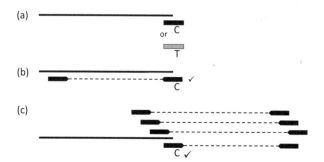

FIGURE 5.14: (a) The top sequence is the template. Two reads (black and gray colors) can be aligned to the 3' end of the template. Based on the black color read, the next base is C. Based on the gray color read, the next base is T. The black read is likely to be correct if the mate of the black read aligns on the template (Figure (b)) or the mate of the black read align with the mate of another read R that aligns on the template (Figure (c)).

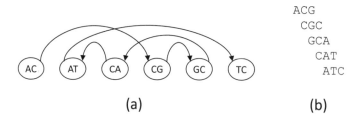

FIGURE 5.15: (a) The de Bruijn graph H_3 for $\mathcal{R} = \{\text{ACGC}, \text{CATC}, \text{GCA}\}$. (b) The overlapping of the 5 3-mers corresponding to the 5 edges of H_3.

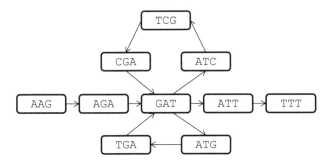

FIGURE 5.16: The de Bruijn graph H_3 for $\mathcal{R}$ = {AAGATC, GATCGAT, CGATGA, ATGATT, GATTT}.

that a Eulerian path of H_k can be computed in $O(n)$ time if H_k has n edges (see Exercise 13). For the example in Figure 5.15(a), a unique path exists from the node AC to the node TC. By overlapping all 3-mers of the edges in the order of the path (see Figure 5.15(b)), we obtain the sequence ACGCATC. The sequence ACGCATC is in fact the superstring formed by overlapping the reads ACGC, GCA, CATC in order. This example suggests that a genome can be recovered from the de Bruijn graph of $\mathcal{R}$.

However, a Eulerian path may not be unique in H_k. For example, consider a set of reads $\mathcal{R}$ = {AAGATC, GATCGAT, CGATGA, ATGATT, GATTT}. Assume $k = 3$. The de Bruijn graph H_3 is shown in Figure 5.16. There are two possible Eulerian paths in H_3. If we traverse the top cycle first, we obtain AAGATCGATGATTT. If we traverse the bottom cycle first, we obtain AAGATGATCGATTT. This example indicates that the Eulerian path may not return a correct sequence. Even worse, a Eulerian path may not exist in some H_k.

Below, we describe the de Bruijn graph assembler when (1) there is no sequencing error and (2) there are sequencing errors. Then, we will discuss the issue of selecting k.

5.3.3.1 De Bruijn assembler (no sequencing error)

As the Eulerian path is not unique and may not exist, we do not aim to obtain the complete genome. Instead, we try to obtain a set of contigs. The contig is a maximal simple path in H_k. Precisely, each maximal simple path is a maximal path in H_k such that every node (except start and end nodes) has in-degree 1 and out-degree 1. Figure 5.17 gives the pseudocode of this simple method.

For the de Bruijn graph H_3 in Figure 5.16, we can construct 4 contigs: AAGAT, GATCGAT, GATGAT, GATTT.

Note that k is an important parameter. We obtain different sets of contigs when we change k. To illustrate this issue, Figure 5.18 gives the de Bruijn graphs for $k = 4$ and $k = 5$.

Algorithm De_Bruijn_Assembler($\mathcal{R}, k$)

Require: $\mathcal{R}$ is a set of reads and k is de Bruijn graph parameter
Ensure: A set of contigs
 1: Generate the de Bruijn graph H_k for $\mathcal{R}$;
 2: Extract all maximal simple paths in H_k as contigs;

FIGURE 5.17: A simple de Bruijn graph assembler.

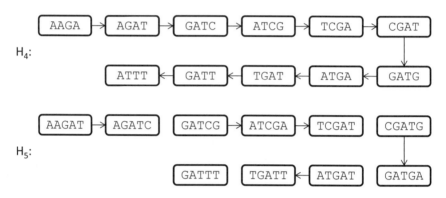

FIGURE 5.18: The de Bruijn graph H_4 and H_5 for $\mathcal{R}$ = {AAGATC, GATCGAT, CGATGA, ATGATT, GATTT}.

When $k = 4$, H_4 is a single path. We can construct one contig AAGATCGATGATTT.

When $k = 5$, H_5 has 5 connected components, and each is a single path. We can construct 5 contigs: AAGATC, GATCGAT, CGATGA, ATGATT, GATTT.

By the above examples, we observe that the de Bruijn graph is good when we know the correct k. When k is small (see H_3 in Figure 5.16), there are many branches due to repeat regions. This results in many short contigs. When k is large (see H_5 in Figure 5.18), some k-mers are missing (especially for regions with low coverage). This results in disconnected components, which also generate many short contigs.

We need to identify k to balance these two issues. We will discuss more on k selection in Section 5.3.3.3.

5.3.3.2   De Bruijn assembler (with sequencing errors)

The previous subsection assumes there are no sequencing errors in the reads, which is unrealistic. When sequencing errors exist, we can try to remove errors by cleaning up the de Bruijn graph. This section presents the solution proposed by Zerbino and Birney [329]. Observe that short reads have low error rates (about 1 error in every 100 bases). Most k-mers contain at most 1 error. These erroneous k-mers will create two possible anomalous subgraphs in the

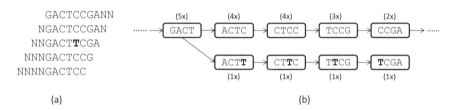

(a)   (b)

FIGURE 5.19: (a) is the multiple sequence alignment of a set of 5 reads where the third read has a sequencing error (shown in bold font). (b) shows the de Bruijn graph corresponding to the set of reads in (a). A tip is formed. The numbers in the brackets are the multiplicity of the 4-mers.

de Bruijn graph: the tip and the bubble. Below, we describe how to clean them up.

A tip is a path of length at most k where all internal nodes are of in-degree 1 and out-degree 1 while one of its ends is of in-degree 0 or out-degree 0. It can produce a potential contig of length at most $2k$. Figure 5.19 illustrates a tip created by a single mismatch in one read. If all nodes on the tip are of low multiplicity (i.e., lower k-mer counts), such a short contig is unlikely to be true. We can remove this tip from the de Bruijn graph. Note that removing a tip may generate more tips. The procedure needs to remove tips recursively.

Bubbles are two paths starting from the same vertex and ending at the same vertex, where the two paths represent different contigs that differ by only one nucleotide. For example, Figure 5.20(a) consists of a set of reads where the third read has a single mismatch. Figure 5.20(b) shows the corresponding de Bruijn graph and one bubble is formed. The top path of the bubble represents GACTCCGAG. The bottom path of the bubble represents GACTTCGAG. When the two paths in the bubble are highly similar, the path with lower multiplicity is likely to be a false positive. We can try to merge the bubble. In the example, the two paths have only one mismatch. Furthermore, since the nodes in the bottom path have lower multiplicity, we merge the bubble and obtain the graph in Figure 5.20(c). More precisely, we can define the weight of a path $w_1 \rightarrow w_2 \rightarrow \ldots \rightarrow w_p$ as $\sum_{i=1}^{p} f(w_i)$ where $f(w_i)$ is the multiplicity of w_i. Then, when we merge two paths in a bubble, we will keep the path with highest path weight.

To merge bubbles, we can use the tour bus algorithm. The tour bus algorithm is a Dijkstra-like breadth-first search (BFS)-based method. The algorithm starts from an arbitrary node s and it visits the nodes u in increasing order of distance from the start node. When the algorithm processes an unvisited node u, it checks all its children v. For each child v, it performs two steps. First, it assigns u as the parent of v in the BFS tree by setting $\pi(v) = u$. Second, if a child v of u is visited, a bubble is detected; the algorithm computes the lowest common ancestor c of u and v in the BFS tree defined by $\pi()$. Then, the two paths $c \rightarrow u$ and $c \rightarrow v$ are compared. If they are similar

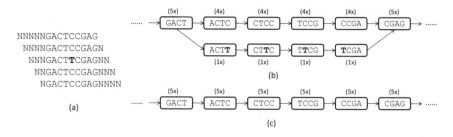

NNNNNGACTCCGAG
NNNNGACTCCGAGN
NNNGACT**T**CGAGNN
NNGACTCCGAGNNN
NGACTCCGAGNNNN

(a)

FIGURE 5.20: (a) is the multiple sequence alignment of a set of 5 reads where the third read has a sequencing error (shown in bold font). (b) shows the de Bruijn graph corresponding to the set of reads in (a). A bubble is formed. (c) is the path obtained by merging the bubble in (b).

enough (with similarity at least 80% in Velvet), the two paths will be merged and we will keep the path with the highest path weight. The algorithm is detailed in Figure 5.21.

Figure 5.22 gives an example to illustrate the steps of the tour bus algorithm. The original de Bruijn graph is shown in Figure 5.22(a). Starting from the node r, BFS traversal is performed to visit descendants of r in increasing order of the distance from r. The BFS traversal stops when it revisits the same node. Figure 5.22(b) shows the BFS tree (in bold) when the node v is revisited from the node u. By identifying the lowest common ancestor c in the BFS tree, we find two paths $c \to u$ and $c \to v$ which form a bubble. We merge the two paths and keep the path with the higher multiplicity. After pruning the bubble, we obtain Figure 5.22(c). Then, the BFS is continued. After the BFS visits u', it revisits the node v'. c' is the lowest common ancestor of u' and v'. We find two paths $c' \to u'$ and $c' \to v'$ which form a bubble. After pruning the bubble, we obtain Figure 5.22(d). Then, we cannot find any bubble. The tour bus algorithm is finished.

After we remove tips and merge bubbles in the de Bruijn graph, we can further filter noise by removing k-mers with multiplicity smaller than or equal to some threshold m (says $m = 1$). For the example in Figure 5.22(d), if $m = 1$, we need to remove the two edges of weight 1. Note that this technique is also used when we perform error correction in Section 5.3.1.

In summary, this subsection introduces three tricks: (1) remove tips, (2) merge bubbles and (3) filter k-mers with low multiplicity. By combining these techniques, we obtain an algorithm that can handle sequencing errors, which is shown in Figure 5.23.

5.3.3.3   How to select k

As discussed in Section 5.3.3.1, the selection of k can affect the performance of the de Bruijn graph algorithm.

A simple solution is to run the algorithm in Figure 5.23 for multiple k's.

Algorithm Tour_Bus(H, s)

Require: H is the de Bruijn graph and s is an arbitrary node in H
Ensure: A graph formed after merging the bubbles
 1: Set Q be a queue with one node s;
 2: **while** $Q \neq \emptyset$ **do**
 3:    $u = $ dequeue(Q);
 4:    **for** each child v of u **do**
 5:      **if** visited[v] = false **then**
 6:        Set $\pi(v) = u$; /* set u as v's parent in the BFS tree */
 7:        Set visited[v] = true;
 8:        enqueue(Q, v);
 9:      **else**
10:        Find the lowest common ancestor c of u and v by $\pi()$;
11:        **if** the paths $c \to u$ and $c \to v$ are similar enough **then**
12:          Merge the two paths and keep the path with the highest path weight;
13:        **end if**
14:      **end if**
15:    **end for**
16: **end while**

FIGURE 5.21: The tour bus algorithm.

Then, the contigs are clustered and merged. This technique is used by some transcriptome assemblers [193, 291].

One issue with this simple solution is that the contigs obtained by different k's have different quality. For contigs obtained from H_k where k is small, they are highly accurate; however, the contigs are short since there are many branches due to repeat regions. For contigs obtained from H_k where k is large, they are longer; however, they may contain many errors.

IDBA [227] suggested that we should not build the de Bruijn graphs independently for different k's. Instead, IDBA builds the de Bruijn graph H_k incrementally from small k to large k. When k is small, we can get high-quality contigs, though they are short. Then, these high-quality contigs are used to correct errors in the reads. Incrementally, we build de Bruijn graphs H_k for larger k. Since the reads in $\mathcal{R}$ are corrected, the noise in H_k is reduced. This allows us to obtain high-quality long contigs from H_k when k is large. Figure 5.24 details the idea proposed by IDBA.

5.3.3.4 Additional issues of the de Bruijn graph approach

In the above discussion, we do not use the k-mers appearing in the reverse complement of the reads. This is not correct. We actually need to consider

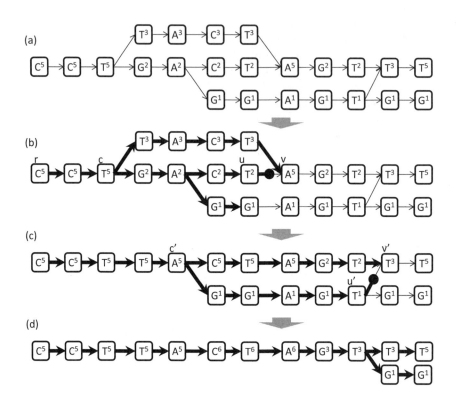

FIGURE 5.22: An example illustrates how the tour bus algorithm prunes the de Bruijn graph. For clarity, every node shows the last base of its k-mer and the corresponding integer superscript is its multiplicity. The algorithm performs breadth-first search (BFS) and the BFS tree is shown as the bold edges. In (a), there are two (nested) bubbles. In (b), we perform BFS starting from r. When we visit u, a child v of u is visited (i.e., within the BFS tree). We identify a bubble and we merge it. Then, we obtain (c). In (c), we continue to perform BFS. When we visit u', a child v' of u' has been visited. We identify another bubble and we merge it. Then, we obtain (d). In (d), we continue to perform BFS. As no bubble is identified, the algorithm stops.

Algorithm Velvet$(\mathcal{R}, k)$

Require: $\mathcal{R}$ is a set of reads and k is de Bruijn graph parameter

Ensure: A set of contigs

1: Generate the de Bruijn graph H_k for $\mathcal{R}$;
2: Remove tips;
3: Merge bubbles;
4: Remove nodes with multiplicity $\leq m$;
5: Extract all maximal simple paths in H_k as contigs;

FIGURE 5.23: A de Bruijn graph assembler Velvet, which can handle sequencing errors.

Algorithm IDBA$(\mathcal{R}, k_{min}, k_{max})$

Require: $\mathcal{R}$ is a set of reads and k_{min} and k_{max} are de Bruijn graph parameter

Ensure: A set of contigs

1: **for** $k = k_{min}$ to k_{max} **do**
2:    Generate the de Bruijn graph H_k for $\mathcal{R}$;
3:    Remove tips;
4:    Merge bubbles;
5:    Remove nodes with multiplicity $\leq m$;
6:    Extract all maximal simple paths in H_k as contigs;
7:    All reads in $\mathcal{R}$ are aligned to the computed contigs;
8:    The mismatch in the read is corrected if 80% of reads aligned to the same position has the correct base;
9: **end for**
10: Extract all maximal simple paths in $H_{k_{max}}$ as contigs;

FIGURE 5.24: An iterative de Bruijn graph assembler (IDBA), which can handle sequencing errors.

FIGURE 5.25: This example gives an arrangement of the contigs and the alignments of paired-end reads on the contigs. (1) is concordant while (2), (3) and (4) are discordant.

the reverse complement of the k-mers when we create the de Bruijn graph. See Exercise 11 for an example.

Note that a single error in the read will create k noisy k-mers. Due to such noisy k-mers, de Bruijn graphs can be huge. Hence, it requires big memory. Alternatively, a compression method [28] has been proposed to reduce the size of a de Bruijn graph.

Another issue is that the de Bruijn graph approach does not use the connectivity information of paired-end reads. To resolve the problem, SPAdes [13] is proposed, which uses a paired de Bruijn graph to capture such information.

5.3.4 Scaffolding

The contig building step in Sections 5.3.2 and 5.3.3 reports a set of contigs. The next question is how to recover the correct ordering and the correct orientation of the contigs in the sample genome. This question can be answered by solving the scaffolding problem. Figure 5.25 gives an example scaffold for 4 contigs A, B, C and D. The scaffold is $(A, -B, -C, D)$. Note that the orientations of contigs B and C are reversed. To verify the scaffold, we can utilize the paired-end reads extracted from the sample genome. Recall that in whole genome sequencing, every paired-end read (i.e., a read and its mate) is expected to map in inward orientation (based on Illumina sequencing) and has its insert size within a short range ($span_{\min}, span_{\max}$). A paired-end read is said to support the scaffold if the paired-end read is aligned inwardly on the scaffold and its insert size does not violate the range ($span_{\min}, span_{\max}$). In such case, the paired-end read is said to be concordant. Otherwise, it is said to be discordant. For the example in Figure 5.25, the paired-end read (1) has a correct insert size and a correct orientation. So, the paired-end read (1) is called concordant. The paired-end reads (2) and (3) have the correct insert size. However, their orientations are not correct. For the paired-end read (4), although it has a correct orientation, the insert size is too long. Hence, paired-end reads (2), (3) and (4) are discordant.

The scaffolding problem is defined as follows. The input is a set of paired-end reads $\mathcal{E}$ and a set of contigs $\mathcal{C}$. The scaffolding problem aims to generate the orientations and the orderings of the contigs so that the number of discordant reads is minimized. Figure 5.26 gives an example. The input consists of three contigs A, B and C, and a set of paired-end reads that are mapped

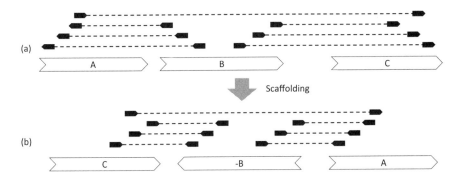

FIGURE 5.26: An example of scaffolding. The input is three contigs A, B and C and a set of paired-end reads mapping on the contigs. The scaffolding problem aims to rearrange the contigs to $(A, -B, C)$ so that the number of discordant reads is minimized.

on the contigs. If we fix the scaffold as (A, B, C), all paired-end reads are discordant. To minimize the number of discordant reads, the scaffold should be $(C, -B, A)$. In this case, there is only one discordant read. Hence, the scaffolding problem aims to report the scaffold $(C, -B, A)$.

The scaffolding problem seems simple. Huson et al. [111] showed that scaffolding is NP-hard. Even for just determining the correct orientations of contigs (i.e., no need to determine their orderings), Kececioglu and Myers [133] showed that such a problem is still NP-hard.

As the scaffolding problem is computationally hard, a number of heuristics or approximation algorithms are proposed. Many of them are internal routines within some assemblers. Some examples include PEAssembler [9], SSAKE [309], ABySS [275] and SOAPdenovo [176]. There are also some standalone scaffolders, including Bambus [235], SSPACE [24], SOPRA [61], MIP Scaffolder [258], Opera [85], Opera-LG [84], and SCARPA [70].

The scaffolding problem can be solved in three steps.

1. Demarcate repeat regions within assembled contigs.

2. Build the contig graph.

3. Identify a linear order of the contigs that minimizes the number of discordant edges.

Step 1: Demarcate all repeat regions within assembled contigs. This step aligns all paired-end reads in $\mathcal{E}$ onto the contigs in $\mathcal{C}$. The median read density is assumed to be the expected read coverage across the genome. Any region with read density higher than 1.5 times the median is considered a repeat region. Demarcate all repeat regions within contigs.

Step 2: Build the contig graph. This step defines the contig graph. For

FIGURE 5.27: This example considers 4 contigs. The gray boxes mark the repeat regions. The figure shows all paired-end reads that are mapped to two different contigs. The contig graph consists of all 4 contigs and all paired-end reads except PE_1 and PE_2.

FIGURE 5.28: This figure gives a scaffold with two paired-end reads $PE_1 = (A, -D, a, e)$ and $PE_2 = (-C, -D, d, e + f)$.

each paired-end read PE in $\mathcal{E}$, it is discarded if (1) both reads of PE align concordantly on the same contig, or (2) either read of PE fails to align to any contig, or (3) either read of PE aligns on a repeat region (see Figure 5.27). Each paired-end read in $\mathcal{E}$ is represented as a tuple (X, Y, x, y) where X and Y are some oriented contigs in $\mathcal{C}$ and x and y are the lengths of the contigs X and Y, respectively, covered by the paired-end read. For the example in Figure 5.28, the paired-end reads PE_1 and PE_2 are represented as $(A, -D, a, e)$ and $(-C, -D, d, e + f)$, respectively.

Step 3: Identify a linear order of the contigs that minimizes the number of discordant edges. This step orientates and orders the contigs of $\mathcal{C}$ to minimize the number of discordant reads. This problem is known as the contig scaffolding problem, which is defined as follows. The input is a contig graph $(\mathcal{C}, \mathcal{E})$ where $\mathcal{C}$ is a set of contigs and $\mathcal{E}$ is a set of paired-end reads. The output is a scaffold, which is defined as a linear ordering of the contigs in $\mathcal{C}$. A paired-end read is said to be discordant if it is not inward and its insert size is bigger than $span_{max}$. For example, in Figure 5.28, the paired-end read PE_2 is discordant if $d + e + f > span_{max}$. The problem aims to find a scaffold S of $\mathcal{C}$ that minimizes the number of discordant paired-end reads.

Here, we present a greedy algorithm [9]. For any scaffold S containing a subset of contigs in $\mathcal{C}$, a score $score(S)$ is defined which equals the total number of concordant paired-end reads minus the total number of discordant paired-end reads that are aligned on S.

The greedy algorithm starts by selecting a random contig as the initial scaffold S. The process then extends the scaffold S by some contig C (denoted as $S \circ C$ or $C \circ S$ depending on whether C is to the right or to the left of S) iteratively until S cannot be extended. A contig C is said to be a right (or left) feasible neighbor of S if S can extend to $S \circ C$ (or $C \circ S$) such

---

Algorithm GreedyScaffolding

Require: A contig graph $(\mathcal{C}, \mathcal{E})$ where $\mathcal{C}$ is the set of contigs and $\mathcal{E}$ is the set of paired-end reads

Ensure: A set of scaffolds S that cover all contigs in $\mathcal{C}$

1: **while** $\mathcal{C}$ is not empty **do**
2:     Denote S be a scaffold consists of a random contig C in $\mathcal{C}$ and remove C from $\mathcal{C}$;
3:     **repeat**
4:         From $\mathcal{E}$, we identify all contigs $C_1, C_2, \ldots, C_r$ that are right feasible neighbors of S and all contigs $C'_1, \ldots, C'_{r'}$ that are the left feasible neighbors of S;
5:         Identify the permutation $C_{i_1}, \ldots, C_{i_r}$ that maximizes $s_R = score(S \circ C_{i_1} \circ \ldots \circ C_{i_r})$;
6:         Identify the permutation $C'_{j_1}, \ldots, C'_{j_{r'}}$ that maximizes $s_L = score(C'_{j_1} \circ \ldots \circ C'_{j_{r'}} \circ S)$;
7:         **if** $s_R > \max\{0, s_L\}$ **then**
8:             Set $S = S \circ C_{i_1}$ and remove C_{i_1} from $\mathcal{C}$;
9:         **else if** $s_L > \max\{0, s_R\}$ **then**
10:            Set $S = C_{j_{r'}} \circ S$ and remove $C_{j_{r'}}$ from $\mathcal{C}$;
11:        **end if**
12:    **until** both ends of S cannot extend;
13:    Report the scaffold S;
14: **end while**

FIGURE 5.29: The GreedyScaffolding algorithm.

that some concordant paired-end read exists whose one end aligns on S and another end aligns on C. Let $C_1, \ldots, C_r$ be all right feasible neighbors of S and let $C'_1, \ldots, C'_{r'}$ be all left feasible neighbors of S. Among all possible permutations of the right feasible neighbors of S, we identify the permutation $C_{i_1}, \ldots, C_{i_r}$ that maximizes the score $s_R = score(S \circ C_{i_1} \circ \ldots \circ C_{i_r})$. Similarly, we also identify the permutation $C'_{j_1}, \ldots, C'_{j_{r'}}$ that maximizes the score $s_L = score(C'_{j_1} \circ \ldots \circ C'_{j_{r'}} \circ S)$. If $s_R > \max\{0, s_L\}$, we extend S to $S \circ C_{i_1}$. Otherwise, if $s_L > \max\{0, s_R\}$, we extend S to $C_{j_{r'}} \circ S$. This process is repeated until both ends are not extendable; then, we obtain one scaffold. After one scaffold is obtained, the entire procedure is repeated on the unused contigs to identify other scaffolds. The detail of the pseudocode is shown in Figure 5.29.

5.3.5 Gap filling

From scaffolding, we identify adjacent contigs. The region between two adjacent contigs is called a gap. For some gaps, we may be able to reconstruct the DNA sequence using local assembly. Below, we discuss one possible way to fill in the gap.

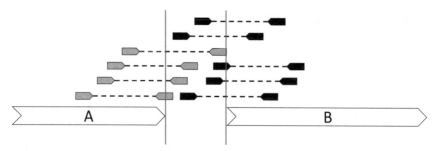

FIGURE 5.30: Suppose contigs A and B are adjacent. This figure shows all paired-end reads where one of the reads maps on contig A or B, but its mate cannot align properly. By collecting these reads, we may reconstruct the gap using local assembly. (Note that this gap cannot be reconstructed using paired-end reads aligned on A only (gray color) or B only (black color). Moreover, the gap can be assembled if we use all paired-end reads aligned on both A and B.)

Let A and B be two adjacent contigs. All paired-end reads that can align to the right end of A or the left end of B are extracted (see Figure 5.30). We assemble these paired-end reads using an assembler, like SSAKE [309], which can still work under low coverage. This approach enables us to fill in some gaps or reduce the size of some gaps.

5.4   Genome assembly for long reads

Previous sections presented genome assemblers for short reads generated from second-generation sequencing. Third-generation sequencing generates long reads. This section describes genome assemblers specialized for long reads, and is organized as follows. Section 5.4.1 discusses techniques to assemble long reads assuming that long reads have a low sequencing error rate. When long reads have a high sequencing error rate, the literature has two solutions. The first solution corrects the sequencing errors of long reads by aligning short reads on them. This approach is called the hybrid approach, and is discussed in Section 5.4.2. Another solution is to correct sequencing errors by overlapping the long reads. This technique is discussed in Section 5.4.3.

5.4.1 Assemble long reads assuming long reads have a low sequencing error rate

When the long reads have a low sequencing error rate, we can use an overlap layout consensus (OLC) assembler like Celera [213] or AMOS [298]. The OLC assembler involves five steps:

1. **Overlapping reads**: This step performs all-versus-all alignments of all long reads. It aims to identify read pairs that are adjacent in the genome assembly.

2. **Forming layouts of reads**: Based on the pairwise overlaps of the reads, this step aims to group reads into layouts such that each layout contains reads coming from one genomic region.

3. **Reconstructing a contig by consensus**: This step overlays the reads in each layout and computes the consensus. Each consensus is a contig.

4. **Scaffolding**: By aligning reads over the contigs, this step aims to link contigs to form scaffolds.

5. **Gap filling**: Gaps may exist between adjacent contigs in each scaffold. This step tries to fill in these gaps by aligning reads over these gaps.

Below, we detail the first three steps. The last two steps were discussed in Sections 5.3.4 and 5.3.5, respectively.

Step 1: Overlapping reads. Given a set of reads $\mathcal{R}$, Step 1 aims to report all read pairs (R, R') that are highly similar. Precisely, a read pair (R, R') is similar if the local alignment score $score(R, R') \geq \theta$ for some threshold θ. The naive solution is to compute the local alignment score of every read pair (R, R') using the Smith-Waterman alignment algorithm; then, (R, R') is reported if $score(R, R') \geq \theta$. Suppose L is the total length of all reads in $\mathcal{R}$. Then, this brute-force solution runs in $O(L^2)$ time.

The brute-force solution is slow. To speed up, one solution is to filter read pairs by k-mers. For every k-mer w, let $H[w]$ be a list of all reads $R \in \mathcal{R}$ such that w is a k-mer in R. Instead of performing local alignment for all read pairs, local alignment will be performed for read pair (R, R') sharing some common k-mer (i.e., $R, R' \in H[w]$ for some k-mer w). This idea significantly reduces the number of pairwise alignments.

Step 2: Forming layouts of reads. Given all read pairs with significant alignments, Step 2 computes the overlap graph. The overlap graph is a graph where each vertex is a read and each edge connects a pair of overlapping reads. (See Figure 5.31(c) for an example.)

We observe that the overlap graph has redundancy. Some edges can be inferred (transitively) from other edges. Precisely, for reads $R_{i_1}, R_{i_2}, \ldots, R_{i_m}$, if the overlap graph contains (1) an edge (R_{i_1}, R_{i_m}) and (2) a path $(R_{i_1}, R_{i_2}, \ldots, R_{i_m})$, then the edge (R_{i_1}, R_{i_m}) is redundant and we can remove

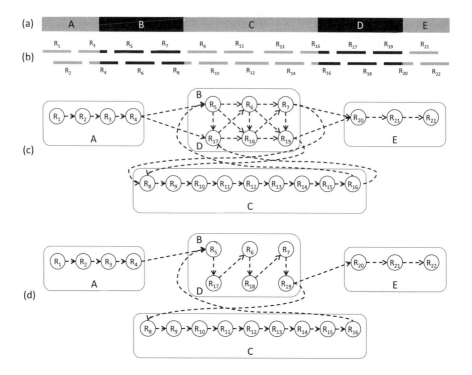

FIGURE 5.31: An illustration of the creation of the overlap graph. (a) is a genome with a repeat (region B is similar to region D). (b) is the reads obtained from whole genome sequencing. (c) is the overlap graph for the reads in (b). The edges are shown as dotted line. Observe that there are edges between reads in regions B and D since they are repeated. The graph in (c) has (transitively) redundant edges. After removing all redundant edges, we obtain the graph in (d).

it. For example, in the overlap graph in Figure 5.31(c), (R_4, R_{17}) is redundant since we have a path (R_4, R_5, R_{17}). By removing all redundant edges in Figure 5.31(c), we obtain the graph in Figure 5.31(d).

After redundant edges are removed from the overlap graph, we can obtain maximal simple paths. (Each maximal simple path is a maximal path such that every node (except start and end nodes) has in-degree 1 and out-degree 1.) These paths form the layouts. For our example in Figure 5.31(d), we obtain 4 simple paths: (R_1, R_2, R_3, R_4), $(R_8, R_9, R_{10}, R_{11}, R_{12}, R_{13}, R_{14}, R_{15}, R_{16})$, (R_{20}, R_{21}, R_{22}) and $(R_5, R_{17}, R_6, R_{18}, R_7, R_{19})$. The first three layouts represent the regions A, C and E, respectively. The last layout represents the repeat region B and D.

Step 3: Reconstructing contigs by consensus. Each layout consists of

```
R₁=GGTAATCT-AC            S₁=GGTAATC-GAC
R₂=    ATC-GACTTA         S₂=    ATC-GACTTA
R₃=        CTGACTTACCG    S₃=        CTGACTTACCG
   GGTAATCTGACTTACCG         GGTAATC-GACTTACCG
        (a)                       (b)
```

FIGURE 5.32: (a) is the multiple sequence alignment of
`GGTAATCTAC, ATCGACTTA, CTGACTTACCG`. (b) is the multiple sequence alignment
of `GGTAATCGAC, ATCGACTTA, CTGACTTACCG`. The bottom sequences in (a) and
(b) are the consensus sequences.

a set of reads $\mathcal{R}$ coming from one genomic region. The third step reconstructs
the genomic region by overlapping the reads in $\mathcal{R}$.

This problem can be solved by computing the multiple sequence alignment
(MSA) of all reads in $\mathcal{R}$; then, from the MSA, a consensus sequence is called
by reporting the most frequent base in each aligned column. This consensus
sequence is called the contig. Figure 5.32 gives two examples to illustrate the
idea.

However, computing MSA is NP-hard. Here, we describe a greedy merging
method (used in Celera [213] and AMOS [298]), which is as follows. First,
we compute the pairwise similarity for every pair of reads in $\mathcal{R}$. (In fact, the
pairwise similarities are known from Step 1.) Then, we greedy merge two reads
with the highest similarity first. The merging is repeated until we obtain the
alignment of all reads in $\mathcal{R}$. Figure 5.33 illustrates the steps of greedy merging
for the set of reads in Figure 5.32(a). Greedy merging runs in $O(\sum_{R\in\mathcal{R}} |R|)$
time. After we obtain the MSA of all reads in $\mathcal{R}$, we can extract the consensus
sequence by reporting the most frequent character in each column of the MSA.

However, greedy merging may not compute the optimal MSA, which leads
to an incorrect consensus sequence. Figure 5.34 gives an example. Greedy
merging reports the MSA in Figure 5.34(b). However, the optimal MSA is in
Figure 5.34(c). The corresponding consensus sequences are also different.

Anson and Myers [8] observed that the MSA generated by greedy merging
can be improved by locally rearranging reads one by one. They proposed a
round-robin realignment algorithm called ReAlign. ReAlign tries to iteratively
improve the MSA as follows. The algorithm iteratively checks if the MSA
can be improved by realigning every read in $\mathcal{R}$ one by one. This process is
repeated until the MSA cannot be further improved. Figure 5.36 details the
algorithm ReAlign. Figure 5.35 illustrates how to apply the algorithm ReAlign
to improve the MSA in Figure 5.34(b).

5.4.2 Hybrid approach

Observe that long and short reads have different properties. The hybrid
approach aims to use the advantages of both long and short reads to improve

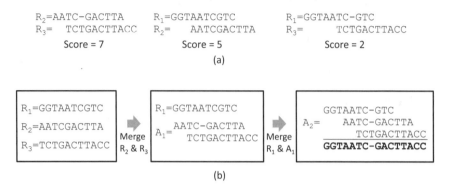

```
R₂=AATC-GACTTA        R₁=GGTAATCGTC        R₁=GGTAATC-GTC
R₃=  TCTGACTTACC      R₂=  AATCGACTTA      R₃=    TCTGACTTACC
     Score = 7             Score = 5              Score = 2
```
(a)

```
┌──────────────────┐       ┌──────────────────┐       ┌────────────────────────┐
│ R₁=GGTAATCGTC    │       │ R₁=GGTAATCGTC    │       │   GGTAATC-GTC          │
│                  │  ➤    │                  │  ➤    │ A₂=  AATC-GACTTA       │
│ R₂=AATCGACTTA    │       │ A₁=  AATC-GACTTA │       │      TCTGACTTACC      │
│                  │ Merge │      TCTGACTTACC │ Merge │ GGTAATC-GACTTACC       │
│ R₃=TCTGACTTACC   │ R₂&R₃ │                  │ R₁&A₁ │                        │
└──────────────────┘       └──────────────────┘       └────────────────────────┘
```
(b)

FIGURE 5.33: Consider three reads R_1 = GGTAATCGTC, R_2 = AATCGACTTA, R_3 = TCTGACTTACC. (a) shows the pairwise alignment of the three reads. We assume match scores 1, mismatch/indel scores −1. (b) illustrates the merging process. Since R_2 and R_3 are more similar, we first merge R_2 and R_3 and obtain the alignment A_1. Then, between $\{R_1\}$ and $\{R_2, R_3\}$, R_1 and R_2 are more similar. We merge R_1 and A_1 according to the pairwise alignment between R_1 and R_2 and obtain the alignment A_2. By reporting the most frequent character in each column of the alignment, we generate the consensus sequence GGTAATCTGACTTA.

	R₂	R₃	R₄	R₅
R₁	R₁=TAAAGCG R₂=TAAAGTGG Score=5	R₁=TAAAG-CG R₃= AGTCGG Score=3	R₁=TAAAGCG R₄= TCGG Score=1	R₁=TAAAGCG R₅= AGCG Score=4
R₂		R₂=TAAAGT-GG R₃= AGTCGG Score=4	R₂=TAAAGT-GG R₄= TCGG Score=2	R₂=TAAAGTGG R₅= AGCG Score=2
R₃			R₃=AGTCGG R₄= TCGG Score=4	R₃=AGTCGG R₅=AG-CG Score=3
R₄				R₄= TCGG R₅=AGCG Score=1

(a)

```
R₁=TAAAGC-G
R₂=TAAAGT-GG
R₃=    AGTCGG
R₄=      TCGG
R₅=    AGC-G
TAAAGT-GG
```
(b)

```
R₁=TAAAG-CGG
R₂=TAAAGT-GG
R₃=    AGTCGG
R₄=      TCGG
R₅=    AG-CG
TAAAGTCGG
```
(c)

FIGURE 5.34: Consider 5 reads: $\mathcal{R}$ = $\{R_1$ = TAAAGCG, R_2 = TAAAGTGG, R_3 = AGTCGG, R_4 = TCGG, R_5 = AGCG$\}$. Assume match scores 1 and mismatch/indel scores −1. (a) shows the pairwise alignments among them. (b) shows the MSA of $\mathcal{R}$ generated by greedy merging. The corresponding consensus sequence is TAAAGTGG. (c) shows the optimal MSA of $\mathcal{R}$. It gives another consensus sequence TAAAGTCGG.

FIGURE 5.35: Given the MSA in Figure 5.34(b), after we realign R_1 and R_5, we obtain the optimal MSA in Figure 5.34(c).

Algorithm ReAlign

Require: An alignment A for the set of reads $\mathcal{R} = \{R_1, \ldots, R_N\}$

Ensure: An alignment A' such that $score(A') \geq score(A)$

1: **repeat**
2:     **for** $i = 1$ to N **do**
3:         Let A_{-i} be the subalignment of A excluding the read R_i;
4:         Update A to be an alignment between R_i and A_{-i} that improves the alignment score;
5:     **end for**
6: **until** the alignment score of A does not increase;

FIGURE 5.36: The round-robin realignment algorithm ReAlign.

genome assembly. We first recall the advantages and disadvantages of long and short reads.

- Second-generation sequencing generates short reads (of length < 300bp) which are highly accurate (sequencing error rate equals ~1%). However, the insert size of their paired-end reads is at most 1,000 bp.

- Mate-pair sequencing is also a kind of short read sequencing. It generates paired-end short reads with longer insert size (2,000 bp to 10,000 bp). The sequencing error rate is also low (~1%). However, to prepare mate-pair reads, we need more input DNA.

- The third-generation sequencing sequences long reads (of length ~10,000 bp). However, their reads have a high sequencing error (the error rate is 15%−18%).

This section discusses methods to reconstruct a genome using a combination of these sequencing technologies. There are two approaches:

- Use mate-pair reads and long reads to improve the assembly from the short reads.

- Use short reads to correct the errors in long reads.

The following subsections cover these two approaches.

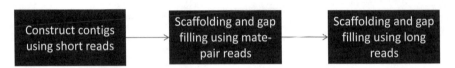

FIGURE 5.37: The pipeline for improving the genome assembly using short reads, mate-pair reads and long reads.

5.4.2.1   Use mate-pair reads and long reads to improve the assembly from short reads

Observe that short reads are more accurate than mate-pair reads while mate-pair reads are more accurate than long reads. One approach is to use a short read assembler to generate a set of accurate contigs first. Then, mate-pair reads and long reads are used to improve the assembly. ALLPATHS-LG [244] and PBJelly [72] use this approach.

Precisely, the steps are as follows (see Figure 5.37). First, we assemble short reads using a short read assembler (like a de Bruijn graph assembler) and obtain a set $\mathcal{C}$ of contigs. Then, if mate-pair reads are available, we can use mate-pair reads to link the contigs in $\mathcal{C}$ through scaffolding (see Section 5.3.4); then some gaps can be closed by gap filling (see Section 5.3.5) and we obtain a set $\mathcal{S}$ of scaffolds. Finally, using long reads, the scaffolds in $\mathcal{S}$ can be further refined through scaffolding and gap filling.

This approach assumes the contigs generated from the short read assemblers are accurate. The errors introduced by the short read assemblers are difficult to correct in the subsequence steps.

5.4.2.2   Use short reads to correct errors in long reads

Long reads generated by third-generation sequencing have a high sequencing error rate. On the other hand, short reads have fewer sequencing errors. One idea is to use short reads to correct errors in long reads. The basic solution consists of three steps.

(1) Short reads are aligned on the long reads.

(2) Errors in long reads are corrected by consensus of the short reads.

(3) Assemble the long reads to reconstruct the genome.

This idea is used by PBcR [147].

Below, we detail the three steps. The first step performs all-versus-all overlaps between paired-end short reads and the long PacBio reads. To avoid misalignment, PBcR proposed the following criteria. First, only paired-end short reads which entirely aligned on the long reads are considered. Second, although each short read is allowed to align to multiple long reads, the short read is permitted to align to exactly one position on each long read, which is the position with the highest alignment score. Third, each short read is permitted to align to at most C long reads with the highest alignment scores, where C is the expected coverage of the long reads on the genome. This constraint is to reduce the chance of aligning the short reads to the repeat regions.

The first step is time consuming. To improve efficiency, the hashing heuristics (similar to the hashing idea used in Step 1 of Section 5.4.1) is applied. Instead of performing all-versus-all overlaps, PBcR identifies pairs of short reads and long reads such that they share some k-mer (by default $k > 14$ bp); then, alignments are performed only for this set of pairs. This filter improves the efficiency.

After short reads are aligned on long reads, the second step corrects the errors on the long reads. For each long read, its corrected version can be estimated by the consensus of its short read alignments, which can be computed using the AMOS consensus module [298] (i.e., Step 3 of Section 5.4.1), DAGcon [45] or Falcon Sense (see Section 5.4.3.2).

Finally, given all the corrected long reads, the third step runs the OLC assembler in Section 5.4.1 to reconstruct the genome.

5.4.3 Long read approach

Third-generation sequencing has a high sequencing error rate ($\sim$15%–18%). The OLC assembler cannot directly assemble such long reads. Existing solutions try to correct the long reads first before they reconstruct the genome using an OLC assembler. FALCON, HGAP [45] and MHAP [20] use this approach, which has four steps.

1. **Overlapping reads**: This step aims to report all read pairs with good alignment scores. Although it is similar to Step 1 in Section 5.4.1, this step is more difficult since long reads have a high error rate.

2. **Correcting reads by consensus**: For each read S, Step 1 gives a set of reads overlapping with S. Based on the alignments, this step aims to correct the errors in the read S.

3. **Reconstructing the genome**: Given the corrected long reads, the genome is reconstructed using an OLC assembler.

4. **Refining the genome using the base quality**: Every base in the third-generation sequencing reads has a quality score. This step refines the genome utilizing the quality scores.

Below, we detail the four steps.

Step 1: Overlapping reads. This step performs pairwise alignments among all long reads. Recall that long reads have 15%–18% sequencing errors. So, the number of differences between two long reads can be as big as 30%–36%. This step cannot be solved by simply applying the method in Step 1 of Section 5.4.1. HGAP, FALCON and MHAP use different methods for this step. HGAP uses BLASR [39] to perform the alignment (BLASR is an aligner tailored for aligning long reads with a high error rate). FALCON uses Daligner (an aligner by Gene Myer tailored for all-versus-all pairwise alignment of long reads with a high error rate). Both BLASR and Daligner are slow ("overlapping reads" is actually the slowest step among the four steps). MHAP proposed a probabilistic method that uses minHash to speed up this step. Section 5.4.3.1 details the method. In addition, other read overlappers are available, including GraphMap and minimap[166].

Step 2: Correcting reads by consensus. For every read S, Step 1 tells us all reads significantly overlapping with S. Let $\mathcal{R}$ be such a set of reads. Step 2 corrects the errors in S by computing the consensus of all reads in $\mathcal{R}$. Apparently, this step can be solved using Step 3 in Section 5.4.1. However, since the error rate is high, we need some better method. HGAP uses PBDAG-con. PBDAG-con first uses BLASR to align all reads in $\mathcal{R}$ on S; then, it builds a directed acyclic graph (DAG) to describe the alignments between the reads in $\mathcal{R}$ and S; finally, from the DAG, it extracts the consensus sequence. FALCON and MHAP use Falcon Sense. Section 5.4.3.2 details the algorithm for Falcon Sense.

Step 3: Reconstructing the genome. Given the set of corrected reads, Step 3 simply runs an OLC assembler (see Section 5.4.3.3) to reconstruct the genome.

Step 4: Refining the genome using base quality. Since the third-generation sequencing has a high error rate, the reconstructed genome generated in Step 3 may still have many mistakes. Step 4 realigns all long reads on the reconstructed genome to refine it. In this step, the quality scores of the bases are also used. The detail is stated in Section 5.4.3.3.

5.4.3.1   MinHash for all-versus-all pairwise alignment

Given a set of reads $\mathcal{R}$, this section describes an algorithm for finding all read pairs (R_1, R_2) such that R_1 and R_2 are significantly similar.

This step is the same as Step 1 in Section 5.4.1. We can perform Smith-Waterman alignment to align reads. However, it requires quadratic time. To speed up, we compute a simpler similarity score instead. One possible score is the Jaccard similarity of the k-mers of two reads (see Section 3.4.4). For every read $R \in \mathcal{R}$, let $kmer(R)$ be its list of k-mers. Consider two reads R and S; their Jaccard similarity is $J(R_1, R_2) = \frac{|kmer(R_1) \cap kmer(R_2)|}{|kmer(R_1) \cup kmer(R_2)|}$. Note that Jaccard similarity is between 0 and 1. When $J(R_1, R_2) = 1$, we have $R_1 = R_2$.

For example, consider $R_1 = $ ACTGCTTACG and $R_2 = $ ACTCCTTATG. We have $kmer(R_1) = \{$ACT, CTG, TGC, GCT, CTT, TTA, TAC, ACG$\}$ and $kmer(R_2) = \{$ACT, CTC, TCC, CCT, CTT, TTA, TAT, ATG$\}$. Then, the set of shared 3-mers is $\{$ACT, CTT, TTA$\}$; hence, $J(R_1, R_2) = \frac{3}{13}$.

Given a set of reads $\mathcal{R}$, our aim is to find $\{(R_1, R_2) \in \mathcal{R} \times \mathcal{R} \mid J(R_1, R_2) \geq \delta\}$. This problem can be solved efficiently using hashing. The detail of the algorithm is shown in Figure 5.38. The algorithm first builds a hash table $H[]$

Algorithm JaccardSimilarity

Require: A set of reads $\mathcal{R}$
Ensure: A set of pairs (R, R') such that $J(R, R') \geq \delta$

1: Build a hash table $H[]$ for all k-mers such that $H[w] = \{R \in \mathcal{R} \mid w \in kmer(R)\}$.
2: **for** every read $R \in \mathcal{R}$ **do**
3:     Set $count[S] = 0$ for $S \in \mathcal{R}$;
4:     **for** every $w \in kmer(R)$ **do**
5:         For every read $S \in H[w]$, increment $count[S]$ by 1;
6:     **end for**
7:     For every read S with $\frac{count[S]}{|kmer(R)|+|kmer(S)|-count[S]} \geq \delta$, report (R, S);
8: **end for**

FIGURE 5.38: The JaccardSimilarity algorithm.

such that $H[w] = \{R \in \mathcal{R} \mid w \in kmer(R)\}$. Then, for every read $R \in \mathcal{R}$, the algorithm computes $count[S] = |kmer(R) \cap kmer(S)|$ for every $S \in H[w]$, for every $w \in kmer(R)$ (see Steps 2−8 in Figure 5.38). Finally, we report all pairs (R, S) if $J(R, S) = \frac{count[S]}{|kmer(R) \cup kmer(S)|} = \frac{count[S]}{|kmer(R)|+|kmer(S)|-count[S]} \geq \delta$.

Although the hashing approach is fast, its running time still depends on the length of the reads. For long reads, the hashing approach is still slow.

Below, we discuss the minHash approach (see Section 3.4.4). Its running time is independent of the length of the reads. MinHash is a technique that allows us to estimate Jaccard similarity $J(R, S)$ for any reads R and S. It requires K functions $h_1(), \ldots, h_K()$ that map $\{0, \ldots, 2^K - 1\}$ to its random permutation. For each read R, minHash creates a length-K signature $sig(R) = (\omega_{h_1}(R), \ldots, \omega_{h_K}(R))$, where $\omega_{h_i}(R) = \min_{w \in kmer(R)} h_i(w)$. Figure 5.39 illustrates the process to generate the signature of a read. By Lemma 3.3, for any two reads R_1 and R_2, the Jaccard similarity $J(R_1, R_2)$ is approximately the same as $sim(sig(R_1), sig(R_2))$ where $sim((a_1, \ldots, a_K), (b_1, \ldots, b_K)) = \frac{|\{i \mid a_i = b_i\}|}{K}$.

By the above discussion, finding all $(R_1, R_2) \in \mathcal{R} \times \mathcal{R}$ such that $J(R_1, R_2) \geq \delta$ can be approximated by finding all $(R_1, R_2) \in \mathcal{R} \times \mathcal{R}$ such that $sim(sig(R_1), sig(R_2)) \geq \delta$. Since comparing the minHash signatures of a pair of reads takes $O(K)$ time where K is a constant, the minHash approach speeds up the all-versus-all comparison of all long reads in $\mathcal{R}$.

5.4.3.2 Computing consensus using Falcon Sense

Given a template read S and a set of reads $\mathcal{R} = \{R_1, \ldots, R_N\}$, Falcon Sense corrects errors in S by computing the consensus of all reads in $\mathcal{R}$. Below, we detail the method. For every read $R \in \mathcal{R}$, we first obtain the semi-global alignment between S and R. Since a mismatch may actually be an indel, we replace every mismatch with a local deletion-insertion pair (for

R$_1$: GATCAACGGACCCA   R$_2$: TCACGACCCATGTC
    GAT AAC GAC   TCA GAC CAT
     ATC ACG ACC   CAG ACC ATG
      TCA CGG CCC   ACG CCC TGT
       CAA GGA CCA   CGA CCA GTC

h_1	h_2	h_3	h_4	
53	5	26	21	GAT
8	12	33	15	ATC
14	**2**	21	31	**TCA**
31	18	10	51	CAA
22	21	13	**9**	AAC
28	7	9	27	**ACG**
7	26	16	32	CGG
25	39	53	33	GGA
27	34	19	63	**GAC**
17	24	**5**	11	**ACC**
9	27	45	39	**CCC**
23	42	12	61	**CCA**

	h_1	h_2	h_3	h_4
TCA	14	**2**	21	31
CAG	36	42	28	13
ACG	28	7	9	27
CGA	34	30	61	7
GAC	27	34	19	63
ACC	17	24	**5**	11
CCC	**9**	27	45	39
CCA	23	58	12	61
CAT	41	61	18	57
ATG	10	53	36	18
TGT	43	9	58	23
GTC	12	45	7	**1**

   7 **2** **5** 9   9 **2** **5** 1
    sig(R$_1$)   sig(R$_2$)

sim(sig(R$_1$), sig(R$_2$))=2/4=0.5

R$_1$: GA**TCA**ACGG**ACC**CA
       | | | | | | | | | |
R$_2$: **TCA**-CG-**ACC**CATGTC

FIGURE 5.39: An example demonstrates the computation of $sim(R_1, R_2)$, where R_1 = GATCAACGGACCCA and R_2 = TCACGACCCATGTC. Note that R_1 and R_2 share 6 3-mers. Hence, the Jaccard similarity $J(R_1, R_2) = \frac{6}{12+12-6} = \frac{1}{3}$. At the same time, $sig(R_1) = (7, 2, 5, 9)$ and $sig(R_2) = (9, 2, 5, 1)$. Hence, $sim(sig(R_1), sig(R_2)) = \frac{2}{4} = 0.5$. Note that $J(R_1, R_2) \approx sim(sig(R_1), sig(R_2))$.

example, see the deletion-insertion pair in the alignment between R_1 and S in Figure 5.41(a)). Moreover, there are multiple ways to place gaps in an alignment. To ensure a consistent result, all gaps are moved to the rightmost equivalent positions. After that, each aligned position is either (1) a match that aligns $S[j]$ and $R[i]$, (2) a deletion that aligns $S[j]$ with $-$ or (3) an insertion that aligns $-$ with $R[i]$. For (1), we include a tuple $(j, 0, R[i])$. For (2), we include a tuple $(j, 0, -)$. For (3), we include a tuple $(j, d, R[i])$ if it is the dth insertion after $S[j]$. Figure 5.41(a) illustrates the computation of the tuples (p, d, b).

Each tuple (p, d, b) is a vote. For $d = 0$, each $(p, 0, b)$ is a support that $S[p]$ should be b. For $d > 0$, each (p, d, b) is a support that b should be inserted at the dth position after $S[p]$. The next step is to compute consensus from all tuples. The steps are as follows. Let $count(p, d, b)$ be the number of occurrences of the tuple (p, d, b). We sort all distinct tuples (p, d, b) by increasing order of j, d and followed by the alphabetic order of b; then, b is a consensus base if $count(p, d, b) > \frac{1}{2} \sum_{x \in \{A,C,G,T,-\}} count(p, 0, x)$. For example, Figure 5.41(b) gives the sorted list of distinct tuples, the corresponding counts and the corresponding consensus bases. For this example, the consensus for the template S is TAAGTCAG. The detail of the algorithm is given in Figure 5.40.

5.4.3.3 Quiver consensus algorithm

After we obtain a genome from the OLC assembler, this section discusses the Quiver consensus algorithm that refines the genome using the base-call information of the reads.

Given the base-call information, for any read R and any reference T, we can compute the likelihood function $Pr(R|T)$. Let κ be any single base substitution/insertion/deletion. Let $T \circ \kappa$ be the genome formed after the mutation κ. If $Pr(R|T \circ \kappa) > Pr(R|T)$, R supports $T \circ \kappa$ better than T. Let $\mathcal{R}$ be the set of all reads. Our aim is to find a set of mutations $\{\kappa_1, \dots, \kappa_m\}$ that maximizes $\prod_{R \in \mathcal{R}} Pr(R|T \circ \kappa_1 \dots \circ \kappa_m)$.

This problem is NP-hard. One heuristic solution is to iteratively improve the genome. More precisely, the genome T is partitioned into subregions. For each subregion W, we iteratively find mutation κ such that $\prod_{R \in \mathcal{R}} Pr(R|W \circ \kappa) > \prod_{R \in \mathcal{R}} Pr(R|W \circ \kappa)$. Then, we update W to $W \circ \kappa$. We repeat the updates until it cannot be further improved. The algorithm is known as the Quiver consensus algorithm. Its detail is shown in Figure 5.42.

Algorithm FalconSense

Require: A set of reads $\mathcal{R}$ that are aligned on the template S

Ensure: A corrected sequence for S based on the consensus of the reads in $\mathcal{R}$

1: **for** every read $R \in \mathcal{R}$ **do**
2: 　　Compute the alignment A between R and S allowing matches or indels (no mismatches);
3: 　　**for** each aligned position in A **do**
4: 　　　　If it is a match aligning $S[j]$ with $R[i]$, add tuple $(j, 0, R[i])$;
5: 　　　　If it is a deletion aligning $S[j]$ with $-$, add tuple $(j, 0, -)$;
6: 　　　　If it is an insertion aligning $-$ with $R[i]$, suppose this is the dth insertion after $S[j]$; add tuple $(j, d, R[i])$;
7: 　　**end for**
8: **end for**
9: For each distinct tuple (p, d, b), let $count(p, d, b)$ be the number of occurrences of the tuple in the list.
10: Sort all distinct tuples (p, d, b) by increasing order of p, d and followed by the alphabetic order of b;
11: **for** each distinct tuple (p, d, b) in sorted order **do**
12: 　　**if** $count(p, d, b) > \frac{1}{2} \left(\sum_{x \in \{A, C, G, T, -\}} count(p, 0, x) \right)$ **then**
13: 　　　　Output b;
14: 　　**else**
15: 　　　　Output $-$;
16: 　　**end if**
17: **end for**

FIGURE 5.40: The algorithm FalconSense.

R₁=TAA-G-CA
S =TAAAGT-AG
p =12345667
d =00000010

R₂=   AGTCAG
S =TAAAGT-AG
p =   456678
d =   000100

R₃=   GTCAG
S =TAAAGT-AG
p =   56678
d =   00100

R₄=TAA--TAG
S =TAAAGTAG
p =12345678
d =00000000

(a)

Tuple (p, d, b)	Count	Consensus
(1, 0, T)	2	T
(2, 0, A)	2	A
(3, 0, A)	2	A
(4, 0, A)	1	-
(4, 0, -)	2	-
(5, 0, G)	3	G
(5, 0, -)	1	-
(6, 0, T)	3	T
(6, 0, -)	1	-
(6, 1, C)	3	C
(7, 0, A)	4	A
(8, 0, G)	3	G

(b)

S =TAAAGT-AG
R₁=TAA-G-CA
R₂=   AGTCAG
R₃=   GTCAG
R₄=TAA--T-AG
Consensus: TAA-GTCAG

(c)

FIGURE 5.41: Consider a set of 4 reads $\mathcal{R} = \{R_1 = \text{TAAGCA}, R_2 = \text{AGTCAG}, R_3 = \text{GTCAT}, R_4 = \text{TAATAG}\}$ and a template $S = \text{TAAAGTAG}$. (a) shows the alignment (allowing match and indel only) between R_k and S for $k = 1, 2, 3, 4$. For each alignment, we list the p and d values below each aligned base. Each aligned position gives one tuple. (b) shows the sorted list of tuples (p, d, b), their counts and their consensus bases. (c) shows the final consensus sequence and the corresponding MSA of $\mathcal{R}$.

Algorithm Quiver

Require: A predicted reference genome T and a set of reads $\mathcal{R}$, where every base in the reads is associated with a quality score

Ensure: A corrected sequence for T

1: **for** every subregion W of T **do**
2:     Identify a subset of reads S of $\mathcal{R}$ that can align on the subregion W;
3:     **while** there exists mutation κ of W such that $\left(\prod_{R \in S} Pr(R|W \circ \kappa) > \prod_{R \in S} Pr(R|W)\right)$ **do**
4:         Set $W = W \circ \kappa$;
5:     **end while**
6: **end for**

FIGURE 5.42: The algorithm Quiver.

5.5 How to evaluate the goodness of an assembly

Different assemblers generate different assemblies. We need some ways to evaluate the goodness of an assembly. The goodness of an assembly can be evaluated by (1) the assembly completeness and (2) the assembly accuracy.

The assembly completeness checks if the assembly is too fragmented. It is usually accessed by statistics including the number of contigs, number of contigs longer than 1000 bp, maximum contig length, combined total length, N50, and N90. They are defined as follows.

- The number of contigs and number of contigs longer than 1000 bp are self-explanatory.

- The maximum contig length is the maximum among the set of contigs in the assembly while the combined total length is the total length of all the contigs in the assembly.

- The N50 length of the assembly is defined as the maximum contig length such that contigs of length equal or longer than that length account for 50% of the combined total length of the assembly. The N90 length is defined similarly. Note that N90 is always shorter than N50.

If the assembly is near complete, the number of contigs should be as small as possible while the maximum contig length, combined total length, N50 and N90 should have values close to the actual genome size.

For example, suppose you reconstructed 10 contigs of lengths 125, 250, 480, 700, 950, 1273, 1435, 1855, 1927 and 2100. The number of contigs is 10 while the number of contigs longer than 1000 bp is 5. The maximum contig length and combined total length are 2100 and 11,095, respectively. N50 equals 1855 since $1855 + 1927 + 2100 = 5882 > 11,095 * 50\%$ while $1927 + 2100 = 4027 < 11,095 * 50\%$. Similarly, N90 equals 700.

For the assembly accuracy, if a trustable reference genome is available, the accuracy can be measured by aligning the contigs or scaffolds to the reference genome. When there is no reference genome, there is no standard measurement to evaluate the goodness of the assembly. Some possible measurements include the number of discordant reads, the GC content of the contigs, etc.

5.6 Discussion and further reading

This chapter discussed different techniques for genome assembly. We cover read error correction, contig building (base-by-base extension approach, de Bruijn graph approach, OLC assembly approach), scaffolding and gap filling.

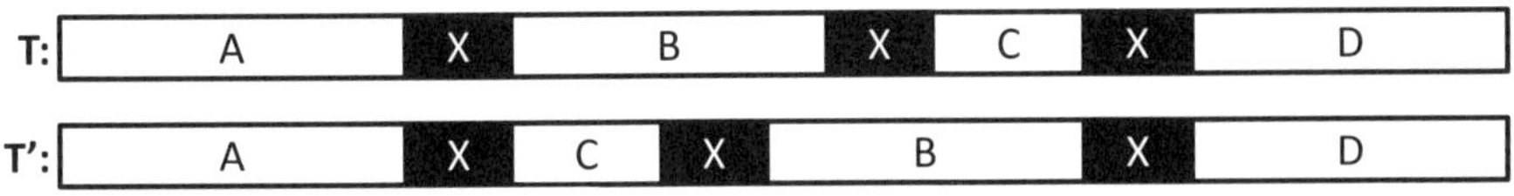

FIGURE 5.43: The top figure visualizes a genome with a repeat X. The bottom genome is formed by swapping fragments B and C.

For the contig building step, apart from the base-by-base extension approach, de Bruijn graph approach and OLC assembly approach, another approach is based on the string graph. Methods that use string graphs include Read-joiner [88], fermi [164], [212] and SGA [273, 274].

Although the current genome assemblers are more advanced, repeat is still a tough problem. Repeats appear frequently and cover more than 50% of our genome. Even worse, many repeats are segmental duplications and they are long. Figure 5.43 illustrates why repeats affect genome assembly. Suppose we obtain a set of reads $\mathcal{R}$ from the genome T. If the length of the repeat X is longer than the length of the insert size, $\mathcal{R}$ cannot distinguish the genomes T and T'. Hence, no genome assembler can reconstruct the correct genome without ambiguity if the insert size is shorter than the length of the longest repeat.

Apart from standard genome assembly, other types of assembly problems exist. They include single-cell genome assembly, meta-genomic assembly and transcriptome assembly.

Single-cell genome assembly is based on the technology of single-cell whole genome sequencing. This technology can sequence the genome of a single cell. It tries to perform a genome-wide amplification of a set of short fragments from a single cell using multiple displacement amplification (MDA); then, paired-end sequencing is performed on those short fragments. As a first glance, it seems that this problem can be solved using normal genome assemblers. How-ever, since the genome-wide amplification has bias, the paired-end reads do not evenly cover the whole genome. This creates technical problems for genome assemblers. A number of methods have been proposed to solve such problems, including the E+V-SC assembler [47], SPAdes [13], and IDBA-UD [230].

Meta-genomic assembly aims to sequence a mixture of genomes, each with different frequency. The difficulty of this problem is that the coverage of each genome is unknown initially. A number of assemblers are proposed to per-form de novo meta-genomic assembly. They include MetaVelvet [214], Meta-IDBA [228], IDBA-UD [230], MAP and Genovo [156]. MetaVelvet, Meta-IDBA, IDBA-UD use the de Bruijn graph approach. Both MAP and Genovo are designed for long reads.

Transcriptome assembly aims to reconstruct all transcripts from the RNA-seq dataset. The length of transcripts is 7000 bp on average, which is much shorter than genomes. Hence, it seems that it is an easier problem. However, there are two difficulties: (1) different transcripts have different expression levels and (2) transcripts may share a number of exons due to splicing. A number of methods have been proposed to address these two difficulties. They include Rnnotator [193], Multiple-k [291], Trans-ABySS [248], T-IDBA [229], Trinity [89], Oases [268], SOAPdenovo-Trans [321] and EBARDenovo [49].

5.7   Exercises

1. Can you explain why the mate-pair sequencing reports two reads at the two ends of the DNA fragment, both from the forward template?

2. Suppose $\mathcal{T} = \{\text{ACT}, \text{CTG}, \text{TGT}, \text{GTC}, \text{TCT}, \text{CTA}\}$ and $R = \text{TGTTA}$. Can you compute the minimum edit distance between R and a $\mathcal{T}$-string? Please illustrate the steps.

3. Suppose $\mathcal{T} = \{\text{ACT}, \text{CAC}, \text{CTG}, \text{CTT}, \text{TCA}, \text{TTC}\}$ and $R = \text{TCACTGCA}$. Can you compute the minimum edit distance between R and a $\mathcal{T}$-string? Please illustrate the steps. (You can assume there is no indel error in TCA.)

4. Show that the dependency graph defined in Section 5.3.1.1 does not have negative cycle.

5. Show that Lemma 5.2 is correct.

6. Consider the weighted directed graph G constructed based on the recursive equation in Lemma 5.1. Let P be a shortest path from v_s to some node $(|R|, t)$ where $t \in \mathcal{T}$. Can you give an algorithm to determine if it is unique? What is the time complexity?

7. Consider an undirected graph $G = (V, E)$. (Assume it has one connected component.) We aim to compute the shortest paths between u and all nodes $v \in V - \{u\}$. Can you propose a solution which runs in $O(|V|+|E|)$ time? (Hint: You can use breadth-first search.)

8. Give an $O(|\mathcal{T}||R|)$-time algorithm for computing the minimum edit distance between R and any $\mathcal{T}$-string.

9. For the following two cases, using the base-by-base extension approach (with the pairing information), can you guess what the next bases are?

   (a) Consider the following set of paired-end reads (in Illumina format): $\{(\text{GCAAC}, \text{CTCGA}), (\text{ACTCG}, \text{CAACG}), (\text{AGCAC}, \text{GCTCG})\}$. Suppose the template is TGCAACGTACGTACGGTCGAG. Can you guess what the next base is?

(b) Consider the following set of paired-end reads (in Illumina format): {(CGTAC, CGTAT), (TACGG, TCGAA), (ACGGA, GTCGA), (CGGAA, TGTCG), (CGGAG, CTCAC)}. Suppose the template is CGTACGTCAGCCACGATACGGA. Can you guess what the next base is?

10. Consider the following set of seven paired-end reads extracted from a genome of length 14 (with no error, the insert size is exact (i.e., 8 bp) and all reads are in positive strand): ACG..GTG, CGT..GTA, CGT..TGC, GTC..GCG, GTG..TAG, TCG..CGT, and TGC..AGA. Perform genome assembly to reconstruct the genome. If we ignore the pairing information (i.e. we assume the input consists of fourteen 3-mers), can we reconstruct the genome? Please explain your answer.

11. Consider the set of reads {AACCTCA, ACCTCAT, GGGGTTC, TATGGAA, TGGAACC}.

    (a) Let TGGAACC be the initial template. Can you perform base-by-base extension to reconstruct the genome? (When you align reads on the template, please assume you only trust exact alignment of length at least 3. Note that the read may be reverse complemented.)

    (b) Can you build the de-Bruijn graph H_4 based on 4-mers in the reads and their reverse complements? After pruning the tips and merging the bubbles, can you reconstruct the genome?

12. Consider the set of reads {ACCTCC, TCCGCC, CCGCCA}. For $k = 2$ or 3, can you build a de Bruijn graph H_k? Can you get the Eulerian path from H_k? Is the Eulerian path unique?

13. Given a de Bruijn graph H with n edges, can you describe an $O(n)$-time algorithm to determine if a Eulerian path exists? If the Eulerian path exists, can you give an $O(n)$-time algorithm to recover the Eulerian path?

14. Given a de Bruijn graph H with n edges, suppose a Eulerian path exists in H. Can you give an algorithm to determine if there is a unique Eulerian path?

15. Consider $S =$ ACTGTTCGCAGTTCGAAGTTCAAGG. Let H_k be the de Bruijn graph of all k-mers of S. What is the smallest k such that there is a unique Eulerian path in H_k? (Assume there is no error.) Please also explain why we don't have a unique Eulerian path from the de Bruijn graph H_{k-1}.

16. Consider $S =$ CCTCGCTTCGTCA. Let H_k be the de bruijn graph of all k-mers of S. What is the smallest k such that there is a unique Eulerian path in H_k? (Assume there is no error.) Please also explain why we don't have a unique Eulerian path from the de bruijn graph H_{k-1}.

17. Please show that the scaffolding problem is NP-complete.

18. Prove or disprove: The scaffolding problem always have a unique solution.

19. Assume the insert size is 400 bp. Suppose we run the greedy scaffolder in Figure 5.29 starting from contig C. Can you reorder the following three contigs?

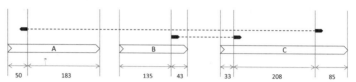

20. Let $\mathcal{R}$ be a set of paired-end reads generated from the following genome. You perform deep sequencing. So, you have many paired-end reads in $\mathcal{R}$. Also, assume fragments A, B, C, D and E have little homology.

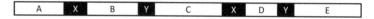

- Suppose the insert size is shorter than the size of X and Y. Is it possible to reconstruct the whole genome as one contig? If not, can you guess the list of contigs?

- Suppose the insert size is shorter than the size of X but not Y. Is it possible to reconstruct the whole genome as one contig? If not, can you guess the list of contigs?

21. Recall that the minHash approach computes a length-K signature $sig(R)$ for every read R given K hash functions $h_1(), h_2(), \ldots, h_K()$. Precisely, let $kmer(R)$ be all k-mers in R. The signature is $sig(R) = (\omega_1(R), \omega_2(R), \ldots, \omega_K(R))$ where $\omega_i(R) = \min_{w \in kmer(R)} h_i(w)$. For any two reads R_1 and R_2, it is known that the Jaccard similarity $J(R_1, R_2) \approx sim(sig(R_1), sig(R_2))$, where $sim((a_1, \ldots, a_K), (b_1, \ldots, b_K)) = \frac{|\{i | a_i = b_i\}|}{K}$.

- If we change $sig(R)$ to $sig'(R) = (\omega'_1(R), \omega'_2(R), \ldots, \omega'_K(R))$ where $\omega'_i(R) = \max_{w \in kmer(R)} h_i(w)$, is it true that $J(R_1, R_2) \approx sim(sig'(R_1), sig'(R_2))$? Why?

- If we change $sig(R)$ to $sig''(R) = (\omega_1(R), \ldots, \omega_K(R), \omega'_1(R), \ldots, \omega'_K(R))$, is it true that $J(R_1, R_2) \approx sim(sig''(R_1), sig''(R_2))$? Why?

- Among the three scores $sim(sig(R_1), sig(R_2))$, $sim(sig'(R_1), sig'(R_2))$ and $sim(sig''(R_1), sig''(R_2))$, which one gives the best approximation of $J(R_1, R_2)$? Why?

22. Consider the template sequence $S = $ GCACTCGTGCAC. Consider the set of reads $\mathcal{R} = \{$ TCGTTGCA, GTCGTTGAC, GCAGTCGT, ACTTTGA $\}$. Can you deduce the consensus sequence using Falcon Sense?

Chapter 6

Single nucleotide variation (SNV) calling

6.1 Introduction

Different individuals have slightly different genomes. These differences are called genome variations. They include single nucleotide variations (SNVs), small indel variations, copy number variations (CNVs) and structural variations (SVs). Figure 6.1 illustrates various genome variations in our genome.

These variations affect phenotypes and how we look. They also cause diseases. Hence, it is important to call these variations for each individual. With the advances in next-generation sequencing, genome variations can be discovered on a genome-wide scale. In this chapter, we focus on discussing techniques to call SNVs and small indels from NGS data. The next chapter covers methods for calling CNVs and SVs.

6.1.1 What are SNVs and small indels?

Single nucleotide variation (SNV) is a point mutation that occurs at some location (called locus) in our genome. It is the most frequent genome variation. Each individual is expected to have 3×10^6 SNVs, i.e., one SNV in every 1000 nucleotides [122, 1]. Figure 6.1(a) shows one SNV where A is mutated to C.

SNVs can occur in protein-coding regions or non-coding regions. A protein coding region is a sequence of triplets of nucleotides called codons, each of which determines a single amino acid. If the SNV does not change the amino acid type, it is called a synonymous SNV; otherwise, it is called a non-synonymous SNV. The non-synonymous SNV is further partitioned into missense SNV and nonsense SNV. The missense SNV changes the amino acid type. The nonsense SNV changes an amino acid codon to a stop codon. The nonsense and missense SNVs can significantly alter the 3D structures and functions of proteins.

When SNVs occur in non-coding regions, they are mostly neutral. However, non-coding SNVs may affect the expressions of genes if they occur on functional sites like transcription factor binding sites or splice junctions.

A small indel is an insertion/deletion of a short segment (of size at most

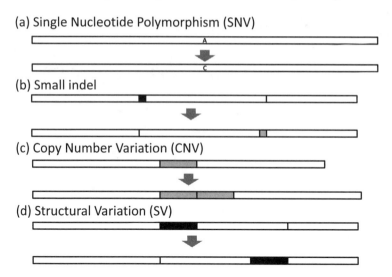

FIGURE 6.1: Four types of variations: (a) SNV, (b) small indel, (c) CNV and (d) SV. For (b), the black region is deleted while the gray region is inserted.

50 bp) from the genome. Figure 6.1(b) shows one insertion and one deletion. It is the second most frequent variation. Bhangale et al. [21] showed that indels represent 7%−8% of human polymorphisms. Recently, some evidence showed that the number of indels may be under-estimated. For example, in a Yoruban genome (NA18507), we expect approximately 1 indel per 3000 nucleotides, where most of the indels (up to 98.5%) are of size 1−20 bp [122]. Most indels (43%−48%) are located in 4% of the genome (called indel hotspots). 75% of small indels are caused by polymerase slippage [206]. (Polymerase slippage is a process occurring during replication. It is thought to occur in a section with repeated patterns of bases (e.g. CAG repeats). In such a region, a bubble is formed in the new strand. Then, subsequence replications create the indel. See Figure 6.2.)

When an indel occurs in a protein coding region, it can cause frameshift variation. If the indel is of a size that is a multiple of 3, it will cause deletion or insertion of a few codons (called in-frame indel), which may or may not affect the property of the gene. If the indel's size is not a multiple of 3, a frameshift indel occurs which can totally change/destroy the whole protein. Similar to SNVs, when indels occur in non-coding regions, many are neutral unless they occur in some functional sites like transcription factor binding sites or splice junctions.

The effect of SNVs and indels is further complicated by the fact that the human genome is diploid, that is, every chromosome has two copies, one from the father and the other one from the mother. For each locus (i.e., each position

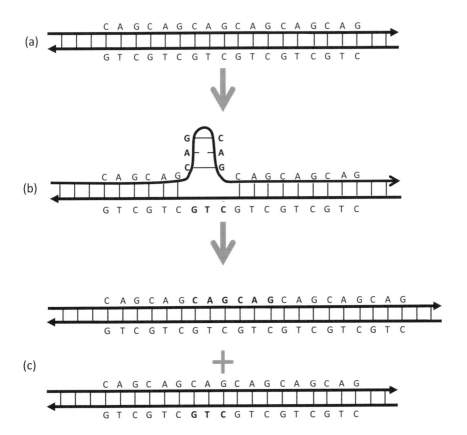

FIGURE 6.2: Illustration of polymerase slippage. (a) is a DNA sequence with repeats of CAG. (b) During replication, polymerase slippage occurs in the upper strand, which creates a bubble. (c) Then, subsequence replication of the top strand creates a small insertion.

...ACGTCATG...
...ACGCCATG...

FIGURE 6.3: Consider a chromosome pair. The 4th position has two different alleles C and T. This locus has a heterozygous genotype. For the rest of the loci, each has a single allele; hence, they have homozygous genotypes.

on the genome), the nucleotide that appears in each copy of the chromosome is called its allele. The pair of alleles that appears at a locus is called its genotype. If the two alleles of a genotype are the same, it is a homozygous genotype; otherwise, it is a heterozygous genotype. (See Figure 6.3 for an example.) Different alleles have different effects or phenotypes. Some alleles mask the effect of other alleles, which are called dominant alleles. Alleles that are masked by others are called recessive. So, depending on whether SNVs and indels affect dominant or recessive alleles, different effects will be observed.

Increasing evidence shows that SNVs and indels are involved in a wide range of diseases[95, 324]. For example, SNVs in TP53 and CTNNB1 genes have been discovered to recurrently occur in hepatocellular carcinoma [129]. Indels that appear in microsatellites have been linked to more than 40 neurological diseases [226]. The deletion in intron 2 of the BIM gene has been associated with the resistance to tyrosine kinase inhibitors in CML patients [218].

6.1.2   Somatic and germline mutations

Mutations can be classified into two types: somatic mutations and germline mutations. Germline mutations are mutations that are transmitted from parents to offspring through the germline cells (gametes). These mutations are present in virtually every cell in an individual. Somatic mutations are mutations that occur in only a subgroup of cells of an individual (except the germ cells). Hence, these mutations are not passed to the children. Somatic mutations may cause diseases. Cancer is a typical example where cells acquired somatic mutations and developed into a tumor.

6.2   Determine variations by resequencing

To call SNVs and indels in a sample, the first step is to sequence the sample's genome by next-generation sequencing. Depending on whether we want to obtain SNVs and indels in the whole genome or in some targeted regions, we can sequence the sample by whole genome sequencing or targeted sequencing. For whole genome sequencing, we use the wet-lab protocol described in

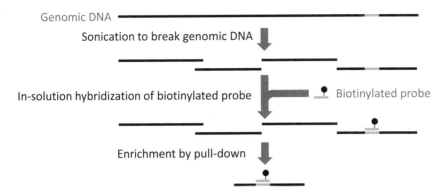

FIGURE 6.4: Illustration of the biotinylated target enrichment workflow. This illustration aims to enrich the region with the gray-colored probe.

Section 5.2.1. Below, we describe targeted sequencing and how to generate datasets for calling somatic SNVs and indels.

6.2.1 Exome/targeted sequencing

Whole genome sequencing is expensive since we need to sequence reads that cover the whole genome. Instead, we can study variations in some targeted regions. The sequence-capturing technologies enable us to capture targeted regions from the sample genome; then, we can study variations in these targeted regions by high-throughput sequencing. Targeted sequencing is useful since it is cheaper. One popular targeted sequencing is exome sequencing (see [292] for a survey). Exome sequencing only sequences regions containing promoters and protein coding genes. Since the regions are much shorter than the whole genome, exome sequencing costs $\sim 1/6$ of whole genome sequencing. Furthermore, although protein coding sequences are short (constitute approximately 1%–2% of the human genome), they harbor $\sim 85\%$ of the mutations with large effects on disease-related traits [48]. Hence, exome or targeted sequencing provides a cost-effective solution to identify causal mutations of different genetic diseases.

There are two general approaches for targeted sequencing: (1) target enrichment workflow and (2) amplicon generation workflow.

The target enrichment workflow enriches the selected target regions. Figure 6.4 illustrates the idea of Biotinylated target enrichment workflow. Biotinylated probes are designed so that each targets one specific region. After the genomic DNA is sheared by sonication, the sheared DNA fragments are hybridized with the biotinylated probes in solution. Through affinity enrichment, the targeted DNA fragments are captured.

Amplicon generation workflow amplifies the selected target regions. A set of oligo pairs are designed, where each oligo pair flanks one target region. The

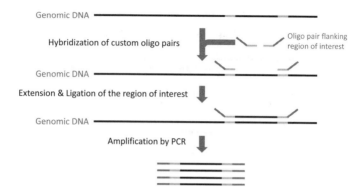

Genomic DNA

Hybridization of custom oligo pairs

Oligo pair flanking
region of interest

Genomic DNA

Extension & Ligation of the region of interest

Genomic DNA

Amplification by PCR

FIGURE 6.5: Illustration of the amplicon generation workflow. This illustration aims to amplify the region flanking the gray-colored oligo pair.

oligo pairs are hybridized on the genomes. Then, by extension and ligation, we obtain an amplicon of the target region for each oligo pair. Finally, through PCR, the target regions are amplified. Figure 6.5 illustrates the amplicon generation workflow.

6.2.2   Detection of somatic and germline variations

By resequencing the genome of a sample, we can identify a set of mutations with respect to the reference genome. These mutations include both somatic and germline mutations.

These two types of mutations cannot be differentiated by just sequencing the targeted sample. We also need to sequence normal tissue that represents the germline. For example, for a solid tumor like in the liver, we can sequence the tumor and its adjacent normal. Mutations that appear in both the tumor and normal tissue are assumed to be germline mutations. Somatic mutations are mutations that appear in the tumor but do not appear in normal tissue.

6.3   Single locus SNV calling

A single nucleotide variant (SNV) is the modification of a single reference nucleotide by another base. SNVs are known to be the drivers of tumorigenesis and cellular proliferation in many human cancer types. Next-generation sequencing becomes a high-throughput and cost-effective sequencing method to discover them on a genome-wide scale.

Given a set of reads obtained from whole genome sequencing or exome/targeted sequencing, the basic SNV calling approach is as follows.

1. Align all reads on the reference genome (see Chapter 4) and obtain a BAM file.

2. Some reads may be aligned incorrectly. This step filters alignments with a low mapQ score and alignments with many mismatches.

3. Alleles appearing in each locus are collected. Suppose that some alleles that appear on a particular locus are different from the reference allele. An SNV may exist in such locus. Figure 6.6 gives an example. There are three loci whose non-reference alleles occur at least once. They appear at positions 5, 23 and 28. These alleles may be SNVs or sequencing errors.

4. For each locus with a non-reference allele, different statistics can be applied to determine if an SNV exists in such locus.

5. After SNVs are called, the called SNVs are stored in VCF files [60] (see Section 2.5).

This section focuses on Step 4, which is called the single locus SNV calling problem. Precisely, the problem is as follows. Consider a particular locus j with a reference base r. Suppose there are n reads whose alignments cover this locus and let $\mathcal{D} = \{b_1, \ldots, b_n\}$ be the set of n bases on the n reads that cover this locus. Furthermore, let $q_1, \ldots, q_n$ be the corresponding PHRED quality scores of the n bases. (This means that $e_i = 10^{-\frac{q_i}{10}}$ is the sequencing error probability of b_i.) This single locus SNV calling problem aims to determine if the locus j has an SNV.

A number of methods are proposed to solve this problem including:

- identifying SNVs by counting alleles,

- identifying SNVs by binomial distribution,

- identifying SNVs by Poisson-binomial distribution, and

- identifying SNVs by the Bayesian approach.

The following subsections will cover these methods. Calling single locus somatics SNVs will be covered in Section 6.4.

6.3.1 Identifying SNVs by counting alleles

The basic approach to call SNVs is based on counting alleles. Let $\mathcal{D}'$ be the subset of bases in $\mathcal{D}$ that are of high confidence. Usually, we keep bases with a quality score ≥ 20. Hence, $\mathcal{D}' = \{b_i \in \mathcal{D} \mid q_i \geq 20\}$. Then, among all bases in $\mathcal{D}'$, the number of occurrences of each allele is counted.

- If the proportion of the reference allele in $\mathcal{D}'$ is below θ_{low} (says, 20%), it is called a homozygous non-reference allele;

- If the proportion of the reference allele in $\mathcal{D}'$ is above θ_{high} (says, 80%), it is called a homozygous reference allele;

- Otherwise, it is called a heterozygous genotype.

Figure 6.6 gives an example to illustrate the idea. There are three loci whose non-reference alleles occur at least once. For position 5, the reference base T occurs less than 20% of the time. We predict the genotype of this locus to be AA (homozygous for non-reference allele). For position 23, the reference base A occurs more than 80% of the time. We predict the genotype of this locus to be AA (homozygous for reference allele). For position 28, 75% of reads contain the reference base T. We predict the genotype of this locus to be GT (heterozygous site).

This method is used in a number of commercial software programs including Roches GSMapper, the CLC Genomic Workbench and the DNSTAR Lasergene. It works fairly well when the sequencing depth is high ($> 20\times$).

However, this method may under-call heterozygous genotypes. Also, it does not measure uncertainty.

6.3.2   Identify SNVs by binomial distribution

To resolve the limitation of simple counting, we can determine uncertainty by binomial distribution. Recall that $\mathcal{D} = \{b_1, \ldots, b_n\}$ is the set of bases covering a particular locus. The null model is that all non-reference variants in $\mathcal{D}$ are generated by sequencing errors. Let the random variable X be the number of variant bases among the n bases. Denote $Pr_n(X = k)$ as the probability of observing k variants in $\mathcal{D}$ under the null model.

Suppose there are K non-reference variants in $\mathcal{D}$. We declare that this site is an SNV (i.e., reject the null model) if $Pr(X \geq K)$ (which equals $\sum_{k \geq K} Pr_n(X = k)$) is smaller than some user-defined threshold (say 0.05).

Assume that the sequencing errors of the n bases are independent. When the sequencing error probability p is known (say $p = 0.01$), X follows a binomial distribution. Then, we have

$$Pr_n(X = k) = \binom{n}{k} p^k (1 - p)^{n-k}.$$

Figure 6.7(a) shows an example. Three reads cover locus j and the three bases at locus j are $\mathcal{D} = \{A, G, A\}$. Note that two bases are non-reference variants. The p-value of observing two non-reference variants is $Pr_3(X \geq 2) = \binom{3}{2}(0.01)^2(1 - 0.01)^1 + \binom{3}{3}(0.01)^3 = 0.000298$. With p-value threshold 0.05, we call this locus an SNV.

Although this method determines the uncertainty probability, it does not utilize the quality score of each base.

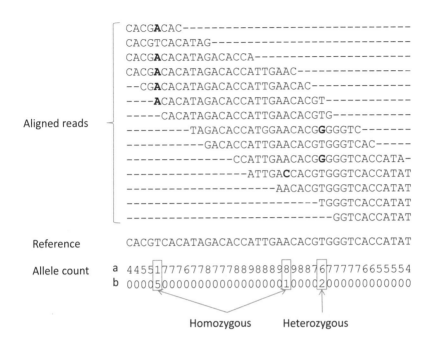

```
            ⎡ CACGACAC-------------------------------
            │ CACGTCACATAG---------------------------
            │ CACGACACATAGACACCA---------------------
            │ CACGACACATAGACACCATTGAAC---------------
            │ --CGACACATAGACACCATTGAACAC-------------
            │ ----ACACATAGACACCATTGAACACGT-----------
Aligned reads ⎨ -----CACATAGACACCATTGAACACGTG----------
            │ ---------TAGACACCATGGAACACGGGGGTC-------
            │ -----------GACACCATTGAACACGTGGGTCAC-----
            │ --------------CCATTGAACACGGGGGTCACCATA-
            │ ----------------ATTGACCACGTGGGTCACCATAT
            │ -------------------AACACGTGGGTCACCATAT
            │ ------------------------TGGGTCACCATAT
            ⎣ --------------------------GGTCACCATAT

Reference     CACGTCACATAGACACCATTGAACACGTGGGTCACCATAT

Allele count  a 4455177767787778898888989887677776655554
              b 0000500000000000000000010000200000000000
```

Homozygous   Heterozygous

FIGURE 6.6: An example illustrates the simple SNV caller. The top part shows all the aligned reads. The middle part is the reference sequence. The bottom part shows the allelic counts. It has two rows. The first row shows the counts of the reference bases while the second row shows the counts of the non-reference bases.

GAACTCGCACGATCAG
GAACTCACAC
ACTCGCACGA
TCACACGATC

Base b_i	Qscore q_i	Err prob e_i
A	20	10^{-2}
G	10	10^{-1}
A	50	10^{-5}

(a)   (b)

FIGURE 6.7: (a) shows the alignments of three reads on the reference genome. For locus j, the reference base is G while there are two non-reference variants. (b) shows the quality scores and the sequencing error probabilities for these three bases.

6.3.3   Identify SNVs by Poisson-binomial distribution

Section 6.3.2 assumes the sequencing error rate is the same for every base aligned on a locus. However, the sequencing error rates for different bases are actually different. Luckily, the sequencing error rate for each base can be estimated by its PHRED quality score. Given the PHRED quality scores, can we improve the sensitivity of SNV calling? Based on Poisson-binomial distribution, we have a positive answer.

The null model is that all variants are generated by sequencing errors. Let the random variable X be the number of variant bases among the n bases. Denote $Pr_n(X = k)$ as the probability of observing k variants in $\mathcal{D} = \{b_1, \ldots, b_n\}$ under the null model.

We generalize the binomial distribution to a Poisson-binomial distribution, where the sequencing error probabilities are different for different bases. Then, we have

$$Pr_n(X = k) = \sum_{b_1 \ldots b_n} \left\{ \left(\prod_{b_i = r}(1 - e_i) \right) \left(\prod_{b_i \neq r} e_i \right) \mid \text{the number of } (b_i \neq r) \text{ is } k \right\}.$$

For example, consider the example in Figure 6.7(a,b). $Pr_3(X = k)$ for $k = 0, 1, 2, 3$ can be computed as follows.

$$
\begin{aligned}
Pr_3(X = 0) &= (1 - e_1)(1 - e_2)(1 - e_3) = 0.89099109 \\
Pr_3(X = 1) &= (e_1)(1 - e_2)(1 - e_3) + (1 - e_1)(e_2)(1 - e_3) + (1 - e_1)(1 - e_2)(e_3) \\
&= 0.10800783 \\
Pr_3(X = 2) &= (1 - e_1)(e_2)(e_3) + (e_1)(1 - e_2)(e_3) + (e_1)(e_2)(1 - e_3) \\
&= 0.00100107 \\
Pr_3(X = 3) &= (e_1)(e_2)(e_3) = 0.00000001
\end{aligned}
$$

Since the observed number of non-reference variants in Figure 6.7(a,b) is 2. The p-value is $Pr_3(X \geq 2) = 0.00100108 < 0.05$. We reject the null hypothesis and accept that this locus is an SNV. This approach is sensitive and it can detect low-frequency mutations.

However, computing $Pr_n(X = k)$ using the above equation takes exponential time. When n is large, it is inefficient to compute $Pr_n(X = k)$. LoFreq [317] proposed a dynamic programming solution, whose recursive equation is stated in the following lemma.

Lemma 6.1 *When $k = 0$ (base case), $Pr_n(X = k) = Pr_n(X = 0) = \prod_{i=1}^{n}(1 - e_i)$. When $k > 0$ (recursive case), we have*

$$Pr_n(X = k) = (1 - e_n)Pr_{n-1}(X = k) + e_n Pr_{n-1}(X = k - 1). \qquad (6.1)$$

Algorithm LoFreq

Require: n is the number of bases at the locus and K is the number of non-reference bases, $\{q_1, \ldots, q_n\}$ is the set quality scores.

Ensure: $Pr_n(X \geq K)$

1: $Pr_0(X = 0) = 1$
2: **for** $i = 1$ to n **do**
3:    Set $Pr_i(X = 0) = (1 - e_i)Pr_{i-1}(X = 0)$, where $e_i = 10^{-\frac{q_i}{10}}$;
4: **end for**
5: **for** $i = 1$ to n **do**
6:    **for** $k = 1$ to $\min\{i, K - 1\}$ **do**
7:       Compute $Pr_i(X = k)$ by Equation 6.1;
8:    **end for**
9: **end for**
10: Report $1 - \sum_{k=0}^{K-1} Pr_n(X = k)$;

FIGURE 6.8: The dynamic programming algorithm to estimate $Pr_n(X \geq K)$.

Proof The base case is simple.

For the recursive case, note that we have either $b_n = r$ or $b_n \neq r$. When $b_n = r$, this means that there are k variants in $\{b_1, \ldots, b_{n-1}\}$; thus, the probability is $Pr_{n-1}(X = k)(1 - e_n)$.

When $b_n \neq r$, this means that there are $k - 1$ variants in $\{b_1, \ldots, b_{n-1}\}$; thus, the probability is $Pr_{n-1}(X = k - 1)(1 - e_n)$. The lemma follows. ∎

LoFreq aims to compute the exact p-value by Equation 6.1. We can compute $Pr_n(X \geq K)$ using $O(Kn)$ time by dynamic programming. The algorithm is detailed in Figure 6.8. Steps 1−4 compute the base case in Lemma 6.1 while Steps 5−8 compute the recursive case in Lemma 6.1.

6.3.4 Identifying SNVs by the Bayesian approach

We can call SNVs by the Bayesian approach. Let $\mathcal{G}$ be the set of all possible genotypes, i.e., $\mathcal{G} = \{\text{AA}, \text{CC}, \text{GG}, \text{TT}, \text{AC}, \text{AG}, \text{AT}, \text{CG}, \text{CT}, \text{GT}\}$. The Bayesian approach aims to compute $Pr(G|\mathcal{D})$ for every genotype $G \in \mathcal{G}$, where $\mathcal{D} = \{b_1, \ldots, b_n\}$ is the set of bases. Then, we report G, which maximizes $Pr(G|\mathcal{D})$, as the genotype at this locus.

The issue is how to compute $Pr(G|\mathcal{D})$. By the Bayesian approach, $Pr(G|\mathcal{D}) \propto Pr(G)Pr(\mathcal{D}|G)$, where $Pr(G)$ is the prior probability and $Pr(\mathcal{D}|G)$ is the likelihood.

First we discuss how to compute the likelihood $Pr(\mathcal{D}|G)$. Suppose the genotype $G = A_1 A_2$. Since the bases in D are extracted from different reads, they are expected to be independent. Then,

$$Pr(\mathcal{D}|G) = \prod_{i=1..n} Pr(b_i|\{A_1, A_2\}) = \prod_{i=1..n} \frac{Pr(b_i|A_1) + Pr(b_i|A_2)}{2}.$$

Recall that each base b_i is associated with a PHRED score q_i. By the definition of the PHRED score, the error probability of b_i is $e_i = 10^{-q_i/10}$. Then, for $j = 1, 2$, we have $Pr(b_i|A_j) = 1 - e_i$ if $b_i = A_j$; and $e_i/3$ otherwise.

Next, we discuss how to estimate the prior probability $Pr(G)$ for every genotype $G \in \mathcal{G}$. Typically, we assume the heterozygous SNV rate (altHET) is 0.001 (since we expect 1 SNV per 1000 bp) and the homozygous SNV rate (altHOM) is 0.0005. For example, suppose the reference allele is G and the alternative allele is A. Then, we set $Pr(\text{AA}) = 0.0005, Pr(\text{GG}) = 0.9985, Pr(\text{AG}) = 0.001$ while $Pr(G) = 0$ for $G \in \mathcal{G} - \{\text{AA}, \text{GG}, \text{AG}\}$.

Given the prior probability and the likelihood, we can compute the posterior probability $Pr(G|\mathcal{D})$ for every $G \in \mathcal{G}$. For the example in Figure 6.7, the following equations illustrate the computation of $Pr(\text{AG}|\mathcal{D})$.

$$Pr(b_1 = \text{A}|\text{AG}) = \frac{1}{2}\left((1 - 10^{-2}) + \frac{10^{-2}}{3}\right) = 0.49667$$

$$Pr(b_2 = \text{G}|\text{AG}) = \frac{1}{2}\left(\frac{10^{-1}}{3} + (1 - 10^{-1})\right) = 0.466667$$

$$Pr(b_3 = \text{A}|\text{AG}) = \frac{1}{2}\left((1 - 10^{-5}) + \frac{10^{-5}}{3}\right) = 0.499997$$

$$
\begin{aligned}
Pr(\mathcal{D}|\text{AG}) &= Pr(b_1 = \text{A}|\text{AG})Pr(b_2 = \text{G}|\text{AG})Pr(b_3 = \text{A}|\text{AG}) \\
&= 0.49667 * 0.466667 * 0.499997 = 0.115888 \\
Pr(\text{AG}|\mathcal{D}) &= Pr(\mathcal{D}|\text{AG})Pr(\text{AG}) = 0.115888 * 0.001 = 0.000116
\end{aligned}
$$

By the same approach, we can compute $Pr(\text{AA}|\mathcal{D}) = 5.5 \times 10^{-11}$, $Pr(\text{GG}|\mathcal{D}) = 9.985 \times 10^{-9}$ while $Pr(G|\mathcal{D}) = 0$ for $G \in \mathcal{G} - \{\text{AA}, \text{GG}, \text{AG}\}$. Among $Pr(G|\mathcal{D})$ for $G \in \mathcal{G}$, $Pr(\text{AG}|\mathcal{D})$ has the maximum value. We report AG as the genotype for this locus.

In the above discussion, we use a simple prior probability. The prior probability can be estimated more accurately using knowledge in the literature. For example, there are two types of mutations: transition (Ti) and transversion (Tv). Transition mutates either purine to purine (A $\leftrightarrow$ G) or pyrimidine to pyrimidine (C $\leftrightarrow$ T). Transversion mutates purine to or from pyrimidine (A $\leftrightarrow$ C, A $\leftrightarrow$ T, G $\leftrightarrow$ C and G $\leftrightarrow$ T). From the literature, people observed that transition is more frequent than transversion. Based on this fact, SOAP-snp [169] improved the estimation of the prior probability. Another knowledge is that we know some loci have higher mutation rate. For example, dbSNP is a database that provides the mutation frequency of each locus. We can use such information to improve the prior probability estimation.

6.4 Single locus somatic SNV calling

Suppose we sequence both the tumor sample and its matching normal sample. One question is to identify somatic SNVs in the tumor sample (i.e., SNVs that appear in the tumor sample but not in the matching normal sample).

The simple approach is to first identify SNVs in both the tumor and normal samples; then, somatic SNVs are defined as SNVs that appear in the tumor sample but not the normal sample. However, this approach may miss results. Consider the example in Figure 6.9(a). In the tumor sample, seven out of eight reads contain the alternative allele. It should be an SNV. In the normal sample, two out of eleven reads contain the alternative allele. (Note: the two alternative alleles may be the contamination of tumor samples.) Some SNV caller may also call this locus an SNV in the normal sample. In such case, the simple approach will not call this locus a somatic SNV. Below, we describe techniques for calling somatic SNVs.

6.4.1 Identify somatic SNVs by the Fisher exact test

Somatic SNVs can be identified by the Fisher exact test. This approach is used by VarScan2 [144] and JointSNVMix [250].

Consider a particular locus j. Let r and m be the reference allele and the alternative allele, respectively, at the locus j. Let c_t and c_n be the observed number of alleles in the tumor sample and the normal sample, respectively. Let c_r and c_m be the observed number of reference alleles and alternative alleles, respectively, in both the tumor and normal samples. Denote $c_{t,r}$ as the observed number of reference alleles in the tumor sample. For the example in Figure 6.9(a), we have $c_t = 8$, $c_n = 11$, $c_r = 10$, $c_m = 9$ and $c_{t,r} = 1$. The contingency table is in Figure 6.9(b).

Under the null hypothesis that the reference alleles and alternative alleles are independently distributed in the tumor and normal samples, the probability of observing x reference alleles in the tumor sample can be estimated by hypergeometric distribution, that is, $Pr(x|c_t, c_n, c_r, c_m) = \frac{\binom{c_t}{x}\binom{c_n}{c_r-x}}{\binom{c_t+c_n}{c_m}}$.

To test if the reference allele is under-represented in the tumor sample, we perform an one-tailed Fisher exact test. Precisely, the p-value for the one-tailed Fisher exact test is:

$$\sum_{x=0}^{c_{t,r}} Pr(x|c_t, c_n, c_r, c_m).$$

In VarScan2 [144], if the p-value is smaller than 0.1, locus j is called somatic if the normal matches the reference; otherwise, locus j is called LOH (loss of heterozygosity) if the normal is heterozygous; if the p-value is bigger than 0.1, locus j is called a germline variant.

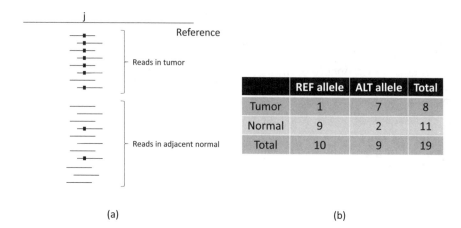

(a)   (b)

FIGURE 6.9: (a) shows all reads in the tumor and matching normal samples covering the locus j. The black squares are alternative alleles appearing in the locus j. (b) shows the corresponding 2×2 contingency table.

For the example in Figure 6.9, the one-tailed Fisher exact test p-value is $\frac{\binom{8}{1}\binom{11}{9}}{\binom{19}{10}} + \frac{\binom{8}{0}\binom{11}{10}}{\binom{19}{10}} = 0.0049 < 0.1$, which indicates this locus is a somatic SNV.

6.4.2 Identify somatic SNVs by verifying that the SNVs appear in the tumor only

Somatic SNVs can also be called as follows. First, we find all SNVs appearing in the tumor sample. Then, each candidate SNV is verified if it does not appear in the normal sample. A number of methods use this approach including MuTect [50] and LoFreq [317]. Below, we detail the method used by MuTect.

Consider a locus j whose reference base is r. Let $\mathcal{D}_T = \{b_1, \ldots, b_n\}$ and $\mathcal{D}_N = \{b'_1, \ldots, b'_{n'}\}$ be the sets of bases in the tumor dataset and the normal dataset, respectively, covering the locus.

MuTect [50] performs two steps. First, it finds candidate somatic SNVs in the tumor sample (see Section 6.4.2.1). Second, it verifies if these candidate somatic SNVs are real by checking if these SNVs appear in the normal sample (see Section 6.4.2.2).

6.4.2.1 Identify SNVs in the tumor sample by posterior odds ratio

Consider the set of alleles $\mathcal{D}_T = \{b_1, \ldots, b_n\}$ covering the locus in the tumor dataset. Suppose the reference base is r. This section discusses how to use the posterior odds ratio to determine if the locus is likely to be SNV.

The proportion of true variant allele m in $\mathcal{D}_T$ may not be 100%. For instance, if the locus is heterozygous, the proportion of allele m in $\mathcal{D}_T$ is expected to be 50%. Under some situations, the proportion of the variant allele may not be 50% and 100%. For example, for Down syndrome samples, there are three copies of chromosome 21. The proportion of variant alleles in chromosome 21 can be $1/3$ or $2/3$.

Instead of fixing the proportion of true variant allele m in the sample, MuTect [50] introduces a parameter f, that is, the proportion of the variant allele m at the locus (i.e., it assumes the proportion of reference alleles is $(1-f)$). Then, MuTect detects if the locus contains the variant allele m at some unknown allele fraction f.

Precisely, MuTect explains $\mathcal{D}_T$ using two models: M_0 and M_f^m.

- Model M_f^m: A variant m exists and the frequency of m is f.

- Model M_0 ($= M_0^m$): There is no variant at this locus. The observed non-reference bases are due to random sequencing errors.

Note that $M_0 = M_0^m$. The following equation describes the likelihood of the model M_f^m given that we observe the set $\mathcal{D} = \{b_1, \ldots, b_n\}$ covering the locus.

$$L(M_f^m|\mathcal{D}) = Pr(\mathcal{D}|M_f^m) = \prod_{i=1}^{n} Pr(b_i|M_f^m) = \prod_{i=1}^{n} Pr(b_i|e_i, r, m, f)$$

Assume that sequencing errors of the observed bases are independent. Then, we have the following lemma.

Lemma 6.2

$$Pr(b_i|e_i, r, m, f) = \begin{cases} f\frac{e_i}{3} + (1-f)(1-e_i) & \text{if } b_i = r \\ f(1-e_i) + (1-f)\frac{e_i}{3} & \text{if } b_i = m \\ \frac{e_i}{3} & \text{if } b_i \neq r, m \end{cases}$$

Proof When $b_i = r$, there are two cases: (1) this locus is a mutant (i.e., this locus is m) and (2) this locus is not a mutant (i.e., this locus is r). For (1), the probability is $Pr(\text{this locus is a mutant}) \cdot Pr(\text{the site is mutated to } r)$, which is $f \cdot \frac{e_i}{3}$. For (2), the probability is $Pr(\text{this locus is not a mutant}) \cdot Pr(\text{this site is not mutated})$, which is $(1-f) \cdot (1-e_i)$.

Similarly, we can derive the equations for $b_i = m$ and for $b_i \neq r, m$. ∎

Given the likelihood, the posterior probabilities $Pr(M_f^m|\mathcal{D})$ and $Pr(M_0|\mathcal{D})$ can be computed by the Bayesian approach as follows:

$$Pr(M_f^m|\mathcal{D}_T) \propto Pr(M_f^m)Pr(\mathcal{D}_T|M_f^m)$$
$$Pr(M_0|\mathcal{D}_T) \propto Pr(M_0)Pr(\mathcal{D}_T|M_0).$$

We declare m as a candidate variant with frequency f if the posterior odds ratio of the two models is bigger than some threshold δ_T (by default, 2). Precisely, the posterior odds ratio test is as follows.

$$\frac{Pr(M_f^m|\mathcal{D}_T)}{Pr(M_0|\mathcal{D}_T)} = \frac{Pr(M_f^m)Pr(\mathcal{D}_T|M_f^m)}{Pr(M_0)Pr(\mathcal{D}_T|M_0)} = \frac{Pr(M_f^m)Pr(\mathcal{D}_T|M_f^m)}{(1 - Pr(M_f^m))Pr(\mathcal{D}_T|M_0)} \geq \delta_T.$$

(6.2)

In the equation, $Pr(M_f^m)$ is the prior probability that the mutant is m and the mutation frequency is f. We assume m and f are independent and $Pr(f)$ is uniformly distributed (i.e., $Pr(f) = 1$). Hence, $Pr(M_f^m) = Pr(m) \cdot Pr(f) = Pr(m)$.

$Pr(m)$ is one third of the expected somatic mutation frequency. We expect there are 3 somatic SNV out of 1 million bases. Hence, MuTect sets $Pr(M_f^m) = Pr(m) = 1 \times 10^{-6}$. (Note that this is the somatic mutation rate. Don't mix it up with the SNV rate, which is 0.001.)

Define the log odds score as $LOD(m, f) = \log_{10} \frac{L(M_f^m|\mathcal{D}_T)}{L(M_0|\mathcal{D}_T)}$. By Equation 6.2 and set $\delta_T = 2$ and $Pr(M_f^m) = 10^{-6}$, we predict m as a candidate somatic SNV if

$$\max_f LOD(m, f) = \max_f \log_{10}\left(\frac{L(M_f^m|\mathcal{D}_T)}{L(M_0|\mathcal{D}_T)}\right) \geq \log_{10}\left(\frac{1 - Pr(M_f^m)}{Pr(M_f^m)}\delta_T\right) \approx 6.3.$$

It is time consuming to compute f that maximizes $LOD(m, f)$. MuTect estimates f to be $\hat{f} = \frac{\text{number of mutants in } \mathcal{D}}{\text{total number of bases in } \mathcal{D}}$.

In other words, a mutant m is declared as a candidate variant if $LOD(m, \hat{f}) \geq 6.3.$.

For the example in Figure 6.7(a,b), out of three bases, two of them equal the variant allele A. Hence, we estimate $\hat{f} = \frac{2}{3}$. Then, we have:

$$Pr(b_1 = \mathtt{A}|e_1 = 10^{-2}, r = \mathtt{G}, m = \mathtt{A}, f = \frac{2}{3}) = \frac{2}{3}(1 - 10^{-2}) + (1 - \frac{2}{3})\frac{10^{-2}}{3} \quad = 0.661111$$

$$Pr(b_2 = \mathtt{G}|e_2 = 10^{-1}, r = \mathtt{G}, m = \mathtt{A}, f = \frac{2}{3}) = \frac{2}{3}\frac{10^{-1}}{3} + (1 - \frac{2}{3})(1 - 10^{-1}) \quad = 0.322222$$

$$Pr(b_3 = \mathtt{A}|e_3 = 10^{-5}, r = \mathtt{G}, m = \mathtt{A}, f = \frac{2}{3}) = \frac{2}{3}(1 - 10^{-5}) + (1 - \frac{2}{3})\frac{10^{-5}}{3} \quad = 0.666661.$$

So, we have $L(M_f^m|\mathcal{D}) = 0.661111 * 0.322222 * 0.666661 = 0.142015$. By the same method, we have $L(M_0|\mathcal{D}) = 1 \times 10^{-8}$. Then, we have $LOD(m = \mathtt{A}, f = \frac{2}{3}) = \log_{10}\frac{0.142015}{1 \times 10^{-8}} = 7.15 \geq 6.3$. We predict this locus is a candidate somatic SNV.

The above approach is sensitive. However, since the tumor sample may have a small amount of cross-individual contamination, some mutations from the cross-individual sample may be called by checking if $LOD(m, \hat{f}) \geq 6.3$. These mutations are treated as false variants since they are not the variants in tumor. MuTect observed that even 2% contaimination can give rise to 166 false positive calls per megabase and 10 false positive calls per megabase when excluding known SNP sites.

To avoid these false positive calls, MuTect estimates the amount of cross-individual contamination by tools like ContEst [51]. Let f_{cont} be the estimated contamination frequency. Then, we make a variant call by replacing M_0 with $M_{f_{cont}}^m$ in Equation 6.2. Precisely, m is declared as a candidate variant if $\log_{10} \frac{L(M_f^m)}{L(M_{f_{cont}}^m)} \geq 6.3$. Using this trick, we can reduce the number of false candidate SNVs.

6.4.2.2 Verify if an SNV is somatic by the posterior odds ratio

Section 6.4.2.1 predicts a list of candidate SNVs in the tumor sample. For each locus j with candidate SNV, the next step verifies if the variant allele m also appears at locus j in the matching normal sample. If yes, this locus is a germline SNV; otherwise, the locus is declared as a somatic SNV.

The input is the set of alleles $\mathcal{D}_N$ that covers locus j in the normal sample. To determine if m appears in locus j in the matching normal, we first compute the probabilities $Pr(\text{locus } j \text{ is reference}|\mathcal{D}_N)$ and $Pr(\text{locus } j \text{ is mutated}|\mathcal{D}_N)$ as follows.

$$
\begin{aligned}
Pr(\text{locus } j \text{ is reference}|\mathcal{D}_N) &\propto Pr(\text{reference in normal})Pr(\mathcal{D}_N|\text{locus } j \text{ is reference}) \\
&= Pr(\text{somatic})Pr(\mathcal{D}_N|M_0) \\
&= Pr(\text{somatic})L(M_0|\mathcal{D}_N) \\
Pr(\text{loucs } j \text{ is mutated}|\mathcal{D}_N) &\propto Pr(\text{mutated in normal})Pr(\mathcal{D}_N|\text{locus } j \text{ is mutated}) \\
&= Pr(\text{germline})Pr(\mathcal{D}_N|\text{locus } j \text{ is mutated}) \\
&\approx Pr(\text{germline})Pr(\mathcal{D}_N|M_{0.5}^m) \\
&= Pr(\text{germline})L(M_{0.5}^m|\mathcal{D}_N)
\end{aligned}
$$

The likelihood $L(M_0|\mathcal{D}_N)$ and $L(M_{0.5}^m|\mathcal{D}_N)$ can be computed using the method in Section 6.4.2.1.

We expect that there are 3 somatic SNVs out of 1 million bases. Hence, we set the prior probability $Pr(\text{somatic}) = 3 \times 10^{-6}$. The prior probability $Pr(\text{germline})$ is determined depending on whether the locus is a dbSNP site. There are 30×10^6 dbSNP sites. Each individual genome is expected to have $\sim 3 \times 10^6$ SNVs and 95% of them are in dbSNP sites. Hence, we expect that 0.05×10^6 SNVs are in non-dbSNP sites. Since our genome is of size 3×10^9, we set $Pr(\text{germline}|\text{non-dbSNP}) = \frac{0.05*3\times 10^6}{3\times 10^9} = 5\times 10^{-5}$. For the dbSNP site, we expect that $0.95*3\times 10^6$ SNVs are in dbSNP sites. Since there are 30×10^6 dbSNP sites, we set $Pr(\text{germline}|\text{dbSNP}) = \frac{0.95*3\times 10^6}{30\times 10^6} = 0.095$.

We declare that a locus j is a somatic SNV if the odds ratio between $Pr(\text{locus } j \text{ is somatic SNV}|\mathcal{D}_N)$ and $Pr(\text{locus } j \text{ is germline SNV}|\mathcal{D}_N)$ is bigger than the threshold δ_N (by default, 10).

$$
\frac{Pr(\text{locus } j \text{ is reference}|\mathcal{D}_N)}{Pr(\text{locus } j \text{ is mutated}|\mathcal{D}_N)} = \frac{Pr(\text{somatic})L(M_0|\mathcal{D}_N)}{Pr(\text{germline})L(M_{0.5}^m|\mathcal{D}_N)} \geq \delta_N
$$

FIGURE 6.10: Given a set of raw reads, SNV and indel calling involves 5 steps.

Define the log odds score LOD_N as $\log_{10} \frac{L(M_0|\mathcal{D}_N)}{L(M_{0.5}^m|\mathcal{D}_N)}$. If locus j is a non-dbSNP, it is called a somatic SNV if $LOD_N \geq 2.2$. If locus j is a dbSNP, it is called a somatic SNV if $LOD_N \geq 5.5$.

6.5   General pipeline for calling SNVs

The previous two sections describe the basic single-locus SNV calling methods. However, single-locus SNV callers give many false callings. The reasons are: (1) systematic errors exist in base calling and (2) read alignment may not be accurate. To reduce false callings, we need to correct the systematic errors and the alignment issues prior to the single-locus SNV calling. In practice, the SNV calling pipeline involves 5 steps (see Figure 6.10 for the standard pipeline to call SNVs). Below, we give an overview of these techniques.

- **Local realignment.** The alignments of reads may be incorrect. This technique corrects the alignments by realigning the reads locally (see Section 6.6).

- **Duplicate read marking.** Some reads may be PCR duplicates and they affect the SNV calling. Techniques for resolving duplicate reads is discussed in Section 6.7.

- **Base quality score recalibration.** This technique recalibrates the base quality scores in the reads to reduce the bias (see Section 6.8).

- **Rule-based filter.** Sequencing machines may have bias in base calling. This technique filters away SNV calls which are likely to be systematic errors (see Section 6.9).

- **Single-locus SNV caller.** Given the alleles appearing in a locus, statistical technique is used to determine if the variant is a sequencing error or a real SNV (This technique has been discussed in Sections 6.3 and 6.4).

We detail local realignment, duplicate read marking, base quality score recalibration and rule-based filtering in Sections 6.6–6.9.

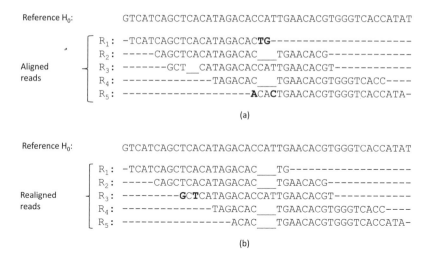

Reference H₀:

Aligned reads

Realigned reads

(a)

Reference H₀:

(b)

FIGURE 6.11: (a) An example alignment of 5 reads on the reference genome. Three alignments contain gaps. (b) An realignment of the 5 reads.

6.6 Local realignment

Although Sections 6.3 and 6.4 describe a number of methods for predicting SNVs, their accuracy depends on whether the reads align correctly. DePristo et al. [65] found that nearly two-thirds of the differences in SNV calling can be attributed to different read mapping algorithms (for HiSeq and exome call sets). This implies that read alignment is a key factor that affects SNV calling. In particular, read mapping near indels and mutation hotspots is difficult.

For example, consider the alignments of 5 reads in Figure 6.11(a). Based on the alignments, we may predict a few SNVs around the indel. Moreover, after the reads R_1, R_3 and R_5 are realigned (see Figure 6.11(b)), we will predict another list of candidate SNVs. The correctness of these SNV calls depends on whether the alignments of the 5 reads are correct or not.

Below, we present the local realignment algorithm of GATK [65]. The algorithm aims to realign reads that are likely to be misaligned. It has four steps: (1) find regions that potentially have misaligned reads, (2) for each selected region, construct a set of haplotypes that are possibly appearing in the sample, (3) realign reads on all constructed haplotypes and check which haplotypes are correct.

For Step 1, the algorithm finds regions that contain either (1) some known indel (e.g., from dbSNP), (2) a cluster of mismatch bases, or (3) at least one read with an indel. (Recall that indel sequencing error is very low for Illumina reads. These regions are selected since reads with indels may be misaligned).

In Step 2, for each region H_0 selected in Step 1, the algorithm constructs haplotypes $H_1, \ldots, H_h$ that are possibly appearing in the sample. Precisely, we create haplotypes from the reference sequence by inserting/deleting either (1) known indels or (2) indels in reads spanning the site from the Smith-Waterman alignment of reads that do not perfectly match the reference genome.

For example, consider the reference H_0 in Figure 6.11(a). The reads R_2 and R_4 span one deletion while R_3 spans another deletion. For these two deletions, we generate two haplotypes H_1 and H_2 (one deletes CA from the reference and another one deletes CAT. The symbol "|" indicates the position with deletion):

$$H_0 = \texttt{TCATCAGCTCACATAGACACCATTGAACACGTGGGTCACCATA}$$
$$H_1 = \texttt{TCATCAGCTCACATAGACAC|TGAACACGTGGGTCACCATA}$$
$$H_2 = \texttt{TCATCAGCT|CATAGACACCATTGAACACGTGGGTCACCATA}$$

In Step 3, among the reference haplotype and all constructed haplotypes in $\{H_0, H_1, \ldots, H_h\}$, GATK checks which haplotypes are likely to be correct. For each haplotype H_i, GATK realigns all reads $R_1, \ldots, R_m$ on H_i without gaps. Suppose the read R_j is aligned at the position p of H_i. Let $e_{j,k}$ be the error probability determined from the quality score of the kth base of the read R_j. Denote $L(R_j|H_i)$ as the probability that R_j is a read from H_i. Then, we have:

$$L(R_j|H_i) = \prod_{k=1}^{|R_j|} \left\{ \begin{array}{ll} (1 - e_{j,k}) & \text{if } R_j[k] = H_i[p + k - 1] \\ e_{j,k} & \text{otherwise} \end{array} \right\}.$$

From all haplotypes, we determine if the haplotype H_i appears in the sample as follows: Denote $L(H_0)$ as the likelihood that H_0 is the only haplotype in the sample. Denote $L(H_0, H_i)$ as the likelihood that H_0 and/or H_i exist in the sample. Then, we estimate $L(H_0, H_i) = \prod_{j=1..m} \max\{L(R_j|H_i), L(R_j|H_0)\}$ and $L(H_0) = \prod_{j=1..m} L(R_j|H_0)$. We accept H_i as the correct haplotype if $\log(L(H_0, H_i)/L(H_0)) > 5$.

As an illustration, Figure 6.12 shows the ungapped alignments of $R_1, \ldots, R_5$ to H_0, H_1, H_2. The figure also shows the score $L(R_j|H_i)$. (In the illustration, we assume the PHRED quality score of every base is 20 and we assume $1 - \epsilon_{j,k} \approx 1$.) Then, we have

$$L(H_0) = L(R_1|H_0)L(R_2|H_0)L(R_3|H_0)L(R_4|H_0)L(R_5|H_0) = 10^{-34}$$
$$L(H_0, H_1) = L(R_1|H_1)L(R_2|H_1)L(R_3|H_0)L(R_4|H_1)L(R_5|H_1) = 10^{-4}.$$

Since $\log(L(H_0, H_1)/L(H_0)) = 30 > 5$, we accept that H_1 is the correct haplotype.

Note that a read R_j is aligned to the haplotype H_i if $L(R_j|H_i) > L(R_j|H_0)$; otherwise, R_j is aligned to the original reference H_0. For our example, R_1, R_2, R_4, R_5 are aligned to H_1 while R_3 is aligned to H_0. Hence, the local alignment step reports the alignment in Figure 6.11(b).

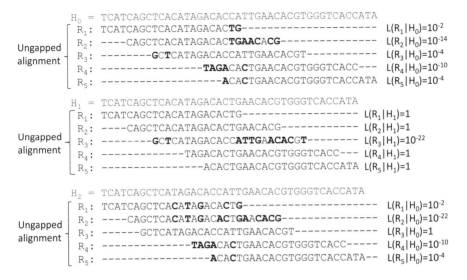

FIGURE 6.12: The ungapped alignment of the reads on the two haplotypes H_1 and H_2. For each read R_j, the figure also shows the score $L(R_j|H_i)$.

6.7 Duplicate read marking

Due to PCR amplification, duplicate reads may be generated. These duplicate reads may bias the SNV calling. For example, if a read containing a sequencing error is duplicated 10 times, these 10 duplicated reads may bias the SNV caller to accept the sequencing error as an SNV. Hence, it is important to mark these duplicated reads.

Many methods like SOAPsnp [169] and GATK [65] perform duplicate read marking. The basic idea is to identify paired-end reads whose endpoints are aligned to the same positions on the genome. These paired-end reads are marked as duplicates.

6.8 Base quality score recalibration

Recall that every base has a quality score. If the error probability of the base is p, then the quality score of the base is expected to be $-10\log_{10} p$. Due to technology limitations, the quality scores may be inaccurately estimated. Even worse, the inaccuracy and covariation patterns differ strikingly among sequencing technologies. Recall that some SNV callers in Sections 6.3 and

6.4 utilize the PHRED base quality score to improve SNV calling. Incorrect estimation of quality scores will affect the correctness of SNV calling. Hence, we need to recalibrate the quality scores.

Below, we briefly describe possible sources of errors when we perform sequencing using Illumina machines. The sequencing process is performed within the flow cell of the sequencing machine. Each flow cell consists of 8 lanes. The sequencing machine cannot read all DNA sequences on the entire lane. Instead, it reads the DNA sequences tile by tile. Different machine models contain a different number of tiles per lane. For instance, GA II has 100 tiles per lane while HiSeq 2000 has 68 tiles per lane. For each tile, there are hundreds of thousands to millions of DNA clusters. Each DNA cluster contains one DNA template. Through bridge amplification, one DNA template in the DNA cluster is amplified to approximately 1000 identical DNA templates. The amplification is to increase the intensity level of the emitted signal since it is difficult to detect a single fluorophore. To read the DNA templates in all DNA clusters in the tile, the Illumina machine applies fluorescent bases and imaging. To capture reads of length ℓ, it requires ℓ cycles. In the ith cycle, the machine will read the ith bases of the DNA templates of all DNA clusters in the tile in parallel. Precisely, in the ith cycle, the machine performs 3 steps. First, it extends the ith base of the nascent strand of each DNA cluster by a base attached with fluorophores. By reading the fluorescent signals of all DNA clusters through imaging, we read the ith bases of all clusters. Then, the machine cleaves the fluorophores from the bases. These 3 steps will be iterated for ℓ cycles.

In this process, four possible biases will contribute to the mis-calling of the bases. They are:

- Fluorophore cross talk: The emission spectra of the fluorophores for the four bases $\{A, C, G, T\}$ may be overlapping. This may generate incorrect base calling.

- Phasing: In a perfect situation, in the ith cycle of the signal reading, we should extend the nascent strand by the ith base. However, because the chemistry is imperfect, the extension of some strands may lag behind or lead forward. These lagging and leading strands contaminate the signal from the ith base.

- Fading: During the sequencing process, the DNA strands in each DNA cluster are subjected to laser emissions and excessive washing. Hence, the DNA fragments will disappear gradually and lead to a decrease in fluorescent signal intensity. This process is called fading.

- Insufficient fluorophore cleavage: In each cycle, we need to wash away the fluorophores. A small number of fluorophores are left behind. Hence, there is an increase in fluorescent signal as the number of cycles increase. The "stickiness" of the fluorophores is base dependent. This leads to a bias.

TABLE 6.1: The table shows the statistics of all bases at position 2 of the reads such that the base is A with quality score 40 under different dinucleotide context. The last column is the empirical score $\epsilon(x\text{A}, 2, 40)$ where $x\text{A}$ is the dinucleotide.

Dinucleotide	count	error	ϵ
AA	3239	8	$-10\log_{10}\frac{8+1}{3239+1}$
CA	4223	5	$-10\log_{10}\frac{5+1}{4223+1}$
GA	3518	2	$-10\log_{10}\frac{2+1}{3518+1}$
TA	4032	20	$-10\log_{10}\frac{20+1}{4032+1}$

A number of methods are proposed to recalibrate the quality score, including SOAPsnp [169], GATK [65] and ReQON. Below, we briefly describe the idea.

Let $b_1, \ldots, b_n$ be all bases in the reads. Let $q_1, \ldots, q_n$ be the corresponding quality scores. Note that the error probability of the base b_i is $e_i = 10^{-\frac{q_i}{10}}$. The average error probability is $\frac{1}{n}\sum_{i=1}^{n} e_i$. Let q_{global} be the phred score for the average error probability, i.e., $q_{global} = -10\log_{10}\left(\frac{1}{n}\sum_{i=1}^{n} e_i\right)$. Denote ϵ as $-10\log_{10}\left(\frac{\text{number of true error}}{\text{total number of bases}}\right)$. The recalibration process adjusts the quality score of each base so that the average recalibrated error rate equals ϵ. This means that, for $i = 1, \ldots, n$, the recalibrated quality score for base b_i is $q_i + (\epsilon - q_{global})$. However, ϵ is unknown. In practice, the recalibration algorithm assumes that any SNV that is not in dbSNP128 is an error. (dbSNP128 is the list of known SNPs, which contains 12 million known SNPs.) Thus, ϵ is estimated to be the proportion of bases that are different from the reference bases but are not known SNPs.

For example, suppose we have 1000 reads, each of length 100 bp. This means that we sequenced 100,000 ($= 1000 \times 100$) bases. Let $b_1, \ldots, b_{100000}$ be these 100,000 bases. Assume 100 of them are different from the reference bases. Out of these 100 bases, suppose 95 of them are dbSNP. Then, we estimate $\epsilon = -10\log_{10}\frac{100-95}{100000} = 43$. Suppose $q_{global} = 45$. If a particular base b_i with quality score $q_i = 30$, we recalibrate its quality score to $30 + 43 - 45 = 28$.

The above recalibration algorithm is simplified. Many factors affect the base qualities. The three main factors are (1) the position of the base in the read (different machine cycles have different error rates), (2) the substitution bias (which gives bias to certain types of miscalls) and (3) the dinucleotide context (the error rate depends on the preceding character).

To incorporate the bias into the recalibration process, we collect and classify every base according to its position, quality score and dinucleotide context. Note that we exclude bases on known dbSNP sites and assume all mismatches are sequencing errors. Table 6.1 gives example statistics of the bases at position 2 of all reads such that the bases are A and the corresponding quality scores are 40. The table has 4 rows corresponding to the 4 dinucleotide contexts. Each row shows the number of bases that match the reference (which are assumed to be correct) and the number of bases that are different from the reference (which are assumed to be incorrect). From the table, we can compute the empirical score for each base under different dinucleotide context. Precisely, for each base b_i with quality score q_i, the empirical score $\epsilon(prev(b_i)b_i, pos(b_i), q_i)$ equals

$$-10\log_{10} \frac{error(prev(b_i)b_i, pos(b_i), q_i) + 1}{count(prev(b_i)b_i, pos(b_i), q_i) + 1},$$

where $prev(b_i)$ is the preceding base of b_i, $pos(b_i)$ is the position of b_i in the read, $count(prev(b_i)b_i, pos(b_i), q_i)$ is the number of bases that have the same property of $(prev(b_i)b_i, pos(b_i), q_i)$ and $error(prev(b_i)b_i, pos(b_i), q_i)$ is the number of bases that have the same property of $(prev(b_i)b_i, pos(b_i), q_i)$ and they are the non-reference and non-dbSNP bases. For the example in Table 6.1, if the base b_i is A, at position 2 with dinucleotide AA and quality score $q_i = 40$, then the empirical score $\epsilon(\text{AA}, 2, 40)$ is $-10\log_{10}\frac{8+1}{3231+8+1} = 25.56$.

Given the empirical score $\epsilon(prev(b_i)b_i, pos(b_i), q_i)$, the recalibrated quality score for the base b_i is $(\epsilon - q_{global}) + (\epsilon(prev(b_i)b_i, pos(b_i), q_i) - q_i)$.

6.9   Rule-based filtering

Even after we perform realignment and base quality score recalibration, a few studies showed that the SNV callers may still give false positive SNVs. For instance, Reumers et al. [242] studied the SNV callings using the whole genome sequencing data from monozygotic twins. Their genomes are expected to be nearly identical. So, the discordant SNVs represent errors in SNV calling. Among the discordant SNVs, 19.7% are near an indel, 25.3% are in SNV-dense regions, 51.9% are in regions with low or extremely high read coverage regions, and 37.9% are in repeat regions like microsatellites. This suggested that the false SNVs are not randomly distributed.

To remove these artifacts, researchers proposed some rules to filter them. This idea is used by a number of SNV callers. Some example rules are as follows. If the SNV is supported by only a few reads, it is likely that they are caused by sequencing errors and we may want to filter it. As another example, if the reads that support the SNV also contain indels, it is possible that the reads are misaligned. In this case, the SNV is not trustable and we may want to filter it. Table 6.2 highlights the rules used by SAMtools [170] (or MAQ [172]) and MuTect [50].

Furthermore, suppose we have the SNVs of a panel of normal samples. If we want to find disease-causing SNPs, we can filter SNVs that appear in all normal samples since they are expected not to cause disease.

TABLE 6.2: Rules used to filter SNPs used by MAQ and MuTect.

Samtools (or MAQ) rule-base filter:
- Discard SNPs near indels (within 3 bp flanking region of a potential indel).
- Discard SNPs with low coverage (covered by 3 or fewer reads).
- Discard SNPs covered by reads with poor mapping only (mapping quality lower than 60 for all covered reads).
- Discard SNPs in SNP dense regions (within a 10 bp region containing 3 or more SNPs).
- Discard SNPs with consensus quality smaller than 10.

MuTect rule-base filter:
- Discard SNPs near indels (false positives caused by misaligned small indel events).
- Discard SNPs covered by reads with poor mapping.
- Discard SNPs on triallelic sites.
- Discard SNPs covered by reads with strand bias.
- Discard SNPs covered by reads mapped to similar location.
- Discard SNPs in tumor if some reads in normal also contain the SNPs.

6.10   Computational methods to identify small indels

More and more evidence shows that indels (insertions and deletions of size < 50 bp) occur frequently in our genomes [122]. Furthermore, they are known to be the causes of a number of diseases [324]. Accurate calling of small indels is important. (Note that calling long indels is also important and they will be described in Chapter 7.)

A number of small indel calling methods have been proposed. Roughly speaking, they can be classified into four approaches: (1) the realignment-based approach (GATK [65], Dindel [4]), (2) the split-read approach (Pindel [327], microindels [149], Splitread [131]), (3) the span distribution-based clustering approach (MoDIL [159]) and (4) the local assembly approach (SOAPindel [177]). We described the realignment-based approach in Section 6.6. Below, we describe the other three approaches one by one.

6.10.1   Split-read approach

The split-read approach is the most popular technique to identify small indels. Methods applying the split-read approach include Pindel [327], Dindel [4], and microindels [149]. Below, we detail Pindel [327].

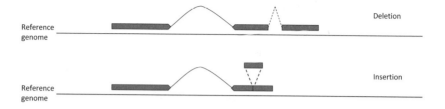

FIGURE 6.13: Consider every paired-end read, where only one read R (i.e., the read on the left) is uniquely aligned on the reference genome while another read R' is unaligned. Pindel tries to align the 3' end of R' within $2\times$ the average insert size from the read R. If the 3' end of R' is uniquely aligned, Pindel tries to align the 5' end of R' nearby. Then, a candidate indel is found.

Pindel utilizes paired-end reads to identify indels of size at most s. Candidate indels are identified by finding paired-end reads that support the indels. Figure 6.13 illustrates the idea. A paired-end read (R, R') is said to support a candidate indel if (1) a read R is uniquely aligned on the reference genome, (2) its mate R' is uniquely aligned on an indel as a split alignment, and (3) the distance between the mapping locations of R and R' is shorter than the maximum insert size d (d is controlled by the size selection step in the wet-lab protocol). This is known as the anchored split mapping signature.

To simplify the method, Pindel does not allow a mismatch when it performs split read alignment for R'. (Pindel finds that such an assumption will not reduce the sensitivity by a lot.) The anchored split mapping signatures can be found in three steps. First, Pindel finds the shortest and longest prefixes of R' such that it is uniquely aligned within distance d from R. Let p be the position of the alignment. Second, Pindel finds the shortest and longest suffixes of R' such that it is uniquely aligned within distance s from p. Let q be the position of the alignment. Third, Pindel verifies if R' can form a split read alignment at positions p and q. If yes, an anchored split mapping signature is identified. The anchored split mapping signature can be found in $O(s + d)$ time if we build a suffix tree (see Exercise 8). Pindel also suggested using the pattern growth approach to find the anchored split mapping signature.

Finally, an indel event is reported if an indel event is supported by at least two anchored split mapping signatures.

6.10.2   Span distribution-based clustering approach

Indels can also be identified by studying the span distribution of the paired-end reads. The idea is that if an indel exists, the span of the paired-end reads covering the indel is expected to derivate from the expected insert size. This enables us to detect indels.

As an example, suppose we have a library of paired-end reads of a diploid

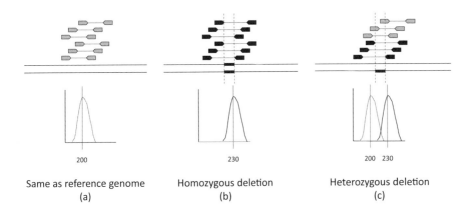

FIGURE 6.14: (a) The distribution of the insert size for a normal region. (b) The distribution of the insert size for a region with a homozygous 30 bp deletion. (c) The distribution of the insert size for a region with a heterozygous 30 bp deletion.

genome whose expected insert size is 200 bp. After the paired-end reads are aligned on the reference genome, there are three cases as shown in Figure 6.14. Figure 6.14(a) illustrates the case where the sample genome is the same as the reference genome. In this case, the span distribution peaks at 200 bp. Figures 6.14(b,c) illustrate two regions that have homozygous and heterozygous deletions, respectively, of 30 bp. If we check the span distribution of the paired-end reads in Figure 6.14(b), we observe that the mean of the insert size is shifted from 200 bp to 230 bp. By detecting the shift of insert size, we can detect the small homozygous deletion. The span distribution of the paired-end reads in Figure 6.14(c) showed that it is a mixture of two insert sizes: 200 bp and 230 bp. This implies that there is a heterozygous deletion.

MoDIL [159] is a method that uses the insert size distribution to detect indels. Below, we detail the method.

MoDIL's algorithm consists of three steps. The first step aligns paired-end reads to the reference genome. Any read mapper in Chapter 4 can be used in this step. The second step finds clusters of aligned reads that potentially overlap with some putative indels. Precisely, for each aligned paired-end read that has an abnormal insert size, one cluster is formed by grouping all paired-end reads covering the same locus. The last step verifies if each cluster contains an indel. The rest of this subsection details the last step.

Let $\mathcal{Y} = \{Y_1, \ldots, Y_{|\mathcal{Y}|}\}$ be the set of insert sizes of all aligned paired-end reads in the dataset. The distribution of $\mathcal{Y}$ represents the insert size distribution for regions with no indel. Denote $\mu_{\mathcal{Y}}$ as the mean of $\mathcal{Y}$. Denote $Pr(y \in \mathcal{Y})$ as the probability that the insert size is y.

Consider a genomic region with an indel of size s (see Figure 6.14(b)). Let

$\mathcal{Z}$ be the set of insert sizes of the aligned paired-end reads covering such an indel. We expect the mean $\mu_{\mathcal{Z}}$ is shifted by s, that is, $\mu_{\mathcal{Z}} = \mu_{\mathcal{Y}} + s$. The shape of the distribution of $\mathcal{Z}$ is expected to be the same as the shape of $\mathcal{Y}$. Precisely, we have $Pr(z \in \mathcal{Z}|\mu_{\mathcal{Z}}) = Pr((z - \mu_{\mathcal{Z}} + \mu_{\mathcal{Y}}) \in \mathcal{Y})$. To check if $\mathcal{Y}$ and $\mathcal{Z}$ have the same shape, it is equivalent to check if $\{Z_i - \mu_{\mathcal{Z}} \mid Z_i \in \mathcal{Z}\}$ and $\{Y_i - \mu_{\mathcal{Y}} \mid Y_i \in \mathcal{Y}\}$ have the same distribution. This can be tested by Kolmogorov-Smirnov goodness of fit statistics. Precisely, let $F_{\mathcal{Z}}(v)$ be the proportion of $Z_j \in \mathcal{Z}$ such that $Z_j - \mu_{\mathcal{Z}} \leq v$ (i.e., $F_{\mathcal{Z}}(v) = \frac{1}{|\mathcal{Z}|} \left(\sum_{j=1}^{|\mathcal{Z}|} I(Z_j - \mu_{\mathcal{Z}} \leq v) \right)$ where $I(Z_j - \mu_{\mathcal{Z}} \leq v) = 1$ if $Z_j - \mu_{\mathcal{Z}} \leq v$; and zero otherwise). The KS statistics compute $D_{\mathcal{Z}} = \max_v |F_{\mathcal{Z}}(v) - F_{\mathcal{Y}}(v)|$. If $D_{\mathcal{Z}}$ is significantly small, this means $\mathcal{Z}$ and $\mathcal{Y}$ have the same shape.

Consider a genomic region with two haplotypes A and B, each with its own indel (see Figure 6.14(c)). Let $\mathcal{X} = \{X_1, \ldots, X_n\}$ be the set of paired-end reads extracted from this genomic region. Note that the paired-end reads in $\mathcal{X}$ are extracted from both haplotypes A and B. Since $\mathcal{X}$ is a mixture of paired-end reads from two haplotypes, the shape of the insert size distribution of $\mathcal{X}$ looks different from that of $\mathcal{Y}$. Suppose we can partition $\mathcal{X}$ into two sets $\mathcal{X}^A$ and $\mathcal{X}^B$ such that $\mathcal{X}^A$ and $\mathcal{X}^B$ are the sets of paired-end reads extracted from haplotypes A and B, respectively. Then, the insert size distributions of $\mathcal{X}^A$ and $\mathcal{X}^B$ are expected to have the same shape as $\mathcal{Y}$. Let μ_A and μ_B be the means of $\mathcal{X}^A$ and $\mathcal{X}^B$, respectively. Then, we can infer the expected sizes of the indels in the two haplotypes to be $\mu_A - \mu_{\mathcal{Y}}$ and $\mu_B - \mu_{\mathcal{Y}}$, respectively.

However, we only observe the set $\mathcal{X}$ in real life. $\mathcal{X}^A$ and $\mathcal{X}^B$ are hidden information. $\mathcal{X}^A$ and $\mathcal{X}^B$ can be estimated to be the bipartition of $\mathcal{X}$ that minimizes $\lambda_A D_{\mathcal{X}^A} + \lambda_B D_{\mathcal{X}^B}$, where $\lambda_t = \frac{|\mathcal{X}^t|}{n}$ is the fraction of reads in $\mathcal{X}^t$ for $t \in \{A, B\}$. (Note that $\lambda_A + \lambda_B = 1$.)

The above minimization problem can be solved by enumerating all possible bipartitions of $\mathcal{X}$. However, it runs in exponential time. Instead, MoDIL applies the EM algorithm (Section 3.3.2) to partition $\mathcal{X}$ into $\mathcal{X}^A$ and $\mathcal{X}^B$.

MoDIL first initializes μ_A, μ_B and λ_A. (By default, we initialize $\mu_A = \mu_B = \mu_{\mathcal{Y}}$ since we expect the majority of regions do not have an indel. We also set $\lambda_A = 0.5$.)

Then, MoDIL iteratively executes two steps: the E-step and the M-step. Given the estimation of μ_A, μ_B, λ_A, the E-step of MoDIL applies Equation 6.3 to determine $\gamma_{jt} = Pr(X_j \in \mathcal{X}^t|\mu_A, \mu_B, \lambda_A)$, that is, the likelihood that the j-th read is generated from the haplotype t. This is the formula for computing γ_{jt}:

$$\gamma_{jt} = \frac{\lambda_t Pr(X_j \in \mathcal{X}^t|\mu_t)}{\lambda_A Pr(X_j \in \mathcal{X}^A|\mu_A) + \lambda_B Pr(X_j \in \mathcal{X}^B|\mu_B)}. \quad (6.3)$$

The M-step of MoDIL estimates λ_A, μ_A and μ_B, given γ_{jt} for $j = 1, \ldots, |\mathcal{X}|$

Algorithm MoDIL

Require: The set $\mathcal{X} = \{X_1, \ldots, X_{|\mathcal{X}|}\}$ of insert sizes in a cluster and the set $\mathcal{Y} = \{Y_1, \ldots, Y_{|\mathcal{Y}|}\}$ of insert sizes in the library.

Ensure: Either report no indel or report the expected indel sizes μ_A, μ_B.

1: Initializes $\mu_A = \mu_B = \mu_{\mathcal{Y}}$, initializes $\lambda_A = 0.5$;
2: **repeat**
3:     **for** $j = 1$ to $|\mathcal{X}|$ **do**
4:         Compute γ_{jA} and γ_{jB} by Equation 6.3;
5:     **end for**
6:     Compute λ_A by Equation 6.4;
7:     Find μ_A, μ_B that minimizes Equation 6.5;
8: **until** the score D in Equation 6.5 reaches local optimal;
9: Discard the cluster if Kolmogorov-Smirnov statistic rejects the null hypothesis at 95% confidence level; otherwise, report the inferred indel sizes to be $\{\mu_A - \mu_{\mathcal{Y}}, \mu_B - \mu_Y\}$;

FIGURE 6.15: The EM algorithm to estimate μ_A, μ_B.

and $t \in \{A, B\}$. Precisely, the M-step sets $\lambda_B = 1 - \lambda_A$ and

$$\lambda_A = \frac{1}{n} \left(\sum_{j=1}^{n} \gamma_{jA} \right). \tag{6.4}$$

Then, it finds μ_A and μ_B, which minimizes

$$D = \sum_{t \in \{A,B\}} \lambda_t \max_{v} \left| \frac{\sum_{j=1}^{n} (\gamma_{jt} I(X_j - \mu_t \leq v))}{\sum_{j=1}^{n} \gamma_{jt}} - F_{\mathcal{Y}}(v) \right|. \tag{6.5}$$

Figure 6.15 details the EM algorithm.

6.10.3 Local assembly approach

All methods described above are good at finding deletions. However, they are not good at detecting insertions. One technique to resolve this shortcoming is by local assembly. The local assembly approach is used by SOAPindel [177] and Scalpel [215]. Below, we briefly describe SOAPindel.

Given the alignments of all paired-end reads, SOAPindel first identifies a set of reads whose mates do not map on the reference genome. Utilizing the insert size, these unmapped reads are piled up at their expected position (see Figure 6.16(a)). These reads are called virtual reads. Then, clusters of virtual reads are identified. For each cluster, the virtual reads are subjected to de novo assembly and generate some contigs. Those contigs are aligned to the reference genome to obtain potential indels (see Figure 6.16(b)).

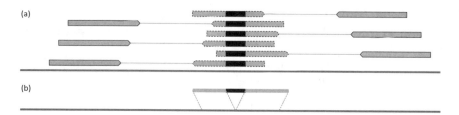

FIGURE 6.16: (a) In this example, there are 6 reads whose mates are not mapped. Those unmapped reads are represented as the dashed arrows and they are piled up at their expected position. (b) Those unmapped reads form virtual clusters and they are subjected to de novo assembly. We obtained a contig. When we realign the assembled contig on the reference genome, we discovered an insertion (represented by the black block).

6.11   Correctness of existing SNV and indel callers

Due to the huge effort in the development of SNV and indel callers, existing methods are highly accurate. It is estimated that the error rate is about 1 in 100,000–200,000 bp. The remaining errors are mostly contributed by (1) low-complexity regions and (2) an incomplete reference genome.

Low-complexity regions comprise about 2% of the human genome. They mostly appear in centromeric regions and acrocentric short arms. Although they are short, they harbor 80%–90% of heterozygous indels and up to 60% of heterozygous SNVs [165]. It is estimated that many of these variation calls are errors. Two major sources of errors are PCR errors and mis-alignment errors. Many heterozygous indels called in low-complexity regions are 1 bp heterozygous indels appearing in long poly-A or poly-T runs. They are most likely contributed by the error of inserting/deleting one A (or T) during the PCR amplification. Another issue is alignment. Mis-alignments of reads happen more frequently in low-complexity regions. Due to the mis-alignments, false positive SNVs/indels are predicted. This problem can be partially solved by the realignment step (see Section 6.6).

Another issue is the incomplete reference genome. Our genome is incomplete. For example, the 1000 Genomes Project identifies an extra 33.4 million bp sequences that appear in other human assemblies but are not present in the primary human reference hg19 (or GRCh37).

These extra sequences are called decoy sequences. The reference formed by hg19 and the decoy sequences is known as hs37d5.

Comparing the number of SNVs and indels called using the references hg19 and hs37d5, the number of heterozygous SNVs/indels called using hg19 is twice as many as that called using hs37d5 [165]. This indicates that many reads which are misaligned in hg19 are aligned to the decoy seuqences in hs37d5; hence, this approach reduces the number of false SNV/indel calls.

6.12 Further reading

This section gives the techniques for calling SNVs and indels. For calling SNVs, we describe the complete pipeline, which includes local realignment, duplicate read marking, base quality score recalibration, statistics for single-locus SNV calling, and rule-based filtering. When the input involves a tumor sample and the matching normal sample, we also discuss how to call somatic SNVs.

We also discuss a few techniques for calling indels. They include the realignment-based approach, the split-read approach, the span distribution-based clustering approach, and the local assembly approach.

Due to the space constraint, we did not cover some other interesting techniques. In this section, we assume the SNVs are sparse. When there are multiple hotspots (2 or more SNVs clustered together), the methods described in this section cannot work. In such case, we use the iterative mapping-based method to call SNVs. The basic idea is to use an SNV caller to call some SNVs with high confidence. Then, the reference genome is updated and we reapply the SNV caller iteratively to call more SNVs. This approach is used by methods like iCORN [221] and ComB [283]. It improves the sensitivity of calling SNVs in mutation hotspots.

This section discusses methods for single sample SNV calling. There are methods for multi-sample SNV callings. In such case, each sample is sequenced using low- to medium-coverage data ($3\times-20\times$). There are three solutions. The first solution is to run a single SNV caller on the pooled sample (consider multiple samples as a single sample). This solution, however, will treat the rare SNVs as noise and miss them. The second solution is to run a single SNV caller on every individual sample. This solution is good for finding SNVs that are unique to one sample. However, it will give many false positives as the sequencing coverage is not high. The third solution is to develop methods that are tailor-made for multi-sample SNV calling. A number of methods have been developed, which include MultiGeMS [211], SNIP-seq [14] and [210].

Evaluation of somatic SNV calling can be found in [246]. Fang et al. [75] performed an extensive evaluation of indel calling and showed that there are still many errors in indel calling, due to library preparation, sequencing biases, and algorithm artifacts.

6.13   Exercises

1. Assume the sequencing error probability is 0.01. Consider a locus with reference base r. Suppose 5 reads cover the locus and two covered bases are non-reference variants. Using binomial distribution, what is the p-value that this locus is an SNV?

2. Consider a locus with reference base G. Suppose $\mathcal{D} = \{b_1, b_2, b_3, b_4, b_5\}$ is the set of 5 bases on the 5 reads covering this locus. The following table shows the bases and the quality scores for $b_1, \ldots, b_5$.

i	1	2	3	4	5
b_i	G	G	G	C	G
q_i	10	30	30	30	30

Using LoFreq, what is the probability that this locus is an SNV?

3. Consider the 5 bases $\mathcal{D} = \{b_1, b_2, \ldots, b_5\}$ in Question 2. By the Bayesian approach, what is the most likely genotype in this locus? You can assume the following prior probability table (which is used by SOAPsnp [169]).

	A	C	G	T
A	3.33×10^{-4}	1.11×10^{-7}	6.67×10^{-4}	1.11×10^{-7}
C		8.33×10^{-5}	1.67×10^{-4}	2.78×10^{-8}
G			0.9985	1.67×10^{-4}
T				8.33×10^{-5}

4. Consider a locus i. Suppose $c_t = 10$, $c_n = 8$, $c_r = 7$, $c_m = 11$, and $c_{t,r} = 2$. Please give the contingency table and determine if it is a somatic SNV by Fisher Exact Test. (Assume the p-value threshold is 0.05.)

5. Consider a locus j whose reference base is G. Also, locus j is a non-dbSNP site. Suppose $\mathcal{D} = \{b_1, b_2, \ldots, b_5\}$ is the set of bases at locus j in a tumor sample. This table shows the bases and the quality scores for $b_1, \ldots, b_5$.

i	1	2	3	4	5
b_i	G	G	C	C	C
q_i	10	30	30	30	30

Suppose $\mathcal{D}' = \{b'_1, \ldots, b'_4\}$ is the set of bases at locus j in the matching normal. This table shows the bases and the quality scores for $b'_1, \ldots, b'_4$.

i	1	2	3	4
b'_i	G	G	C	C
q'_i	30	30	20	30

Can you determine if the locus j is a somatic SNV using MuTect in Section 6.4.2?

6. Consider the following alignments for 4 reads. Assume the quality scores for all bases are 20. Can you perform realignment of these reads using the local realignment algorithm of GATK?

```
Reference H₀:       ATAGACACTGAACACGTGGGTCATCATATATGTGGGTCATCATATG

            ┌ R₁:  -TAGACACTGAACACGTCGGTCA---------------------
Aligned     │ R₂:  ----ACACTGAACACGTCGGTCATCA------------------
reads       │ R₃:  ------ACTGAACACGTCGGTCATCATA----------------
            └ R₄:  -------CTGAAC_____ATGTCGGTCATCATAT-
```

7. Suppose $q_{global} = 40$. Consider 1,000,000 reads, each of length 150 bp. Assume 2,000,000 bases are different from the reference. Suppose 500,000 of them are known dbSNPs. Consider a base b with quality score $q = 35$. Based on the above information, can you recalibrate its quality score?

8. Consider a paired-end read (R, R'). Let d be its maximum insert size. Let s be the maximum size of the deletion. Suppose R is uniquely aligned at position x. We aim to find the anchored split mapping signature of $R'[1..m]$. We aim to check if there is an anchored split mapping signature that supports a deletion of size at most s. (Note: you can assume $d > m$.) Can you give an $O(s + d)$-time algorithm to solve this problem?

9. Pindel can be used to find deletions of size at most s where s is a user-defined parameter. Suppose the length of each read is 20. Do you suggest setting $s = 10^6$? Why?

10. Let $\mathcal{Y} = \{92, 95, 96, 99, 99, 100, 101, 102, 103, 105, 108\}$ be a set of insert sizes for paired-end reads in regions with no indel. Consider the following two sets of insert sizes. Please check if they have the same shape as the distribution of $\mathcal{Y}$.

   - $\mathcal{X}_1 = \{127, 128, 130, 131, 131, 133\}$
   - $\mathcal{X}_2 = \{77, 78, 79, 80, 81, 82, 83, 127, 129, 130, 131, 133\}$

11. Refer to Question 10. Can you determine the size of indels from the distribution of $\mathcal{X}_1$ and $\mathcal{X}_2$?

12. Let ℓ be the insert size of the paired-end read library. For SOAPindel, can it detect an indel of size 3ℓ?

Chapter 7

Structural variation calling

7.1 Introduction

The previous chapter discussed methods for calling SNVs and small indels, which are localized variations in our genome. This chapter studies methods for calling variations that cover longer regions, which are known as structural variations (SVs).

Structural variations (SVs) are large-scale changes in our genome, often more than 50 nucleotides [80]. Figure 7.1(a−f) lists the simple SVs. They include insertion, deletion, reversal, duplication, transposition and translocation. The simple SVs can be classified into balanced SVs and unbalanced SVs. Balanced SVs just rearrange our genomes. They include reversal, transposition and translocation. Unbalanced SVs alternate the number of copies of some DNA segments, which is usually referred to as copy number variation (CNV). They include insertion, deletion and duplication. Complex SVs are formed by combining a number of simple SVs. For example, Figure 7.1(g) is a complex SV formed by a transposition with a deletion.

The above discussion covers SVs within one genome. Some SVs of an individual are inherited from two parents. One example is unbalanced translocation (see Figure 7.2). Suppose parent 2 of the individual has a balanced translocation. If the individual obtains a normal chromosome from parent 1 and a chromosome with translocation from parent 2, the individual has an unbalanced translocation.

We expect each individual has about tens of thousands of SVs[40, 270]. Although the number of SVs is much lower than that of SNVs and small indels per individual, SVs affect more bases. Reference [54] showed that up to 13% of bases are subjected to variations of SVs. SVs not only cover a large number of bases, they are also highly related to disease risks [286] and phenotypic variations [130]. Hence, it is important to understand SVs.

This chapter describes techniques for calling structural variations. The organization is as follows. Section 7.2 describes how SVs are formed. Then, Section 7.3 covers the clinical effects of SVs. Section 7.4 details different wet-lab techniques to determine SVs. The computation techniques for calling CNVs are described in Section 7.5. The computation techniques for calling SVs are described in Sections 7.6−7.9.

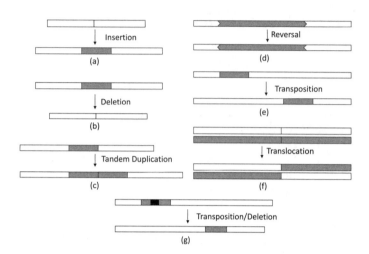

FIGURE 7.1: Different types of structural variations (SVs): (a) insertion, (b) deletion, (c) tandem duplication, (d) reversal, (e) transposition and (f) translocation. Note that (a)–(c) are unbalanced SVs while (d)–(f) are balanced SVs. (f) is a type of SV involving more than one chromosome. (g) is an example of a complex SV formed by two simple SVs (transposition and deletion).

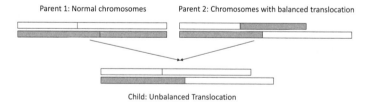

FIGURE 7.2: An illustration of unbalanced translocation.

7.2   Formation of SVs

Since each SV involves at least two breakpoints, assuming uniform distribution, the chance that two individuals share the same SV is very low (1 out of 9×10^{18} individuals). Interestingly, some SVs occur frequently and they are called recurrent SVs. This implies that the SV formation mechanism is not purely random. This section briefly describes a few known mechanisms for forming SVs.

The formation of SVs is mainly driven by inaccurate repair and mistakes in replication. There are five major mechanisms that cause SVs [92, 97]. Below, we briefly describe them.

- **Non-allelic homologous recombination (NAHR):** Due to environmental factors and specific mutagenic exposure (such as smoking, UV light, and chemotherapy), DNA damaging occurs. A double-strand DNA fragment is broken into two pieces, which creates double-strand breaks (DSBs). The DSBs can be repaired through repairing mechanisms.

  One major repairing pathway is homologous recombination. It utilizes the allelic homologous chromosome as a template to help rejoin the DSBs. However, the repair mechanism may misalign the DSBs to another homologous regions (i.e., not the allelic homologous chromosome); then, structural variations are created. This mechanism is called non-allelic homologous recombination (NAHR). NAHR usually occurs between segmental duplication or repeat elements like SINEs, LINEs and LTRs. (Segmental duplication or low copy repeat (LCR) is a DNA region of length > 1,000 bp and its copies have > 90% DNA sequence identity. It appears in about 5% of our genome.)

  Figure 7.3 illustrates the NAHR mechanism (for the formation of translocation). Apart from translocation, NAHR can generate deletion, duplication or inversion (see Figure 7.4). Many SVs formed by NAHR are recurrent.

- **Nonhomologous end joining (NHEJ):** Apart from NAHR, DSBs can be repaired by nonhomologous end joining (NHEJ) and its alternative microhomology-mediated end joining (MMEJ). Neither mechanism requires a homologous region as a template for rejoining the DSBs.

  Figure 7.5(a) illustrates the steps of NHEJ to join two DNA fragments whose single-strand overhangs. NHEJ or MMEJ first join the overhangs of the two DNA fragments. Typically, the joining is guided by some short homologous sequence (of length 1−20 bp) called the microhomology. Then, unaligned or mismatched nucleotides are removed/modified and the gaps are filled by synthesis. After ligation, the SV is formed. The removal/modification of the unaligned or mismatched nucleotides during

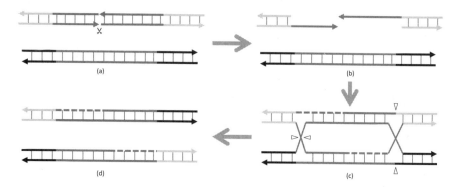

FIGURE 7.3: Illustration of the formation of translocation through the NAHR mechanism. Two chromosomes share a homologous region (in dark gray color). (a) DSBs occur on the chromosome on the top. (b) 5' to 3' resection generates 3' single-stranded overhangs. (c) The single-stranded overhangs on the top chromosome invade the bottom chromosome and form two Holliday junctions. The two Holliday junctions can be resolved by cutting at the positions with the white triangle; (d) Translocation is formed.

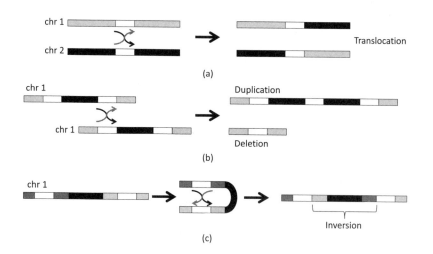

FIGURE 7.4: Possible SVs generated by the NAHR mechanism. The white blocks represent the homologous regions. When DNA damaging occurs at the white blocks, a double strand break (DSB) is created. The repair mechanism may misalign the DSB to non-allelic homologous regions, which leads to an SV. (a) shows the formation of translocation, (b) shows the formation of deletion and tandem duplication and (c) shows the formation of inversion.

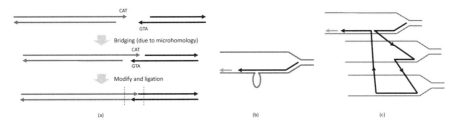

FIGURE 7.5: Illustration of (a) the NHEJ mechanism, (b) replication slippage mechanism and (c) template switching mechanism.

the joining may leave a molecular scar. NHEJ/MMEJ creates simple non-recurrent SVs.

- **Replication-based mechanism**: The above two mechanisms form SVs through the repairing process. SVs can also be generated due to mistakes in the replication process.

  During replication, the duplex DNA sequence is first unzipped. Then, a replication fork is formed. The lagging strand will be synthesized in a direction that is opposite to the direction of the growth of the replication fork. There are two possible mistakes during synthesis: polymerase slippage and template switching.

  In polymerase slippage, the template forms a secondary structure and the synthesis of the lagging strand may skip the DNA segment, which results in a deletion. Figure 7.5(b) illustrates the process.

  In template switching, the lagging strand may disengage from the template and switch to another template in some nearby replication fork. Such close interaction of the two templates is aided by segmental duplications, cruciforms or non-B DNA structures. The switching usually requires micro-homology of 2−5 bp so that the lagging strand can bind on the alternative template. This switching may happen multiple times; then, the replication resumes in the original template. This mistake is known as fork stalling and template switching (FoSTeS) [158] and microhomology-mediated break-induced repair (MMBIR) [96]. Figure 7.5(c) illustrates the process.

  The polymerase slippage and template switching can be interleaved, which results in complex rearrangement of the genome. It is estimated that the percentage of such complex SVs is 5%−16% [238]. Unlike the SVs generated by NAHR, the SVs formed by replication error are usually non-recurrent.

- **Mobile element insertion:** Mobile element insertion is a mechanism that allows some DNA sequences (called transposable elements) to change their positions. Some common transposable elements are Alu sequences and L1 sequences. The movement of transposable elements are mediated by retrotransposons, DNA transposons and endogenous retroviruses. Mobile element insertions are usually flanked with some inverted repeats.

- **Chromothripsis:** Chromothripsis is a recently discovered phenomenon [287]. It involves shattering the genome at multiple breakpoints, followed by error-prone DNA repair by NHEJ within a short time interval. The result is a complex genome rearrangement. Chromothripsis has been observed in a number of cancers like bone cancers, liver cancer [77], etc. It occurs in 2%−3% of the cancers studied and about 25% of bone cancers [287].

Kidd et al. [138] studied the relative importance of different SV formation mechanisms. They found that the formation of SVs is driven by three major mechanisms: (1) microhomology-mediated SVs (accounts for 28% SVs), (2) nonallelic homologous recombination (accounts for 22% SVs), and (3) mobile element insertion (accounts for 19% SVs).

7.3   Clinical effects of structural variations

Structural variations have different downstream effects. They cause numerous diseases and are known to be associated with drug resistance. Below, we briefly describe how structural variations affect the phenotypes.

1. **Loss/gain of gene**: Due to unbalanced SVs, we may gain or lose some genes. Some genes are dosage sensitive. For example, NAHR can cause tandem duplication or deletion of ∼1.4M bp genome fragments in 17p12. The duplication increases the copies of the dosage-sensitive gene PMP22 from 2 to 3, which gives Charcot−Marie−Tooth disease type 1A (CMT1A). The deletion reduces the gene PMP22 from 2 copies to 1 copy, which gives hereditary neuropathy with liability to pressure palsies (HNPP). There are many other diseases contributed to copy number variants, including Crohn's disease [195] and various neurodevelopmental disorders. CNVs are also shown to contribute to drug response [83].

2. **Loss/gain of some part of a gene**: Unbalanced SVs can also duplicate or delete some part of a gene (including promoters and introns). Such variation may cause (a) the gain or loss of some functional part (like the protein domain) of a gene or (b) the change of gene expression level. For example, the deletion of intron 2 of BIM has been shown to associate with TKI resistance (TKI is a treatment for CML patients) [218].

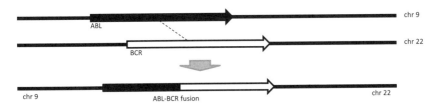

FIGURE 7.6: Due to the translocation event between chromosomes 9 and 22, the genes BCR and ABL1 are fused together and form a BCR-ABL1 fusion. This event happens frequently in chronic myelogenous leukemia.

3. **Creation of fusion gene**: Balanced SVs may link two genes together. In such case, fusion genes are formed. For example, Rowley [252] discovered the translocation of chr9 and chr22 in 1982. This translocation generates the BCR-ABL1 gene fusion (see Figure 7.6). More than 90% of chronic myelogenous leukemias (CMLs) are caused by this fusion. Drugs targeting this fusion exist and significantly improve the survival of CML patients.

4. **Integration of foreign DNAs**: SVs can also integrate some foreign DNA into our genome. An example is the integration of the hepatitis B virus (HBV) into the human genome. HBV integration is commonly observed in liver cancer patients [290]. In fact, HBV integration is one of the main risks that causes liver cancer.

For more information on the phenotypic impact of SVs, please refer to [314] and [184].

7.4 Methods for determining structural variations

To determine the existence of a single SV, we can use qPCR, FISH, comparative genomic hybridization, MLPA [266], etc. To have a rough map of SVs, we can use karyotyping or array-CGH [36]. However, these methods cannot give a high-resolution structural variation map. Furthermore, array-CGH cannot detect balanced SVs.

All the above limitations can be solved by next-generation sequencing. The solution involves a wet-lab phase and a dry-lab phase. The wet-lab phase performs shotgun sequencing to obtain a set of paired-end reads that cover the whole genome. There are two protocols: (1) whole genome sequencing and (2) mate-pair sequencing. Their details are described in Section 5.2. In brief, the sample genome is broken into fragments; then, fragments of a certain insert

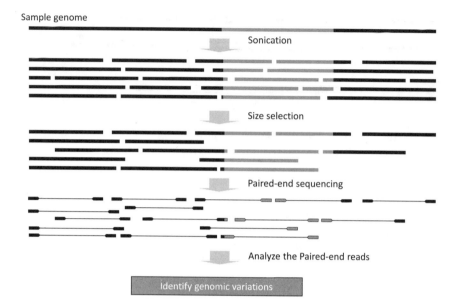

FIGURE 7.7: The illustration of the NGS protocol for whole genome sequencing. In this example, a gray-colored sequence is inserted into the genome. The two boundaries between the black and gray colored sequences are called breakpoints. Paired-end reads that cross the breakpoints are called anomalous paired-end reads. The reads that overlap with the breakpoints are called split-reads. (Split-reads are reads that contain both black and gray colored sequences while anomalous paired-end reads are paired-end reads that contain both black and gray colored sequences.)

size are selected (the insert size is < 1000 bp for whole genome sequencing while the insert size is long (say 10,000 bp) for mate-pair sequencing). Finally, paired-end reads are sequenced from the two ends of these selected fragments by next-generation sequencing. Some fragments cover structural variation, that is, the fragment consists of at least two genomic regions coming from different parts of the reference genome. (See Figure 7.7 for an illustration of the wet-lab step.)

After the wet-lab phase, the dry-lab phase analyzes the paired-end reads to identify structural variations on a genome-wide scale.

Given the whole genome sequencing dataset of a sample, we have two computational problems.

- **CNV calling**: The aim is to compute the copy number variation (CNV) of each genomic region in the reference genome.

- **SV calling**: The aim is to call the structural variations in the sample with respect to the reference genome.

Similar to somatic SNVs (see Section 6.1.2), somatic CNVs/SVs are genomic variations that occur in a subgroup of cells. To discover somatic CNVs/SVs, we need the whole genome sequencing datasets of two different tissue types of an individual. There are two computational problems:

- **Somatic CNV calling**: The aim is to find genomic regions such that the copy number variations are different for the two datasets.

- **Somatic SV calling**: The aim is to find SVs that appear in one tissue type but not in the other.

The somatic CNV/SV calling problems are commonly used to study solid tumor. In this scenario, the two input datasets are the whole genome sequencing datasets for the tumor and the adjacent normal.

The rest of this chapter is organized as follows. Section 7.5 covers CNV calling and somatic CNV calling. Sections 7.6−7.9 cover SV calling and somatic SV calling.

7.5 CNV calling

The number of copies of each genomic region is referred as its copy number variation (CNV). CNVs can be computationally estimated using the read depth approach, clustering approach, split-mapping approach and assembly approach. Among these four approaches, the read depth approach is the only approach which explicitly estimates the copy number. The other three approaches just identify the deleted or duplicated regions. This section will focus on the read depth approach. Other approaches will be covered in Section 7.8.

The read depth approach is applied by many CNV callers, including VarScan2 [144], CNAseg [114], m-HMM [304], SegSeq [44], rSW-seq [142], JointSLM [185], CNV-seq [320], RDXplorer [328] and CNVnator [2]. Its idea is to determine the copy number of a DNA region by measuring the read coverage. The basic principle is as follows. Given a set of paired-end reads generated from the wet-lab protocol, we map them on the reference genome using a read mapper (see Chapter 4). Then, the number of reads mappable to each genomic region can be measured. Assuming reads are uniformly distributed along the genome, the read count is approximately proportional to the copy number of the DNA region. Figure 7.8 illustrates this idea. If the region A is deleted, we expect the region has no read. On the other hand, if the region B is duplicated, we expect the read coverage is double. By computing the read coverage, we can determine if a region is deleted or duplicated.

Precisely, for any DNA region, let C be the number of mappable reads in this region and E be the expected number of reads in this region. If $C >> E$, the DNA region is amplified. Otherwise, if $C << E$, the DNA region is deleted.

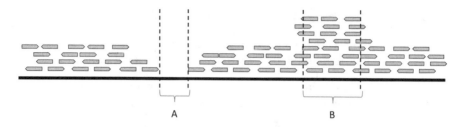

FIGURE 7.8: Different regions have different read depth. In the figure, region A is deleted while region B is duplicated.

The copy number of each DNA region can be estimated as $\frac{2C}{E}$. (Note that the copy number of each DNA region is expected to be 2 since each DNA region has two copies: one from the mother and one from the father.)

However, reads are not uniformly distributed along the genome due to multiple factors like mappability bias and GC content. To avoid the bias, we need to normalize the read counts. Below, we will discuss the general strategy for calling CNVs, which consists of three steps:

1. Computing raw read count: This step maps reads on the reference genome. Then, the genome is partitioned into bins and a read count is computed for each bin.

2. Normalization of read counts: Due to GC bias and mappability, the raw read count may not reflect the actual read coverage. This step normalizes the read counts.

3. Segmentation: This step merges regions with similar normalized read counts. Then, based on the normalized read counts, the copy numbers are determined.

The above procedure is for a single sample. To compute somatic CNVs, the input is a pair of samples. We will run the first two steps for the two input samples. After we obtain the normalized read counts for both samples, we compute the ratio of the normalized counts for all bins. Finally, segmentation (Step 3) is performed on their ratios.

7.5.1 Computing the raw read count

Given the sequencing dataset of a sample, all reads are mapped on the reference genome. For reads with multiple aligned locations, each read is randomly assigned to one mapping location. (Why? See Exercise 2.) After the reads are mapped, the raw read count of each genomic region is computed. Precisely, the length-n genome is partitioned into m equally sized bins (i.e., each bin is of length $\frac{n}{m}$). Denote C_i as the number of mappable reads in the ith bin. The total number of mappable reads is $N = \sum_{i=1}^{m} C_i$.

7.5.2 Normalize the read counts

For the i-th bin, its copy number may not be proportional to its raw read count C_i due to GC bias and mappability issues [19]. GC bias is contributed by the sequencing technology. It is observed that the sequencing machines have fewer reads for DNA regions whose GC content (i.e., percentage of G and C) is too low or too high. Another issue is mappability. Due to repeats, we may fail to map reads of some regions uniquely, which reduces the read coverage of these regions. Hence, it is important to normalize the read counts before we detect the copy numbers.

A number of methods exist to correct GC bias and mappability; please refer to [328, 202, 25, 18]. Here, we describe the method of [25]. For the i-th bin, its GC content is the percentage of G and C in the bin while the mappability is the number of k-mers (for some selected k, say 36) in the bin that can be uniquely aligned. The GC content score and mappability score are scaled to integer values between 0 to 100.

For the ith bin, suppose its GC content score is α_i and its mappability score is β_i. Then, its normalized read count A_i is defined to be

$$\frac{C_i \cdot M}{M_{\alpha_i,\beta_i}},$$

where M is the expected read count for a bin with a normal copy number (i.e., copy number is 2) and M_{α_i,β_i} is the expected read count of a bin with a normal copy number, while its GC content score is α_i and its mappability score is β_i. M is estimated by the median raw read count among all bins while M_{α_i,β_i} is estimated by the median raw read count among all bins that have GC content score α_i and mappability score β_i.

As an illustration, Figure 7.9 shows an example with 12 bins. Rows 2 and 3 show the GC content scores and the mappability scores of the 12 bins, respectively. Row 4 shows the raw read counts C_i, for $i = 1, \ldots, 12$. The median raw read count is $M = 62.5$. By the GC content score and the mappability score, the 12 bins are partitioned into two groups. One group with $\alpha_i = 50, \beta_i = 100$ and another group with $\alpha_i = 90, \beta_i = 60$. Their medians are $M_{50,100} = 100.5$ and $M_{90,90} = 49.5$, respectively. The normalized read counts $A_i = \frac{C_i \cdot M}{M_{\alpha_i,\beta_i}}$ can be found on row 5. The raw read counts C_i and the normalized read counts A_i are also visualized in Figure 7.10.

7.5.3 Segmentation

The previous step computes the normalized read counts (or read count ratios for paired samples) for the n bins. These counts (or ratios) are $A_1, \ldots, A_n$. Segmentation aims to group the bins into intervals so that the normalized read counts of the bins in each interval are approximately the same. Figure 7.10 gives an example. Different methods use different ways to segment the bins. Some methods detect significant change in read depth. Examples in-

bin	1	2	3	4	5	6
GC content α_i	50	50	90	50	90	90
Mappability β_i	100	100	60	100	60	60
C_i	149	151	74	100	50	51
A_i	92.66	93.91	93.43	62.19	63.13	64.39
Actual copy number	3	3	3	2	2	2

bin	7	8	9	10	11	12
GC content α_i	90	50	50	90	50	90
Mappability β_i	60	100	100	60	100	60
C_i	49	97	101	25	51	27
A_i	61.87	60.32	62.81	31.57	31.72	34.09
Actual copy number	2	2	2	1	1	1

FIGURE 7.9: An example that illustrates the computation of normalized read counts.

clude VarScan2 [144] (circular binary segmentation), RDXplorer [328] (EWT approach) and CNVnator [2] (mean-shift approach). Some methods use the HMM approach, including m-HMM [304]. Below, we describe the circular binary segmentation approach.

Circular binary segmentation is a recursive algorithm that partitions $A_1, \ldots, A_n$ into a set of segments such that the values in each segment are similar. The method is as follows. In each step, it identifies an interval $i..j$ such that the values $A_i, \ldots, A_j$ are significantly different from the rest of the values. If we fail to identify such interval, we know that $A_1, \ldots, A_n$ is a segment. Otherwise, three intervals $A_1, \ldots, A_{i-1}$, $A_i, \ldots, A_j$ and $A_{j+1}, \ldots, A_n$ are obtained. The algorithm recursively identifies the segments in these three intervals. The algorithm is detailed in Figure 7.11.

To enable the above solution to work, we need a statistic to determine if $A_i, \ldots, A_j$ are significantly different from the rest of the values. For every interval $i..j$, we compute the t-statistic. Let $n_{i..j} = j-i+1$, i.e., the number of bins in $A_i, \ldots, A_j$. Let $\mu_{i..j} = \frac{A_i+\ldots+A_j}{n_{i..j}}$, i.e., the mean of $A_i, \ldots, A_j$. Let $\overline{\mu}_{i..j} = \frac{(A_1+\ldots+A_{i-1})+(A_{j+1}+\ldots+A_n)}{n-n_{i..j}}$, i.e., the mean of $A_1, \ldots, A_{i-1}, A_{j+1}, \ldots, A_n$. Assume the standard derivation of all A_i's are the same. The t-statistic is proportional to $T_{i,j}$, which is,

$$T_{i,j} = \frac{\mu_{i..j} - \overline{\mu}_{i..j}}{\sqrt{\frac{1}{n_{i..j}} + \frac{1}{n-n_{i..j}}}}.$$

Our statistic is $T_{\max} = \max_{1 \leq i < j \leq n} |T_{i,j}|$.

To determine if $T_{\max}$ is significant, we can use a permutation test. The idea is as follows. We randomly permute all A_i's for 10,000 times. For the kth round of random permutation, denote T^k to be the maximum t-statistic among all intervals. From $T^1, \ldots, T^{10000}$, we determine the $\alpha\%$ confidence

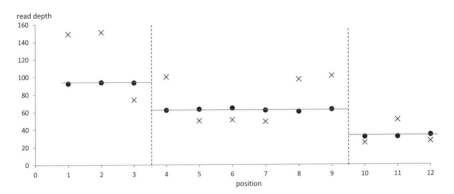

FIGURE 7.10: This figure shows the raw read counts $C_1, \ldots, C_{12}$ (represented by "×") and the normalized read counts $A_1, \ldots, A_{12}$ (represented by "o") for the 12 bins stated in Figure 7.9. Segmentation partitions them into 3 regions. The left, middle and right regions are of copy numbers 3, 2 and 1, respectively.

Algorithm CBS
Require: A list of values $A_p, \ldots, A_q$
Ensure: A set of segments
  1: Find the interval $i..j$ that maximizes the absolute T-statistics $T_{i,j}$;
  2: **if** $T_{i,j}$ is significant by permutation test **then**
  3:    Call CBS($A_p, \ldots, A_{i-1}$), CBS($A_i, \ldots, A_j$) and CBS($A_{j+1}, \ldots, A_q$);
  4: **else**
  5:    Report $p..q$ as a segment;
  6: **end if**

FIGURE 7.11: The Circular Segmentation Algorithm CBS($A_p, \ldots, A_q$) generates the list of segments in the interval $p..q$.

interval. $T_{\max}$ is said to be significant if $T_{\max}$ is not within the $\alpha\%$ confidence interval.

As an example, consider the 12 values $A_1, \ldots A_{12}$ in Figure 7.9. By the above formula, we have $T_{1,3} = 61.32$. Among all $T_{i,j}$, $T_{1,3}$ maximizes the t-statistic. Hence, $T_{\max} = T_{1,3}$. If we perform a permutation test, we can show that $T_{\max}$ is significant. Then, we obtain two intervals $A_1, \ldots, A_3$ and $A_4, \ldots, A_{12}$. We recursively apply the algorithm CSB on these two intervals. Finally, we obtain three segments $A_1, \ldots, A_3$, $A_4, \ldots, A_9$ and $A_{10}, \ldots, A_{12}$.

After we obtain a segment $A_i, \ldots, A_j$, the last step is to determine its copy number. The formula is as follows.

$$\frac{2 \cdot median(A_i, \ldots, A_j)}{median(A_1, \ldots, A_m)} \qquad (7.1)$$

FIGURE 7.12: Steps for calling SVs.

For the example in Figure 7.10, the median of $A_1, \ldots, A_{12}$ is $(62.19 + 62.81)/2 = 62.5$. The medians of $A_1 \ldots A_3$, $A_4 \ldots A_9$ and $A_{10} \ldots A_{12}$ are 93.43, $(62.19 + 62.81)/2 = 62.5$ and 31.72, respectively. Hence, the estimated copy numbers for the three segments are 2.99, 2.00 and 1.01.

7.6   SV calling pipeline

This section discusses methods for calling SVs. The input is an alignment file (i.e., BAM file) of a set of paired-end reads. The basic steps for calling SVs are given in Figure 7.12. First, the insert size is estimated. Then, the reads are filtered and subclassified into different types. Third, candidate SVs are discovered. Finally, each candidate SV is verified.

Somatic SVs can be called using similar steps. For each sample, we just run the first 3 steps to discover the candidate SVs. Then, in the fourth step (SV verification), we verify if the candidate SV appears in one or both samples. If the candidate SV appears in one sample only, we determine a somatic SV.

The following subsections will detail the 4 steps.

7.6.1   Insert size estimation

Recall that the insert size of a DNA fragment is its length. Due to the size selection step (see Section 7.4), the inset size of the DNA fragments are expected to be in a short range, say, $(span_{\min}..span_{\max})$. The insert size range can be determined by wet-lab equipment like a bioanalyzer. However, the range may not be accurate.

Another choice is to estimate the insert size computationally. The basic approach is to identify all paired-end reads (R_1, R_2) that do not cross SV breakpoints (called concordant); then, the difference of the mapping positions of R_1 and R_2 is expected to be its fragment size. For the Illumina sequencing platform, a paired-end read (R_1, R_2) is expected to be concordant if (1) both R_1 and R_2 are mapped on the same chromosome; (2) the two reads point inward to each other. Figure 7.13 shows the histogram of the insert sizes of a set of paired-end reads.

From the insert size distribution, there are two methods to estimate $(span_{\min}..span_{\max})$: (1) the standard derivation and (2) the gradient.

The standard derivation approach first computes the median and the stan-

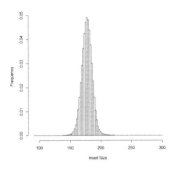

FIGURE 7.13: The distribution of the insert size.

dard derivation of the insert sizes. Precisely, let $L_1, \ldots, L_m$ be the insert sizes of all paired-end reads sorted in increasing order. The median is $L_{m/2}$ and the standard derivation σ is $\sqrt{\frac{1}{m} \sum_{i=1}^{m} (L_i - L_{m/2})^2}$. Then, we estimate $(span_{\min} .. span_{\max}) = (L_{m/2} - k\sigma .. L_{m/2} + k\sigma)$ for some user-defined constant k. (Usually, we set $k = 1, 2$ or 3.)

The gradient approach computes the slope of every point in the span distribution. The $span_{\min}$ and $span_{\max}$ are the positions with gradients approaching 0. For the span distribution in Figure 7.13, by the gradient approach, $span_{\min}$ and $span_{\max}$ are estimated to be 140 and 220.

7.7   Classifying the paired-end read alignments

Consider a set of paired-end reads extracted from a sample genome (say, by Illumina sequencing). When these paired-end reads are mapped onto a reference genome, some paired-end reads cross the SV breakpoint. Figure 7.14 illustrates an example.

With respect to the SV breakpoint, the paired-end read can be classified into 3 types:

(1) a paired-end read that doesn't cross any SV breakpoint;

(2) an anomalous paired-end read that crosses one SV breakpoint; or

(3) an anomalous paired-end read with a split read (i.e., among the two reads of a paired-end read, one read, called a split read, overlaps one SV breakpoint).

Note that it is possible that some anomalous paired-end reads will overlap with more than one SV breakpoint. To simplify the study, we ignore this case.

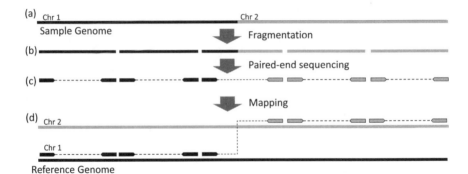

FIGURE 7.14: (a) The line illustrates a sample genome formed by two DNA segments from chromosomes 1 and 2. (b) The sample genome is fragmented. (c) After sequencing the fragments, we extract a set of paired-end reads. (d) When the paired-end reads are mapped on the reference genome, the paired-end read that crosses the breakpoint will be aligned discordantly.

Below, we classify the alignments of the three types of paired-end reads on the reference genome.

(1) For a paired-end read that does not cross any SV breakpoint, it is expected that its two reads R_1 and R_2 are aligned on the same chromosome, R_1 and R_2 are aligned in inward orientation, and the distance between R_1 and R_2 is within $(span_{\min}..span_{\max})$ (see Figure 7.15(a)). Such alignment is called concordant alignment.

(2) For an anomalous paired-end read that crosses one SV breakpoint, it is expected that the two reads do not align concordantly and it is called a discordant alignment (see Figure 7.15(b)).

(3) For an anomalous paired-end read with a split read, the alignment of the split read on the reference genome is soft-clipped due to the SV breakpoint. (A soft-clip is the unaligned portion of the read. Figure 7.16 gives an example.) Depending on whether the 5' or the 3' portion is clipped, we obtain either (i) a discordant alignment with a soft-clip or (ii) a concordant alignment with a soft-clip. When there are more SNVs or sequencing errors near the SV breakpoint, the split read may be unmapped. Then, we obtain (iii) a one-end anchor alignment. See Figure 7.15(c) for an illustration.

The discordant alignments (with or without a soft-clip) can be further classified into six types depending on the orientation of the two aligned reads and the insert size.

(a) inward: Paired-end reads with a correct insert size and the two reads

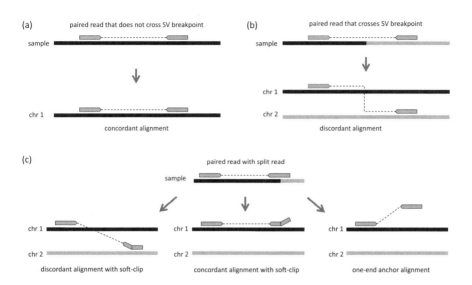

FIGURE 7.15: (a) The alignment of a paired-end read that does not cross any SV breakpoint. (b) The alignment of an anomalous paired-end read that crosses one SV breakpoint. (c) The alignment of an anomalous paired-end read with a split read.

Reference: TCAGTACGGTACAACTGACTTCACGATCAACTGACGTACGGTACCACGTACCAGT

CGATCAACTG TGCCA

FIGURE 7.16: This example gives an alignment of a read CGATCAACTGTGCCA on the reference genome. The 3'-end portion of the read (TGCCA) is not similar to the reference genome and it is not aligned. This portion of the read is called a soft-clip.

pointing inward to each other. They are expected not to cross any SV breakpoint.

(b) long inward: Paired-end reads with a long insert size (longer than $span_{max}$) and the two reads pointing inward to each other. They are expected to be generated from deletion or transposition.

(c) short inward: Paired-end reads with a short insert size (shorter than $span_{min}$) and the two reads pointing inward to each other. They are expected to be generated from insertion.

(d) same orientation: Paired-end reads where both reads align on the same strand. They are expected to be generated from inversion or transposition.

(e) outward: Paired-end reads where the two reads are pointing outward from each other. They are expected to be generated from duplication or transposition.

(f) inter-chromosomal: Paired-end reads where the two reads are aligned on different chromosomes. They are expected to be generated from translocation or transposition.

Figure 7.17 illustrates the 6 types of paired-end read alignments. Except for type (a), all remaining types are discordant alignments (with or without a soft-clip). The positions of these discordant alignments give potential breakpoints for the structural variations.

7.8   Identifying candidate SVs from paired-end reads

From the alignments of the paired-end reads, this section discusses computational methods that infer structural variations. In general, there are 4 approaches to call SVs:

1. Clustering approach (E.g. BreakDancer[43], PEMer[145], DELLY[241] and VariationHunter[106])

2. Split-mapping approach (E.g. SRiC[332], CREST[305] and Socrates[267])

3. Assembly approach (E.g. NovelSeq[94] and TIGRA[42])

4. Hybrid approach (E.g. LUMPY[157], SVMerge[318], MetaSV[204], SoftSV[17])

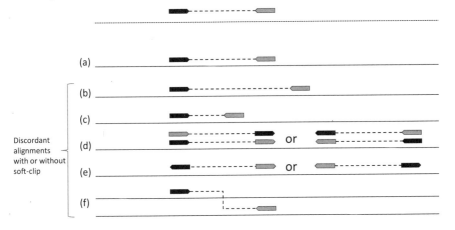

FIGURE 7.17: Six different types of paired-end read alignments: (a) inward, (b) long inward, (c) short inward, (d) same orientation, (e) outward and (f) inter-chromosomal. In the figure, the black and gray reads are the 5' reads and the 3' reads, respectively. The dotted line represents the sample genome while the solid line represents the reference genome.

7.8.1 Clustering approach

Recall that discordant alignment is evidence of structural variation. Hence, we expect an SV occurs when we identify a cluster of discordant alignments. Calling SVs using this idea is called the clustering approach.

Many methods use the clustering approach. They include PEMer [145], VariationHunter [106], BreakDancer [43], MoGUL [160], HYDRA [236], Corona [197], SPANNER [203], GASV [276], GASVPro [277], DELLY [241], Ulysses [87], CLEVER [192], SVMiner [99].

The clustering approach consists of two steps. The first step classifies paired-end reads into different types as stated in Figure 7.17. The second step clusters discordant alignments of the same type to form candidate SVs.

Consider the example in Figure 7.18. There are 12 discordant alignments. By clustering them, 4 clusters are obtained. Each cluster is described by two anchors, where each anchor corresponds to one breakpoint of the SV.

Different methods use slightly different clustering criteria. Below, we cover three ideas.

- Clique-finding approach

- Confidence interval overlapping approach

- Set-cover approach

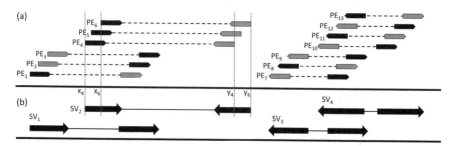

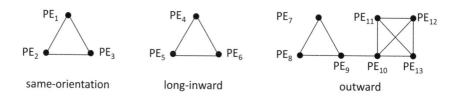

FIGURE 7.18: (a) shows the alignments of a set of 12 paired-end reads. (b) shows that four clusters formed by the 12 paired-end reads. Note that each cluster has two anchors and the sizes of the anchors are at most $span_{\max}$.

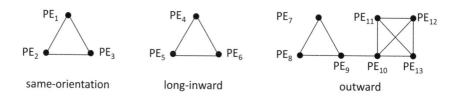

FIGURE 7.19: This figure shows the graph G corresponding to the set of aligned paired-end reads in Figure 7.18.

7.8.1.1   Clique-finding approach

Given a set of paired-end reads $\mathcal{PE} = \{PE_1, \ldots, PE_n\}$, for any i, let the 5'-end mapping positions of the two reads of PE_i be x_i and y_i, respectively. Also, denote the orientations of the left and right reads of PE_i as $sign(x_i)$ and $sign(y_i)$, respectively. For example, for PE_4 in Figure 7.18, $sign(x_4) = +1$ and $sign(y_4) = -1$.

The clustering problem is formulated as a graph problem. We define a graph $G = (\mathcal{PE}, E)$ over the set of vertices $\mathcal{PE}$. For any two paired-end reads PE_i and PE_j, (PE_i, PE_j) is an edge in E if (1) $sign(x_i) = sign(x_j)$, (2) $sign(y_i) = sign(y_j)$ and (3) $|x_i - x_j|, |y_i - y_j| \le span_{\max}$. For example, Figure 7.19 shows the corresponding graph G for the set of aligned paired-end reads in Figure 7.18.

Let $\{PE_{k_1}, PE_{k_2}, \ldots, PE_{k_m}\}$ be a set of discordant alignments crossing an SV breakpoint. For any pair (PE_{k_i}, PE_{k_j}), we expect $sign(x_{k_i}) = sign(x_{k_j})$, $sign(y_{k_i}) = sign(y_{k_j})$ and $|x_{k_i} - x_{k_j}|, |y_{k_i} - y_{k_j}| \le span_{\max}$. Hence, by definition, (PE_{k_i}, PE_{k_j}) is an edge in G. In other words, $\{PE_{k_1}, PE_{k_2}, \ldots, PE_{k_m}\}$ is expected to form a clique in G. By this observation, the SV calling problem can be transformed into a clique-finding problem.

For example, in the graph G of Figure 7.19, there are five maximal cliques: a 3-clique for $\{PE_1, PE_2, PE_3\}$, a 3-clique for $\{PE_4, PE_5, PE_6\}$, a 3-clique for

---

Algorithm DELLY_CliqueFinding

Require: A graph G where each edge (PE_i, PE_j) is assigned a weight, which is the absolute different of the alignment positions of PE_i and PE_j.

Ensure: A set of maximal cliques, each represents an SV

1: **while** edges exist in G **do**
2:     Find an edge (PE_i, PE_j) in G of the minimum weight;
3:     Let $M = \{PE_i, PE_j\}$
4:     **repeat**
5:         Find $PE_k \notin M$ such that (1) $\{(PE_k, PE_h) \mid PE_h \in M\}$ are edges in G and (2) $\min_{PE_l \in M} w(PE_k, PE_h)$ is minimum;
6:         Set $M = M \cup \{PE_k\}$;
7:     **until** PE_k not exists;
8:     Report the clique for M;
9:     Update G removing every edge whose either end is in M;
10: **end while**

FIGURE 7.20: The algorithm DELL_CliqueFinding.

$\{PE_7, PE_8, PE_9\}$, a 4-clique for $\{PE_{10}, PE_{11}, PE_{12}, PE_{13}\}$, and a 2-clique for $\{PE_9, PE_{10}\}$.

However, finding maximal cliques is NP-complete. Here, we present a heuristics solution from DELLY [241]. The heuristic assumes that two paired-end reads PE_i and PE_j are more likely to be in the same cluster if their alignments are close.

For each edge (PE_i, PE_j), its distance $w(PE_i, PE_j)$ is defined as $|x_i - x_j| + |y_i - y_j|$. For example, for PE_4 and PE_6 in Figure 7.18, $w(PE_4, PE_6) = (x_6 - x_4) + (y_6 - y_4)$. DELLY identifies the edge (PE_i, PE_j) in G with the minimum weight and puts it in one cluster M. Then, iteratively, the algorithm finds a paired-end read PE_k such that (1) $M \cup \{PE_k\}$ forms a clique and (2) the distance $\min_{PE_l \in M} w(PE_k, PE_l)$ is minimized. The process is repeated until the cluster forms a maximal clique. The detail of the algorithm is stated in Figure 7.20.

For the example graph in Figure 7.19, we obtain 4 cliques: $\{PE_1, PE_2, PE_3\}$, $\{PE_4, PE_5, PE_6\}$, $\{PE_7, PE_8, PE_9\}$, and $\{PE_{10}, PE_{11}, PE_{12}, PE_{13}\}$. The 2-clique for $\{PE_9, PE_{10}\}$ is not detected since $w(PE_9, PE_{10})$ is longer than the weights of any pairs in the 3-clique $\{PE_7, PE_8, PE_9\}$ and the 4-clique $\{PE_{10}, PE_{11}, PE_{12}, PE_{13}\}$.

7.8.1.2   Confidence interval overlapping approach

Consider any SV whose breakpoint pair is (a, b). For any discordant alignment PE_i that supports the SV, let the alignment positions of the two reads be x_i and y_i while their alignment orientations are $sign(x_i)$ and $sign(y_i)$, re-

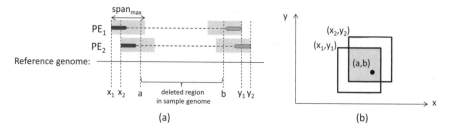

(a)   (b)

FIGURE 7.21: (a) shows two discordant alignments (x_1, y_1) and (x_2, y_2) on the reference genome that potentially captures a deletion between a and b. a and b are unknown. Moreover, by Equation 7.2, we expect the two breakpoints a and b should be within $span_{max}$ from x_i and y_i, respectively. The feasible intervals for the breakpoints a and b are represented by the four transparent boxes, each of length $span_{max}$. (b) shows two squares that correspond to the two discordant alignments (x_1, y_1) and (x_2, y_2).

spectively. Since the insert size is at most $span_{max}$, breakpoint a is expected to be between x_i and $x_i + sign(x_i)span_{max}$ while breakpoint b is expected to be between y_i and $y_i + sign(y_i)span_{max}$ (see Figure 7.21(a)). Hence, the following equation should be satisfied:

$$0 \le sign(x_i)(a - x_i), sign(y_i)(b - y_i) \le span_{max}. \qquad (7.2)$$

This equation can be interpreted from a geometric point of view. All breakpoint pairs (a, b) that satisfy Equation 7.2 form a square on the 2D plane (see Figure 7.21(b)). This square is of length $span_{max}$ and it is called the breakpoint region of (x_i, y_i).

Each discordant alignment corresponds to a breakpoint region. If a set of breakpoint regions overlap, they are expected to support the same SV. For example, Figure 7.21(b) shows two overlapping squares. Their corresponding discordant alignments PE_1 and PE_2 (see Figure 7.21(a)) support the deletion with breakpoint pair (a, b). Hence, SVs can be found by identifying all maximal overlapping breakpoint regions (i.e., squares).

Naively, maximal overlapping squares can be found by enumerating all subsets of squares and checks if they overlap. However, this approach takes $O(2^n)$ time, where n is the number of squares.

This problem can be solved efficiently by the plane-sweeping algorithm. The input is a set of n squares $\{S_1, \ldots, S_n\}$ such that the ith square S_i is represented by its top-left corner (x_i, y_i) and its bottom-right corner $(x_i + span_{max}, y_i - span_{max})$. Without lost of generality, we assume the squares are sorted by x_i (i.e., $x_1 \le x_2 \le \ldots \le x_n$). The plane-sweeping algorithm moves a (conceptual) vertical line from left to right crossing all squares. During the movement of the vertical line, we maintain the set of squares $\{S_{j_1}, \ldots S_{j_h}\}$ overlapping with the line by storing their y-coordinates $\{y_{j_1}, \ldots, y_{j_h}\}$ in a binary search tree (BST) T. Initially, no square overlaps with the vertical line and T is an empty tree. When the line passes through the leftmost and the rightmost ends of every square S_i, the algorithm will insert and delete, respectively, y_i from T. Just before we delete y_i from T, the algorithm reports all maximal lists of squares overlapping with S_i.

Precisely, the algorithm is as follows. For each square S_i, denote i_{start} and i_{end} as the events where the vertical line is at the leftmost and the rightmost position of S_i, respectively (i.e., i_{start} is the event where $x = x_i$ while i_{end} is the event where $x = x_i + span_{max}$). The events i_{start} and i_{end}, for $i = 1, \ldots, n$, are sorted in increasing order of the x-coordinates of the corresponding vertical lines. The processing steps for both the start and end events of the square S_i are as follows.

- For each start event i_{start}, the vertical line starts to overlap with the square S_i and we include y_i in the binary search tree (BST) T.

- For each end event i_{end}, we identify all squares overlapping with S_i by querying the BST T. Precisely, from BST T, we identify squares $S_{j_1}, \ldots, S_{j_h}$ such that $y_i - span_{max} \leq y_{j_1} \leq \cdots \leq y_{j_h} \leq y_i + span_{max}$. Then, we identify and report all maximal intervals $j_\ell..j_{\ell+\Delta}$ such that $y_{j_{\ell+\Delta}} - y_{j_\ell} \leq span_{max}$. After reporting all maximal intervals overlapping with S_i, we remove y_i from T.

The detail of the algorithm is presented in Figure 7.23. It runs in $O(n \log n + t)$ time, where t is the number of maximal lists of overlapping squares.

Figure 7.22 gives an example to illustrate the algorithm. The algorithm handles the events in the following order: $1_{start}, 2_{start}, 3_{start}, 4_{start}, 1_{end}, 5_{start}, 6_{start}, 7_{start}, 8_{start}, 2_{end}, 3_{end}, 4_{end}, 5_{end}, 6_{end}, 7_{end}, 8_{end}$. For events $1_{start}, 2_{start}, 3_{start}, 4_{start}$, we insert y_1, y_2, y_3, y_4 into the BST T and they are ranked by their y-coordinates: $y_1 < y_4 < y_2 < y_3$. To process 1_{end}, we obtain the list of squares overlapping with S_1 from T. The list of squares overlapping with S_1 is S_1 and S_4. Then, we report the maximal set of overlapping squares is $\{S_1, S_4\}$. After that, we remove y_1 from T. Then, we process events $5_{start}, 6_{start}, 7_{start}, 8_{start}$ and insert y_5, y_6, y_7, y_8 to T. The list of y-coordinates in T is $y_8 < y_4 < y_6 < y_2 < y_3 < y_5 < y_7$. To process 2_{end}, we obtain all squares in T overlap with S_2, that is, $\{S_4, S_6, S_2, S_3, S_5, S_7\}$. Then, we try to identify all maximal sets of overlapping squares. For y_4, y_2 is the biggest y-coordinate such that $y_2 \leq y_4 + span_{max}$. We report $\{S_4, S_6, S_2\}$. For y_6, y_3 is the biggest y-coordinate such that $y_3 \leq y_6 + span_{max}$; so, we report $\{S_6, S_2, S_3\}$. For y_2, y_7 is the biggest y-coordinate such that $y_7 \leq y_2 + span_{max}$; so, we report $\{S_2, S_3, S_5, S_7\}$. In a similar way, we can process the rest of the events.

Overlapping squares does not use the constraint that the insert size is at least $span_{min}$. GASV [276] suggested using this information. We first give the key observation. For any paired-end read whose endpoint pair is (x_i, y_i), if it

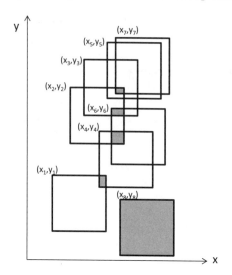

FIGURE 7.22: This figure shows a set of 8 squares $\{S_1, \ldots, S_8\}$ on a plane, where S_i is a length-$span_{max}$ square whose top-left corner is (x_i, y_i). The squares are ordered in increasing order of x_i. The maximal sets of overlapping squares are $\{S_1, S_4\}, \{S_2, S_4, S_6\}, \{S_2, S_3, S_6\}, \{S_2, S_3, S_5, S_7\}, \{S_8\}$. They are highlighted in gray.

Algorithm OverlappingSquare

Require: $\{(x_1, y_1), \ldots, (x_n, y_n)\}$ is a set of top-left corners of squares of size $span_{max}$. (Assume $x_1 \leq x_2 \leq \ldots \leq x_n$.)

Ensure: Report all maximal sets of overlapping squares

1: Create two events i_{start} and i_{end} for each square S_i;
2: Set T be an empty binary search tree;
3: Mark $S_1, \ldots, S_n$ as unused;
4: **for** each event i_ϵ sorted in left-to-right order of the corresponding vertical line **do**
5:   **if** $\epsilon = start$ **then**
6:     Insert y_i into T;
7:   **else**
8:     From T, we identify $y_{j_1}, \ldots, y_{j_h}$ such that $y_i - span_{max} \leq y_{j_1} \leq \ldots \leq y_{j_h} \leq y_i + span_{max}$;
9:     Report all maximal sets $\{j_k, \ldots, j_{k'}\}$ such that $y_{j_{k'}} - y_{j_k} \leq span_{max}$ such that at least one of the square in $S_{j_k}, \ldots, S_{j_{k'}}$ is not used;
10:    Mark $S_{j_1}, \ldots, S_{j_h}$ as used;
11:    Remove y_i from T;
12: **end if**
13: **end for**

FIGURE 7.23: The algorithm OverlappingSquare.

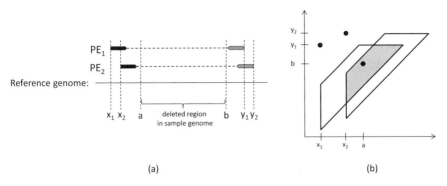

(a)   (b)

FIGURE 7.24: (a) a and b are the breakpoints of a deletion event. For $i = 1, 2$, (x_i, y_i) is the endpoint pair of each paired-end read that crosses this SV. (b) For $i = 1, 2$, (x_i, y_i) can be represented as a dot on the 2-D plane. For each (x_i, y_i), the set of (a, b) that satisfies the constraint in Equation 7.3 corresponds to a trapezoid, which is called the breakpoint region.

corresponds to an SV, the breakpoints a and b of the SV (see Figure 7.24(a)) should satisfy the insert size constraint, which is stated in the following equation,

$$span_{\min} \leq sign(x_i)(a - x_i) + sign(y_i)(b - y_i) \leq span_{\max} \qquad (7.3)$$

where $sign(x_i)$ and $sign(y_i)$ are $+1$ if the read is aligned on a positive strand; and -1, otherwise.

Equation 7.3 can be interpreted from a geometric point of view. Below, we just discuss the case where $sign(x_i)sign(y_i) = -1$. Without lost of generality, we assume $sign(x_i) = +1$ and $sign(y_i) = -1$. (If $sign(x_i) = -1$ and $sign(y_i) = +1$, we swap x_i and y_i.) Each paired-end read whose endpoint pair is (x_i, y_i) can be represented as a point on the 2D plane. Then, all (a, b) pairs that satisfy Equation 7.3 form a trapezoid (see Figure 7.24(b)). This trapezoid is called the breakpoint region B_i of (x_i, y_i). Precisely, the trapezoid is formed by a horizontal line $y = y_i$, a vertical line $x = x_i$ and two parallel lines $span_{\min} = (x - x_i) - (y - y_i)$ and $span_{\max} = (x - x_i) - (y - y_i)$. For paired-end reads crossing the same SV, their breakpoint regions should be overlapping.

Our aim is to find all maximal subsets of overlapping trapezoids. Similar to finding overlapping squares, this problem can be solved efficiently using the idea of plane sweeping (see Exercise 7).

7.8.1.3 Set cover approach

Up until now, the clustering algorithms discussed assume every paired-end read has a unique alignment. Moreover, some paired-end reads may have multiple alignments. If we don't use paired-end reads with multiple alignments, we may miss some SVs. For example, for Figure 7.25(a), PE_1 and PE_5 are multiply mapped on the reference genome. There are 4 possible SVs:

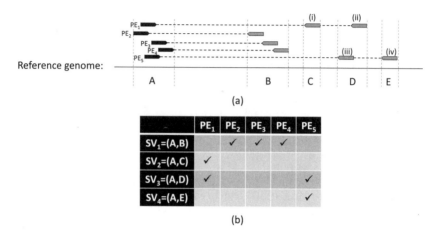

(a)

	PE₁	PE₂	PE₃	PE₄	PE₅
SV₁=(A,B)		✓	✓	✓	
SV₂=(A,C)	✓				
SV₃=(A,D)	✓				✓
SV₄=(A,E)					✓

(b)

FIGURE 7.25: (a) gives an example mapping of 5 paired-end reads on the reference genome. For the paired-end reads PE_1 and PE_5, the black end has a unique alignment while the gray end has two possible alignments at locations. The remaining 3 paired-end reads have unique alignments. (b) is a table where the ith row shows the paired-end reads that support SV_i, i.e., $a(SV_i)$. Note that $a(SV_1) \cup a(SV_3) = \{PE_1, PE_2, PE_3, PE_4, PE_5\}$.

$SV_1 = (A, B)$, $SV_2 = (A, C)$, $SV_3 = (A, D)$ and $SV_4 = (A, E)$. If we only use uniquely mapped reads, we can only discover SV_1. To discover SV_3, we need to use the alignment (ii) of PE_1 and the alignment (iii) of PE_5.

To utilize the information from these multiply aligned paired-end reads, VariationHunter [106] formulated the paired-end read clustering problem as the Maximum Parsimony Structural Variation (MPSV) problem. MPSV is based on the parsimony principle. Consider a set of paired-end reads, where each paired-end read may map to multiple positions. The MPSV problem aims to assign a unique mapping position for each paired-end read such that the number of SVs is minimized. Hormozdiari et al. [105] showed that this problem is NP-hard.

More precisely, the MPSV problem is defined as follows. It requires 3 inputs. First, it requires a set of paired-end reads $\mathcal{PE} = \{PE_1, \ldots, PE_n\}$, where some paired-end reads have multiple alignments. Second, it requires a set of candidate SVs $\mathcal{SV} = \{SV_1, \ldots, SV_m\}$. $\mathcal{SV}$ is computed from all multiple alignments using any algorithm in Sections 7.8.1.1 and 7.8.1.2. Third, it requires, for each SV_j, a set $a(SV_j) \subseteq \mathcal{PE}$ such that all paired-end reads in $a(SV_j)$ support SV_j. The problem aims to find the smallest subset of $\mathcal{SV}$, say $\{SV_{j_1}, \ldots, SV_{j_k}\}$, such that $\bigcup_{p=1}^{k} a(SV_{j_p}) = \mathcal{PE}$.

For the example in Figure 7.25(a), we have $\mathcal{PE} = \{PE_1, \ldots, PE_5\}$. Using the algorithm DELLY_CliqueFinding (see Figure 7.20) over all multiple alignments of $\mathcal{PE}$, we have $\mathcal{SV} = \{SV_1 = (A, B), SV_2 = (A, C), SV_3 = $

Algorithm VariationHunter-SetCover

Require: A set of n paired-end reads $\mathcal{PE}$ and a set of m SVs $\{SV_1, \ldots, SV_m\}$ such that $a(SV_i)$ is a subset of $\mathcal{PE}$ that supports SV_i for $i = 1, \ldots, m$

Ensure: A set $\{j_1, \ldots, j_k\}$ such that $a(SV_{j_1}) \cup \ldots \cup a(SV_{j_k}) = \mathcal{PE}$

1: $ANS = \emptyset$;
2: **while** $\mathcal{PE}$ is not empty **do**
3:   Identify j that maximizes the number of paired-end reads in $a(SV_j)$ that are supported by $\mathcal{PE}$;
4:   Set $ANS = ANS \cup \{j\}$;
5:   Set $\mathcal{PE} = \mathcal{PE} - a(SV_j)$;
6: **end while**
7: Report ANS;

FIGURE 7.26: An $O(\log m)$-approximation algorithm VariationHunter-SetCover for computing the minimum subset of SVs that are covered by the paired-end reads in $\mathcal{PE}$.

$(A, D), SV_4 = (A, E)\}$, $a(SV_1) = \{PE_2, PE_3, PE_4\}$, $a(SV_2) = \{PE_1\}$, $a(SV_3) = \{PE_1, PE_5\}$, and $a(SV_4) = \{PE_5\}$. The MPSV problem reports $\{SV_1, SV_3\}$ since $a(SV_1) \cup a(SV_3) = \mathcal{PE}$ (see Figure 7.25(b)). In other words, for the two multiply aligned reads PE_1 and PE_5, we use alignment (ii) of PE_1 and alignment (iii) of PE_5.

As the MPSV problem is NP-hard, we don't have an efficient algorithm that minimizes the number of SVs. Moreover, this problem is similar to the set cover problem, which has a simple greedy algorithm that ensures an approximation factor of $O(\log m)$. The algorithm iteratively selects some set $a(SV_j)$ until all paired-end reads in $\mathcal{PE}$ are covered. In the pth iteration, the algorithm picks the set $a(SV_{j_p})$ that includes the maximum number of uncovered paired-end reads in $\mathcal{PE}$. Figure 7.26 gives the detail of the algorithm.

The above solution does not consider the confidence of each alignment. Intuitively, a unique alignment should be more trustable than a non-unique alignment. To capture this idea, if a paired-end read has N multiple alignments, we can assign each alignment a confidence weight of $1/N$. Then, for Step 3 of Figure 7.26, instead of maximizing the number of paired-end reads in $a(SV_j)$ that are supported by $\mathcal{PE}$, we maximize the total weight of paired-end reads in $a(SV_j)$ that are supported by $\mathcal{PE}$. Figure 7.27 illustrates the idea. Without weightage, the algorithm reports $\{(A, B), (A, C), (A, D), (A, E)\}$. Moreover, our weightage scheme will assign a weight of $\frac{1}{2}$ for the alignments (i)−(viii) while the remaining alignments have weight 1. Then, the algorithm will report $\{(A, B), (A, C), (A, D)\}$.

The weighting idea is used in a number of methods, including Variation-Hunter [106], HYDRA [236] and Meerkat [325].

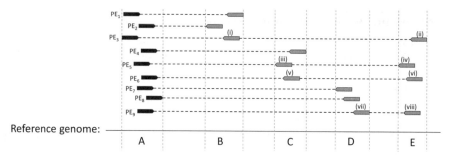

FIGURE 7.27: This example shows the mapping of 5 paired-end reads on the reference genome. For the paired-end reads PE_3 and PE_5, the black end has a unique alignment while the gray end has two possible alignments.

7.8.1.4   Performance of the clustering approach

Based on the evaluation from simulated and real experiments, we observe that the clustering approach is sensitive when the inert size is big. However, the clustering approach is not good for detecting SVs that are short (say, smaller than $span_{max} - span_{min}$.

Another issue is that the clustering approach gives a lot of false positives. It also cannot report the exact breakpoints of the structural variations.

7.8.2   Split-mapping approach

Clustering methods only utilize discordant alignments to identify SVs. However, when the insert size is small, the number of discordant alignments will be small. For example, consider a set of 2×100 bp paired-end reads of insert size 300 bp. Among the paired-end reads covering an SV breakpoint, we expect 66% $(= \frac{2*100}{300})$ of them are paired-end reads with a split read. These paired-end reads may be mapped as concordant alignments with a soft-clip. Then, we only have 33% discordant alignments for identifying the SV. Figure 7.28 illustrates this issue. When these 6 paired-end reads are mapped on the reference genome, only two paired-end reads are confirmed to be discordantly aligned and they enable the clustering methods to locate the SV junction. The remaining 4 paired-end reads may be mapped as concordant alignments with a soft-clip. Hence, instead of using discordant alignments, it may be better to use soft-clip reads to improve SV calling. These methods are called split-mapping methods.

A number of SV callers use split-mapping approach. They include SRiC [332], CREST [305] and Socrates [267]. In particular, this section will detail CREST.

CREST takes advantage of soft-clipping reads for identifying SVs. Recall that a soft-clip of a read refers to its unaligned portion (see Figure 7.16). A soft-clip can be caused by sequencing errors, chimeric reads or errors in

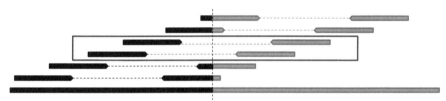

FIGURE 7.28: Consider a region in the sample genome which is formed by connecting two regions in the reference genome. The figure shows a list of six paired-end reads crossing the breakpoint boundary. Only two paired-end reads do not map on the breakpoint. The remaining 4 paired-end reads are called split reads.

the reference genome. Moreover, a soft-clip can also be caused by structural variation where the unaligned portion is actually coming from another part of the genome. When the soft-clips are long enough (say, ≥ 20), we can align the soft-clip portions of the reads to determine the SVs.

Figure 7.29 illustrates the idea of CREST. CREST first identifies a position X where the soft-clip reads are localized (see Figure 7.29(a)). X is the first putative breakpoint of the SV. Then, CREST assembles a contig by piling up the soft-clip portions of the soft-clip reads aligned on position X. Using BLAT [134], CREST tries to align the assembled contig to the reference genome. If the assembled contig can be uniquely aligned to another position Y (see Figure 7.29(b)), we identify another putative breakpoint Y of the SV. To confirm this, CREST extracts the soft-clip portions of the reads at position Y (see Figure 7.29(c)) and assembles them into a contig. If the contig can be uniquely aligned to position X by BLAT (see Figure 7.29(d)), we conclude an SV connecting positions X and Y (see Figure 7.29(e)).

The main advantage of the split read methods is that they can identify the exact breakpoint.

7.8.3 Assembly approach

Paired-end reads may be too short for locating SVs. To resolve this problem, one solution is to assemble paired-end reads that cannot be concordantly aligned on the reference genome into contigs; then, by aligning the contigs, SVs can be identified. This approach is known as the assembly approach. Since contigs are longer than the reads, we expect that the assembly approach improves SV calling near repeats. Furthermore, the assembly approach reconstructs the exact breakpoint sequence. It enables us to detect complex events like inversion occurring together with deletions. It is also useful to recover the novel insertion sequence at the breakpoints (like the microhomology and non-template sequences).

Methods that use the assembly approach include NovelSeq [94] and TIGRA [42].

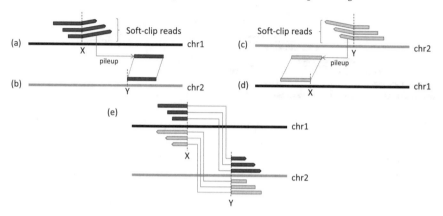

FIGURE 7.29: This is an illustration of the steps in CREST. (a) Identify a set of soft-clip reads co-localized at position X in chromosome 1. (b) Assemble a contig by piling up the soft-clip portions of reads at position X; then, by BLAT, the contig is aligned to location Y in chromosome 2. (c) Identify a set of soft-clip reads that locate around position Y. (d) Assemble a contig by piling up the soft-clip portions of reads at position Y; then, by BLAT, the contig aligns to location X in chromosome 1. (e) Hence, we conclude that there exists an SV connecting positions X and Y.

A number of methods use a de novo assembler to reconstruct the contigs. For example, TIGRA [42] uses a specifically designed assembler to reconstruct the SV breakpoint sequence. Another example is NovelSeq [94]. NovelSeq identifies orphan paired-end reads (i.e., unmappable reads) and then uses a de novo assembler like Euler-SR [38] or ABySS [275] to assemble them into longer contigs. For one-end anchored paired-end reads, NovelSeq developed a local assembler called the micro-read Strand-Aware Assembly Builder to construct the contigs. By aligning these assembled contigs to the reference genome, NovelSeq discovers some SVs.

Some SV callers obtain the contigs by the pile up approach. For example, CREST [305] and Socrates [267] pile up the discordant paired-end reads to form the consensus sequence using voting. Compared with de novo assembly, the pile-up approach is computationally more efficient since it reuses existing alignment information to get the relative positions of the reads.

7.8.4 Hybrid approach

So far the approaches discussed do not fully utilize information from all alignments. The clustering approach uses discordantly aligned reads only. The split read approach uses soft-clip reads only. The assembly approach does not use the read depth information. To fully utilize the information from all reads, a hybrid approach is proposed.

There are a few ways to integrate information. The first way is to use multiple SV callers to integrate the information of all alignments. Examples include SVMerge [318] and MetaSV [204]. SVMerge integrates the SV predictions from BreakDancer [43], Pindel [327], RDXplorer [328], SECluster (unpublished) and RetroSeq [132]. MetaSv integrates the SV predictions from BreakSeq [150], BreakDancer [43], Pindel [327] and CNVnator [2].

Although integrating multiple SV callers can improve sensitivity, each SV caller uses information from one read type (either discordant aligned, soft-clip, one-end anchored reads or read depth). Another choice is to develop an SV single caller that utilizes information from multiple read types. Examples include PRISM [123], LUMPY [157], Meerkat [325] and SoftSV [17].

SoftSV uses different read types to detect different SVs. To detect large SVs, SoftSV uses the clustering approach. Although the clustering approach is good to detect relatively large SVs, it is not effective to detect small SVs (i.e., inversions/tandem duplications of size smaller than the read length and deletions of size smaller than 4σ, where σ is the standard derivation of the insert size). For small SVs, SoftSV uses a split-read approach locally, that is, it aligns the soft-clip portions of reads to the region within 4σ.

PRISM first uses a clustering approach to identify candidate SVs. Then, if there are soft-clip reads in some clusters of discordant alignments, a modified Needleman-Wunsch algorithm is used to align these soft-clip reads as split reads.

Although SoftSV and PRISM use information from different read types, information is still used separately. LUMPY provides a general probabilistic framework to integrate the information of read depth, discordant alignment and split read as the SV breakpoint signal in one step. This allows it to obtain a sufficient SV breakpoint signal more easily. (For example, if an SV is supported by just one discordant alignment and one soft-clip alignment, then the signal is insufficient for each read type. Moreover, for LUMPY, the signal is stronger since it integrates all signals in one step.)

7.9 Verify the SVs

After we call the set of candidate SVs, the next step is to verify them. Different methods use different verification methods. They can be classified as:

- validation by the unused paired-end read information,

- validation by assembling the paired-end reads,

- statistical analysis for filtering the low-confidence SVs, and

- verification using biological features.

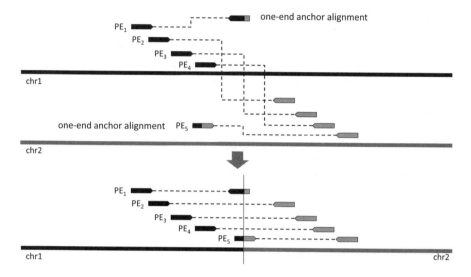

FIGURE 7.30: In the top panel, PE_2, PE_3, PE_4 form a discordant cluster. PE_1 and PE_5 are one-end anchor alignments and they are not used for discovering the candidate SV. In the bottom panel, the validation step tries to align PE_1 and PE_5 to the SV breakpoint. Since both PE_1 and PE_2 can be aligned consistently as a split read, the candidate SV is verified.

Validation by the unused paired-end read information: Note that when we call the candidate SVs, the method may not use all information. The unused information can be utilized to verify the candidate SVs. For example, DELLY [241] is a clustering method. During the clustering step, it does not use the one-end anchored alignments and the concordant alignments with a soft-clip. After the candidate SVs are discovered, these reads are used to verify the SV breakpoints (see Figure 7.30 for an example).

Validation by assembling the paired-end reads: Some methods verify a candidate SV by examining if the paired-end reads in the SV region can be assembled and aligned to the SV region. For HYDRA [236], after the candidate SV is identified by the clustering approach, all reads are assembled and realigned on the reference genome. In SVMerge [318], given a candidate SV, SVMerge extracts all paired-end reads mapped around the breakpoint region of the candidate SV; then, by a de novo assembler like ABySS [275] or Velvet [329], these paired-end reads are assembled into contigs. If the assembled contig can be aligned to the candidate SV, the SV is accepted; otherwise, SVMerge will assume the candidate SV is a false positive. MetaSV [204] and TIGRA [42] use local assembly to improve breakpoint resolution. MetaSV uses SPAdes [13] while TIGRA uses a tailor-made local assembler for constructing the contig.

Statistical analysis for filtering the low-confidence SVs: Some

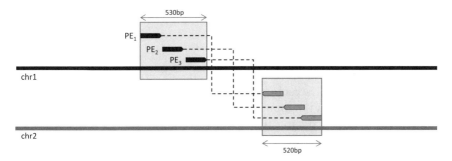

FIGURE 7.31: An example cluster with 3 paired-end reads and the anchor size is $s = 530 + 520 = 1050$. This cluster is a different chromosome cluster. Assume the genome size is $n = 3 \times 10^9$ and the number of different chromosome paired-end reads is $C_t = 10^6$. Then, $c_t \sim Poisson(\lambda_t)$ where $\lambda_t = \frac{sC_t}{n} = 0.35$. Hence, the p-value is $Pr(c_t \geq 3) = 0.005509$.

methods rank the candidate SVs. Then, using a cutoff threshold, a list of high-confidence potential SVs are selected. Examples include PRISM [123], LUMPY [157], Socrates [267], BreakDancer [43], and CREST [305]. A simple score is to rank each cluster by the number of discordant alignments (or split reads) supporting the clusters. We can also estimate the confidence of each cluster. In BreakDancer, a p-value is estimated for each cluster based on a Poisson model. The Poisson model uses the number of paired-end read alignments in the cluster, the size of the two anchor regions and the coverage of the genome. Precisely, let c_t be the random variable of the number of paired-end reads of type t in a cluster. Under the null hypothesis, c_t follows $Poisson(\lambda_t)$ where $\lambda_t = \frac{sC_t}{n}$ and n is the size of the genome, C_t is the total number of paired-end reads of type t in the whole dataset, and s is the sum of the size of the two anchors of all clusters. If k paired-end reads of type t are observed in the cluster, the p-value is defined to be $Pr(c_t \geq k)$. Figure 7.31 gives an example to illustrate the p-value computation. Using a p-value cutoff, we can predict a list of significant structural variations.

Verification using biological features: It is known that SVs predicted near certain genomic regions, like repeat regions, telomere and centromere, are likely to be noisy. Some SV callers verify candidate SVs by these biological features. Examples include FusionSeq [262] and FACTERA [217].

In terms of somatic SV detections in cancer study, the SV verification step will examine the candidate SVs called from both the tumor and the matched normal samples. If a candidate SV appears in the tumor sample while we cannot find any supporting paired-end read for the same candidate SV in the normal sample, such SV is deemed to be a somatic SV.

7.10 Further reading

Although a lot of work has been done on structural variation calling, it is still difficult to call structural variations in repetitive regions due to the ambiguity of short read mapping. In terms of SV types, long insert and complex SVs are still difficult to call since the reads are too short to capture such events.

Sample purity and heterogeneity also affect SV calling. The cancer tumor sample may be contaminated by normal tissue. This reduces the purity of the sample. Due to tumor progression, the tumor sample contains multiple sub-clones, which contribute to heterogeneity [326]. These two factors reduce the number of reads supporting the somatic SVs and somatic CNVs.

This chapter does not cover all methods for calling CNVs and SVs. There exist methods that use information in addition to read depth for CNV calling. For example, CNVer [198] also uses the paired-end reads to determine the boundaries of the CNVs. We also did not cover CNV callers that are tailor-made for whole exome sequencing like EXCAVATOR [233] and ExomeCNV [261].

For SV calling, we did not cover mapping-free SV callers. SMUFIN [205] is one of the mapping-free methods, that uses quaternary tree to compare sequences from two samples (e.g., tumor versus adjacent normal), grouping the reads based on the differences between the normal and tumor reads to identify somatic SVs.

There are methods that can analyze CNVs and SVs for multiple samples. For CNVs, the methods include SegSeq [44], CNV-seq [320], rSW-seq [142] and CNAseg [114]. For SVs, the methods include LUMPY [157], Hydra-Multi [180], etc.

Due to the advance in third-generation sequencing, new methods will be available soon that use long reads to detect SVs. In general, these methods use the split-mapping approach (Section 7.8.2) and the assembly approach (Section 7.8.3) to call SVs.

7.11 Exercises

1. Suppose you sequence a sample using the Illumina sequencer. Consider three regions with the same copy number. Suppose the GC percentage of these regions are 20%, 50% and 80%, respectively. Let A, B and C be the numbers of aligned reads in these three regions, respectively. Can you predict which region has the highest number of aligned raw read? (Hint: Read Section 1.4.3.)

2. To estimate the CNVs, we need to count the number of aligned reads in each region. Some reads may have multiple alignments. There are different ways to handle them. Can you state the advantage or potential problems for the following ways of handling the multiply aligned reads?

   - Ignore all multiply aligned reads.

   - For each multiply aligned read, we always keep the mapping location with the smallest coordinate.

   - For each multiply aligned read, we randomly keep one mapping location.

3. Assume that a set of reads are aligned on a genome, that is partitioned into 10 bins. Suppose the normalized read counts of these 10 bins are $A_1 = 20, A_2 = 19, A_3 = 21, A_4 = 31, A_5 = 28, A_6 = 30, A_7 = 29, A_8 = 10, A_9 = 9, A_{10} = 11$. The circular binary segmentation (CBS) method needs to find the t-statistic of all consecutive intervals. Can you determine the t-statistic of all intervals? Which interval has the highest t-statistic?

4. Segmentation algorithms aim to partition the genome into segments such that the normalized read counts in each segment have similar values. Here, we modify the circular binary segmentation (CBS) algorithm and obtain a new segmentation algorithm, CBS2.

   Algorithm CBS2
   Require: A list of values $A_p, \ldots, A_q$
   Ensure: A set of segments
   1: Find the interval $i..j$ that maximizes the absolute T-statistics $T_{i,j}$;
   2: Report $i..j$ as a segment;
   3: Call CBS2($A_p, \ldots, A_{i-1}$) and CBS2($A_{j+1}, \ldots, A_q$);

   Is CBS a good algorithm for segmentation? Why?

5. Equation 7.1 estimates the CNV of a segment by the median. Why should we use the median? Can we replace the median with the mean? Can you give an example to illustrate the problem?

6. Consider any discordant alignment whose two reads are aligned to positions x_i and y_i while their alignment orientations are $sign(x_i)$ and $sign(y_i)$, respectively. Suppose we define its breakpoint region to be any (a, b) satisfying $span_{\min} \leq sign(x_i)(a - x_i), sign(y_i)(b - y_i) \leq span_{\max}$. Can you propose an algorithm to find all the maximal overlapping breakpoint regions?

7. Please give an efficient algorithm to find all maximal overlapping trapezoids. (Hint: Use plane sweeping where the plane is parallel to the diagonal.)

8. CREST tries to pile up the soft-clip portions of all soft-clip reads to form a contig. Then, by BLAT, the contig is mapped to the whole genome to identify another breakend of the SV. Instead of building a contig, we suggested two alternative modifications to the algorithm.

   - We directly map the soft-clip portions of all soft-clip reads to the genome to identify another breakend of the SV.

   - We identify the longest soft-clip portion S among all soft-clip reads. Then, S is aligned on the genome to identify another breakend of the SV.

   Can you comment on the disadvantages of the above two approaches?

9. Suppose an SV exists that links two genomic regions $A[1..a]$ and $B[1..b]$. Precisely, there exist i and j such that $A[1..i]B[j..b]$ form the SV. Assume a read $R[1..m]$ exists that covers the SV breakpoint. We can identifying i and j that minimizes the edit distance between R and $A[x..i]B[j..y]$ among all $x \leq i$ and $y \geq j$. Can you propose a dynamic programming solution that identifies the breakpoints i and j? What is the running time?

10. You want to discover the SVs in one individual. To reduce sequencing cost, you can only sequence 1 million of 2×100 paired-end reads. You allow sequencing of paired-end reads of insert size (a) 300, (b) 1000 or (c) 10,000. Among the three insert sizes, which one should we choose? (You can assume there are no technical issues in sequencing paired-end reads of different insert sizes.)

11. You want to discover deletions of size 200 bp or above in one individual. You allow sequencing of 2×100 paired-end reads of insert size either (a) 300, (b) 1000 or (c) 10,000. Among the three insert sizes, which one should we choose?

Chapter 8

RNA-seq

8.1 Introduction

Transcription is the process that transcribes genes into transcripts, which is a sequence of RNA nucleotides. There are two types of transcripts: coding transcripts (mRNAs) and non-coding transcripts (ncRNAs). Coding transcripts are transcripts which encode proteins. Non-coding transcripts do not encode proteins. They function in the form of RNA. There are many different types of non-coding RNAs, including rRNAs, tRNAs, miRNAs, siRNA, lncRNAs, etc. The complete set of transcripts that presents in a given cell is called the transcriptome. In human and other multi-cell organisms, different cells express different transcriptomes even though they have the same genome.

The process of transcription is different in prokaryotes and eukaryotes. In prokaryotes, a transcript is an exact copy of each gene in the genome by replacing the DNA bases A, C, G and T in the gene into RNA bases A, C, G and U, respectively, in the transcript. In eukaryotes, the transcript is not an exact copy of the gene. The transcription involves two steps. First, similar to the prokaryote, a pre-mRNA is synthesized which is an exact copy of the gene. Then, splicing occurs during the process. Splicing excludes some RNA segments called introns from the pre-mRNA. The remaining RNA segments (called exons) are joined to form the mature mRNA (see Figure 8.1). The mature mRNA is then translated into a protein sequence.

In humans, the average length of a transcript is 2227 bp [140]. Each transcript consists of one or more exons. The average exon length is 235 bp and about 25% of exons are shorter than 100 bp. Moreover, the smallest exon size can be as small as 2 bp [257]. On the other hand, the lengths of introns vary a lot. In compact genomes like Arabidopsis and fruit fly, the intron sizes are relatively small. In human and mouse, the intron size spans between 50 bp and 100,000 bp. About 93% of introns are of length within [70, 20000]. An extreme example is the gene associated with the disease cystic fibrosis in humans. It has 24 introns of total length approximately 1 million bases, whereas the total length of its exons is only 1 kilo bases.

One gene may transcribe into multiple different transcripts by excluding different sets of introns through the splicing mechanism. This process is called alternative splicing. The different forms of a gene are called its isoforms. A

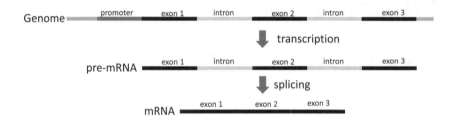

FIGURE 8.1: Illustration of transcription and splicing. First, through transcription, a gene is transcribed into a pre-mRNA. Then, the splicing mechanism cleaves two introns from the pre-mRNA. The remaining three exons are joined to form the mature mRNA.

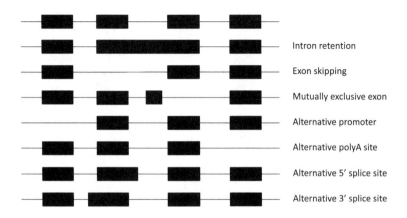

FIGURE 8.2: A list of possible alternative splicing events. The top sequence shows the primary transcript which has 4 exons. The following 7 sequences show possible alternative isoforms of the primary transcript.

genome-wide study showed that about 75% of human genes encode at least two isoforms [125]. Alternative splicing produces different isoforms of a gene by including or excluding different sets of exons. Figure 8.2 gives the list of basic splicing events. They include intron retention, exon skipping, mutually exclusive exon, alternative promoter, alternative polyA site, alternative 5' splice site and alternative 3' splice site. Among the six basic modes of alternative splicing, the most common splicing mode is exon skipping.

Understanding alternative splicing is important since different splice isoforms of a gene are expressed at different stages in cellular development and differentiation [128]. Furthermore, studies showed that dysregulation of alternative splicing events may lead to various human diseases [136, 55].

The splicing mechanism is mostly performed by the spliceosome. It recognizes and cleaves an intron sequence with 5' end (the donor site) equals the consensus motif GTRAGT and the 3' end (the acceptor site) equals YAG. (Note: R stands for A or G and Y stands for G or T.) This intron is called the U2-type intron. However, these motifs are extremely degenerated, leaving just GT − AG as fairly reliable splice sites, found in 98% of known human introns [285]. This type of splice junction is referred as the canonical splice site. For a small number of introns, spliceosomes recognize the splice sites GC − AG and AT − AC. Both GC − AG and GT − AT are referred as the semi-canonical splice sites. Apart from the canonical and the semi-canonical splice sites, non-spliceosomal splicing also occurs. In yeast, the protein Ire1p can splice an RNA. In metazoans, XBP1 is cleaved in a homologous manner. This gives the non-canonical splice sites CA − AG.

Apart from alternative splicing, another interesting transcript is the fusion gene. A fusion gene is formed by merging two different genes. Its formation is the result of structural variations like translocation, inversion and long deletion (see Chapter 7). Recently, it was found that many fusion genes are oncogenes (i.e., a gene that can transform a cell into a tumor cell). One famous example is the BCR-ABL1 fusion (see Figure 7.6). This fusion gene appears frequently in chronic myelogenous leukemia (CML).

In other words, to understand the transcriptome of a sample, we need to learn the expression profile and the isoforms of the transcriptomes. This chapter is devoted to methods that address these issues.

8.2 High-throughput methods to study the transcriptome

To study the expression or the isoforms of a single gene, we can use PCR-based methods. However, this approach cannot give us a global picture of the transcriptome. A number of methods have been proposed to study the transcriptome of a given sample on a genome-wide scale. Below, we give an overview of these methods.

First, the transcriptome can be studied by microarray. A microarray is an array of a set of probes. Each probe is designed so that it hybridizes to a specific transcript. By measuring the hybridization signals of all probes, we can obtain the expression levels of all transcripts. We can also design probes that are specific to the splice junctions. Hence, this approach can also detect splice variants. However, a microarray is biased due to the fact that different probes have different background hybridization noise. Also, a microarray cannot discover novel transcripts or novel alternative splice sites.

Novel transcripts can be discovered using a sequencing approach. One of the earliest sequencing solutions is the expressed sequence tag (EST). This solution clones all transcripts. Then, the 5' ends of the transcripts are sequenced using Sanger sequencing. By counting the number of tags specific to a transcript, we can measure its expression level. We can also discover the splice junctions by mapping the EST to the reference genome. However, it can only discover the splice junctions near the 5' end of the transcript. This approach is relatively expensive due to the high cost of Sanger sequencing.

A cheaper alternative is to sequence short tags like MPSS [31], serial analysis of gene expression (SAGE) indexSAGE[301, 256], CAGE [272] and GIS-ditag [219]. These methods sequence the 5' and 3' ends of all transcripts in the transcriptome. They enable us to measure the expression levels of all transcripts. Furthermore, these methods can detect alternative promoters and alternative polyA sites. However, they cannot detect alternative splicing within the body of the transcripts.

Both EST and short-tag sequencing are cloning-based approaches which are sensitive to cloning biases. RNA-seq [207] is a non-cloning-based sequencing approach. It randomly cleaves the transcriptome into fragments and sequences all fragments by high-throughput sequencing (see Figure 8.3). It is a cost-effective method. It has less noise and does not have the biases introduced by microarrays and does not have cloning biases. Furthermore, it provides a wider range of expression levels. It also can discover and quantify the expression level of transcripts, and provide information to detect allele-specific expression. It can also detect alternative splicing. Experimental validation showed that the RNA-seq protocol is reproducible and sensitive [191, 316]. It is also shown that RNA-seq provides a better estimation of absolute expression level than microarray [82].

Since the RNA fragments are reverse transcribed into cDNAs, the normal RNA-seq protocol cannot report the strand of the RNA fragments. To solve this problem, the strand-specific RNA-seq protocol [162] is developed. This technology enables us to discover anti-sense transcripts, which are common in higher eukaryotes.

8.3   Application of RNA-seq

Different types of cells have different transcriptomes. Hence, by comparing the transcriptomes of different cell types like embryonic stem cells and various adult stem cells, the process of cellular differentiation can be learned.

We can also learn more about cancer by studying the transcriptome. It is known that the transcriptome of the cancer cell is different from its normal form. We can actually use the expression of genes to subgroup cancer patients. For example, the expression of the gene PSA (prostate-specific antigen) is normally low in the healthy prostate. However, its expression is often elevated in prostate cancer. Alternative splicing and fusion genes are often connected with cancer. A few examples were described in [284].

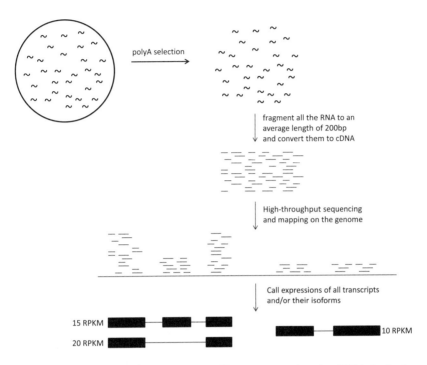

FIGURE 8.3: The RNA-seq process. From a sample, the RNA is first extracted and DNA contamination is removed by DNase. After that, various filters can be applied to remove the highly abundant ribosomal RNA (rRNA). (Note that rRNA typically constitutes over 90% of total RNA. Remove rRNA can enrich the transcripts we are interested in.) We can enrich mRNAs by rRNA depletion or polyA selection. rRNA depletion removes ribosomal RNA only from the sample. It is suitable if you want to sequence all possible RNA molecules. If you just want to analyze mRNA, you can use polyA selection. PolyA selection selects RNAs with polyA tails (as mRNA typically has polyA tails). Then, the extracted RNAs are fragmented into short fragments by magnesium-catalyzed hydrolysis. Afterward, the RNA fragments are reverse transcribed into cDNA by random priming. Sequencing adaptors are ligated and, optionally, the fragments are PCR amplified. Then, RNA fragments of a certain size are selected. All these selected cDNAs are subjected to high-throughput sequencing. The obtained reads are then mapped to the reference genome. Finally, expression and isoforms of all transcripts are called.

8.4   Computational Problems of RNA-seq

This chapter focuses on computational problems related to RNA-seq and we will study 3 computational problems.

- RNA-seq read mapping: Maps RNA-seq reads to a reference genome allowing splice junctions.

- Construction of isoforms: Reconstructs the isoforms for every gene in a transcriptome.

- Quantification of isoforms: Determines the expression level of the isoforms of every gene in a transcriptome.

8.5   RNA-seq read mapping

Due to the existence of introns, RNA-seq reads can be classified as exon reads and junction reads. Exon reads are reads that can be completely aligned within an exon. Junction reads are reads that cross the splice junctions. We could not directly align RNA-seq reads using short read aligners in Chapter 4 since the alignment between the junction read and the reference genome contains some big gaps. Although RNA-seq reads can be aligned by general aligners like BLAT [134] and BLAST [5], these methods are relatively slow.

This section discusses techniques for mapping RNA-seq reads. We first cover features that are used to align RNA-seq reads (see Section 8.5.1). These features include the transcript model and the splice junction signals. Then, we cover two alignment methodologies [86]: (1) the exon-first approach (see Section 8.5.2) and (2) the seed-and-extend approach (see Section 8.5.3).

8.5.1   Features used in RNA-seq read mapping

RNA-seq read mappers generally use two types of features. They are the transcript model and splice junction signals.

8.5.1.1   Transcript model

A transcript model is a list of known RNA transcripts in the literature. In the model, each RNA transcript is represented as a list of exons and splice junction boundaries. A number of methods require the transcript model. They include Cloonan et al. [52], Pan et al. [223], Sultan et al. [288], ERANGE [207], Xmate (old version called RNAMate [53]), RUM [90] and

---

Algorithm AlignWithTranscriptome

Require: A set of RNA-seq reads, the reference genome and the reference transcript model

Ensure: The alignment of each RNA-seq read

1: **for** each read R **do**
2:    Align R on the reference genome and the reference transcript model;
3:    **if** R is aligned to either the genome or the transcript model with little error **then**
4:       Report the alignment
5:    **else if** R is aligned to both the genome and the transcript model and they are consistent **then**
6:       Report the alignment
7:    **else if** R is aligned to both the genome and the transcript model but they are not consistent **then**
8:       Resolve the conflict and return the alignment;
9:    **else**
10:      Use some junction read alignment method to align R on the reference genome;
11:    **end if**
12: **end for**

FIGURE 8.4: The basic algorithm framework for aligning RNA-seq reads with a reference transcript model.

ALEXA-seq [91]. A few methods can optionally use the transcript model. They include TopHat [296, 140], STAR [68] and Mapnext [15].

Figure 8.4 gives the basic algorithm for aligning RNA-seq reads with the help of the transcript model. First, ungapped alignment is performed to align each read R to both the reference genome and the reference transcript model. Depending on the alignment result, we have a few cases.

- If R aligns to the genome only, we report the alignment of R.

- If R aligns to the transcript model only, R should be aligned to some known splice junction. We report the corresponding split alignment of R on the genome.

- If R aligns to both the genome and the transcript model and the alignments are consistent, R should be on some known exon and we report its alignment.

- If R aligns to both the genome and the transcript model but the alignments are not consistent, we need to resolve the conflict; then, we report the alignment.

- If R does not align to both the genome and the transcript model, R may

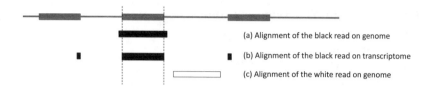

(a) Alignment of the black read on genome

(b) Alignment of the black read on transcriptome

(c) Alignment of the white read on genome

FIGURE 8.5: The gray line is the genome. The gray blocks are the transcript model. (a) is the alignment of the black read on the genome. (b) is the alignment of the black read on the transcript. Suppose the black read and the white read form a paired-end read. (c) is the alignment of the white read on the genome. The white read fails to align on any transcript and it is not shown in the figure.

span across some novel splice junction. We use a junction read alignment method to align R on the reference genome.

The above algorithm is mostly clear. The only case which is not clear is how to resolve the conflict when the alignments of R on the genome and the transcript model are not consistent. Here, we discuss the conflict resolution scheme of RUM [90].

Let A and A' be the alignments of R on the genome and the transcript model, respectively. If the corresponding loci of A and A' are different in the reference genome (i.e., they do not have sufficiently long overlap), RUM will consider this RNA-seq read as a non-unique alignment. If A and A' have a sufficiently long overlap, RUM tries to resolve ambiguities by either giving preference to the transcriptome alignment or by just reporting the overlapping region between the two alignments. For example, in Figure 8.5, the black read has two inconsistent alignments (a) and (b) when it is aligned on the genome and transcriptome, respectively. In this case, RUM will report the alignment in (b) since RUM gives preference to the transcriptome.

Moreover, for the case of a paired-end read, RUM may accept the alignment on the genome instead. For example, suppose the white read is the mate of the black read. Consider the bottom panel in Figure 8.5. Assume the white read maps in the intron between the junction of the black read as shown in Figure 8.5(c). Observe that the alignment (b) for the black read is inconsistent with the alignment for the white read in (c). We know that (b) is incorrect and RUM accepts the alignment in (a).

The transcript model helps improve accuracy. Occasionally, it may create bias and it prevents a read from aligning to a novel splice junction.

8.5.1.2 Splice junction signals

Another feature used by RNA-seq read aligners is the splice junction signal. As stated in the introduction, the 5' end and 3' end of every intron are the donor site and the acceptor site, respectively. They exhibit strong and conserved motifs. 98.12% of sites have the canonical `GT − AG` signal. Semi-canonical junctions occur relatively rarely with 0.76% sites having the `GC − AG` signal and 0.10% the `AT − AC` signal. Such information is used by the RNA-seq read mapper to improve the RNA-seq read alignment accuracy.

Since most of the junctions are canonical and semi-canonical, most RNA-seq aligners assume a fixed junction model and they align RNA-seq reads to these junctions only. Although this approach works well for humans, it may not work for species whose splice junction models are not known. Hence, learning-based methods are also proposed. Learning-based methods perform unbiased splice junction signal discovery. A few methods use this approach. They include QPALMA [62], PALMapper [116] and HMMsplicer [66].

QPALMA [62] and PALMapper [116] require an initial set of splice junctions. Then, it uses a support vector machine (SVM) to predict the junction model. This method is not truly de novo as it requires a training set.

HMMsplicer [66] creates its own training data set. By aligning the RNA-seq reads on the reference genome, it first identifies a set of reads that are not fully aligned. These reads are called initially unmapped (IUM) reads. The IUM reads are split into halves and mapped on the reference genome. A training set is formed by all reads whose halves can be aligned uniquely on the reference genome. Then, using the machine learning method, the splice junction model is learned.

The splice junction model can improve the chance of unique alignment of RNA-seq reads. The techniques will be presented in the following subsections.

8.5.2 Exon-first approach

Recall that there are two approaches for aligning RNA-seq reads: the exon-first approach and the seed-and-extend approach. This section focuses on the exon-first approach, whose idea is illustrated in Figure 8.8(a). The exon-first approach first aligns RNA-seq reads to the exons; then, the remaining RNA-seq reads are aligned to splice junctions. The exon-first approach is used by a number of methods, including Cloonan et al. [52], Pan et al. [223], Sultan et al. [288], ERANGE [207], Xmate (old version called RNAMate [53]), RUM [90], ALEXA-seq [91], TopHat [296], G-Mo.R-Se [63] and PASSion [330].

The basic framework has two phases (see Figure 8.6).

1. Phase 1: Align all exon reads and identify potential exon regions.

2. Phase 2: Align all junction reads.

We will illustrate the basic framework using TopHat. Phase 1 maps the RNA-seq reads on the reference genome allowing no gap. We expect that only exon reads can be mapped. Then, regions, called islands, are identified that are covered by some minimum number of reads. Those islands are candidate exons. Since introns shorter than 70 bp are rare in mammalian genomes, if two islands are close (within 6 bp by default), they will be merged.

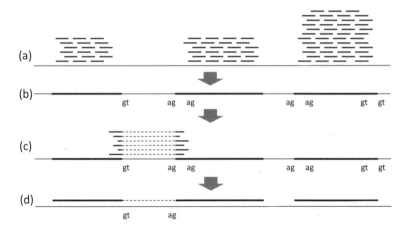

(a)

(b)
        gt   ag ag   ag ag   gt gt

(c)
        gt   ag ag   ag ag   gt gt

(d)
        gt   ag

FIGURE 8.6: Illustration of the two phases of Exon-first approach using TopHat. Phase 1: (a) We first map the reads on the reference genome. (b) From the coverage of the reads, island sequences are formed, each island is a putative exon. Phase 2: (c) We try to recover the splice junctions by mapping reads that span the splice junctions. (d) From each cluster of junction reads, putative intron is formed.

Reads that are not mapped on the genome in Phase 1 are called initially unmapped (IUM) reads. Phase 2 tries to align IUM reads across two candidate exons with the help of the splice signal. It speeds up the alignment of IUM reads on splice junctions by hashing. Precisely, it has two steps. The first step extracts $2k$-mers from every RNA-seq read and puts them in the hash table (by default, $k = 5$). $2k$-mers are only extracted from the first s bases (by default, $s = 28$) of each RNA-seq read. (Note that bases in the 5'-end region in an Illumina read usually have higher quality.) Figure 8.7(a,b) shows an example that illustrates the first step.

The second step checks if each $2k$-mer in the hash table can be aligned to some splice junction. Precisely, it checks if its first k bases and its last k bases represent the segments before and after the splice junction, respectively. Recall that the lengths of 93% of introns are within $[70, 20000]$. For every pair of islands within $[70, 20000]$, TopHat enumerates all possible splice junctions containing the canonical splice signals GT − AG. For each possible splice junction, it generates a length-$2k$ seed S, where k bases are downstream of the acceptor and k bases are upstream of the donor. If the seed S appears in the hash table, we examine the IUM read in $H[S]$ one by one. For each IUM read R, we extend the seed to get the full alignment between R and the two islands allowing a user-specified number of mismatches. If the extension is feasible, the IUM read R is reported to align to the splice junction. Figure 8.7(c) gives an example to demonstrate this step.

After the IUM reads are aligned on the splice junctions, we obtain a set

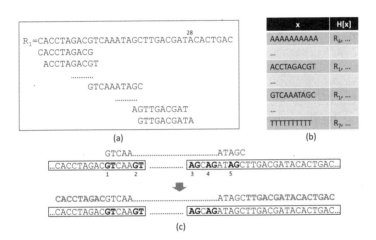

(a)

(b)

(c)

FIGURE 8.7: Illustration of the mapping of IUM read R. (a) We first extract all 10mers from $R_1[1..28]$. (b) For each 10mer x in $R_1[1..28]$, we hash R_1 to $H[x]$. (c) Consider two islands in the genome. The left island has two donor sites while the right island has three acceptor sites. In total, there are 6 combinations of splice site pairs. In particular, from the 5mer upstream of the donor site 2 and the 5mer downstream of the acceptor site 4, we can generate GTCAA − ATAGC. From the hash table in (b), $H[\text{GTCAAATAGC}]$ contains R_1. After we extend the alignment between R_1 and the two islands, we found that R_1 aligns to the splice junction perfectly.

of candidate splice junctions. For each candidate junction, let avL and avR be the average read coverage in the left and right flanking regions of the junction respectively. Let J be the number of alignments crossing the junction. If $J/\max(avL, avR) > 15\%$, TopHat reports this splice junction. (15% is set to be the minimum minor isoform frequency since Wang et al. [303] observed that 86% of the minor isoforms expressed at least 15% of the level of the major isoform.)

In the actual implementation, when an island has high read coverage, TopHat also pairs donor and acceptor sites within the island. Precisely, for an island spanning positions i to j in a genome of length n, its read coverage is defined as $D_{ij} = \frac{\sum_{m=i}^{j} d_m}{j-i} \frac{1}{\sum_{m=0}^{n} d_m}$, where d_m is the depth of coverage at position m. All D_{ij} values are normalized so that it is in the range $[0, 1000]$. TopHat observed that splice junctions tend to occur within an island when D_{ij} is high. By default, TopHat looks for splice junctions within an island spanning positions i to j if its normalized coverage $D_{ij} \geq 300$.

TopHat can work for both short and long reads. (When a read is longer than 75 bp, TopHat version 1.0.7 and later splits the read into three or more segments of approximately equal size and maps them independently.) However, since TopHat is based on read coverage, if the sequencing depth is not high, it will miss solutions. Another disadvantage is that TopHat only finds splice junctions with canonical terminal dinucleotides.

The exon-first approach is usually fast since the time-consuming splice mapping is only performed for a subset of reads. However, this approach is biased to exons and may fail to find some splice junctions. For example, due to pseudogenes (pseudogenes are created when the mRNA transcript of a gene is inserted into a chromosome by spontaneous reverse transcription), some junction reads may map to the pseudogenes instead of the correct splice junctions.

8.5.3   Seed-and-extend approach

The exon-first approach assumes that (1) most reads are exon reads and (2) junction read alignment is difficult. These two points are true only when the RNA-seq reads are short. However, as reads get longer, they may not be valid. For (1), based on the length distribution of exons in humans and assuming RNA-seq reads are uniformly distributed along a transcript, [140] showed that about 33%−38% of the length-100 reads are junction reads (i.e., cover at least one splice junction). As reads get longer, the proportion of junction reads is even higher. For (2), aligning junction reads becomes easy when the reads are long.

This motivates the seed-and-extend approach. The seed-and-extend approach maps seeds (i.e., subsequences) of each read to the reference genome first; then, the seeds are extended to find the splice junction (see Figure 8.8(b)).

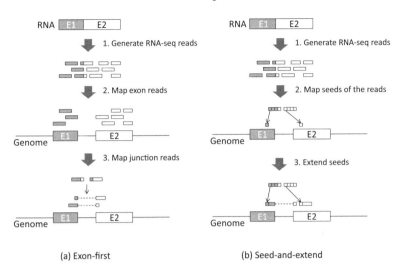

(a) Exon-first   (b) Seed-and-extend

FIGURE 8.8: Two general approaches to map RNA-seq reads. (a) is the exon-first approach and (b) is the seed-and-extend approach.

Many methods have been proposed that apply this idea. They include SplitSeek [6], Mapnext [15], MapSplice [306], Olego [319], SpliceMap [10], Soapsplice [109] and STAR [68]. This section will detail the STAR method.

STAR aims to align every read R on the genome T so that the alignment score is maximized. Given an alignment, its alignment score is defined as follows. For positions where a base is aligned with another base, a match score (+1, default) is given if the two bases are the same; otherwise, a mismatch score (-1, default) is given. A gap smaller than the minimum intron size (21, default) is given an indel score, which equals $indel\_open + indel\_size * indel\_cost$. By default, $indel\_open = -2$ and $indel\_cost = -2$. For a gap of at least the minimum intron size (where the gap appears in the read R), it is assumed to be an intron and is given an intron gap score, which equals $gap\_open + \log_2(gap\_size) * gap\_cost$. By default, $gap\_cost = -0.25$ and $gap\_open = 0$ if the splice signal is GT/AG, -4 if the splice signal is GC/AG and -8 otherwise.

For example, consider the alignment in Figure 8.9, there are 24 matches, 2 mismatches and 3 gaps of sizes $1, 2, 22$. The match score is 24, the mismatch score is -2, the score for the size-1 gap is -4 ($= -2 - 2 * 1$), the score for the size-2 gap is -6 ($= -2 - 2 * 2$), and the score for the size-22 gap is -1.115 ($= 0 - 0.25 * \log_2(22)$) since the splice signal is GT/AG. In total, the alignment score is 10.885.

To find the best alignment of an RNA-seq read R on the reference genome T efficiently, STAR uses a seed-and-extend approach. STAR defines "seeds" of R as all maximum mappable prefixes of R in T. For every position i on R, the maximum mappable prefix $MMP(R, i, T)$ is defined as the longest string

```
      1 2 3 4 5 6 7 8 9    10 11 12 13 14 15 16 17 18 19 20 21 22 23 24 25 26 27 28 29 30 31 32 33 34 35 36 37 38 39 40 41 42 43 44 45 46 47 48 49 50 51 52 53 54 55 56 57
T=...TAACCACGA-GTCGCGTGTCACGGCAAGACCCAAGATAGCGACATAGCAACAGTCGGA...
R=CCGCGACGTCGCGT--------------------CGACAT--CAACCGT
  1 2 3 4 5 6 7 8 9 10 11 12 13 14                                       15 16 17 18 19 20    21 22 23 24 25 26 27
```

FIGURE 8.9: An alignment between a read R and a genome T. The bold gray bases are the mismatches or indels. The bold black bases are the splice signal. The long indel is expected to be the intron gap.

$R[i..i + \ell - 1]$ that appears in T, where ℓ is the maximum mappable length. $MMP(R, i, T)$ can be computed efficiently (see Lemma 8.1). If the number of occurrences of a $MMP(R, i, T)$ is smaller than a certain threshold (50, by default), it is called an anchor.

Lemma 8.1 *Given the suffix array of $T[1..n]$, if $MMP(R, i, T)$ is of length ℓ, $MMP(R, i, T)$ can be computed in $O(\ell \log n)$ time.*

Proof See Exercise 3. ∎

The STAR algorithm is as follows. STAR extracts seeds of R by finding MMPs of R. To save time, STAR does not extract MMPs from every position on R. Instead, the seed-finding algorithm of STAR is as follows.

1. Finds the first MMP starting at position 1 of R.
2. Iteratively finds more MMPs on the unmapped portion of R.

For the example read R in Figure 8.9, the seed-finding algorithm will report six seeds as follows.

1. Initially, $MMP(R, 1, T)$ is CC, which occurs at positions $4, 29, 30$ in T.
2. Second, $R[3..27]$ is the unmapped portion. $MMP(R, 3, T)$ is GCGAC, which occurs at position 38 in T.
3. Third, $R[8..27]$ is the unmapped portion. $MMP(R, 8, T)$ is GTCGCGT, which occurs at position 10 in T.
4. Fourth, $R[15..27]$ is the unmapped portion. $MMP(R, 15, T)$ is CGACAT, which occurs at position 39 in T.
5. Fifth, $R[21..27]$ is the unmapped portion. $MMP(R, 21, T)$ is CAAC, which occurs at position 47 in T.
6. Sixth, $R[25..27]$ is the unmapped portion. $MMP(R, 25, T)$ is CGT, which occurs at position 14 in T.

After the seeds of R are identified, we retain seeds occurring at most 50 times. These seeds are called the anchor seeds. Then, STAR finds alignment windows in the reference genome T that have potential hits of R. T is partitioned into equal size bins (default bin size = 2^{16}). If a bin contains an anchor seed of R, it is an alignment window of R. If two nearby bins are alignment windows of R, they are merged. By default, two alignment windows are nearby if there exist at most 9 bins between them. Furthermore, if there are two alignment windows where one is for R while another is for the mate of R, we can also merge the alignment windows.

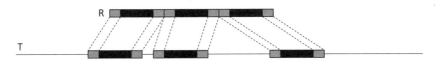

FIGURE 8.10: Consider the set of hits of anchor seeds (black segments) in an alignment window of a read R. This figure illustrates the stitching process that generates the potential hit of R.

Algorithm STAR(T, R)
Require: The suffix array $SA[1..n]$ of T. An RNA-seq read R.
Ensure: A list of hits of the read R
  1: Find the seeds of R (i.e., find MMPs with the help of suffix array);
  2: For seeds with at most 50 hits in the genome, they are called anchors;
  3: Partition T into bins. For bins with hits of anchor seeds, they are called the alignment windows of R;
  4: **for** each alignment window of R **do**
  5:   Generate alignments between R and the alignment window by stitching the seeds;
  6:   We score the alignment window by the best alignment score;
  7: **end for**
  8: Report alignments in all windows with alignment scores bigger than the user-defined threshold;
  9: If the best alignment cannot cover the entire read, connect the main window to other windows to form chimeric alignments.

FIGURE 8.11: The algorithm STAR.

For each alignment window of the read R, STAR generates the alignment of R by stitching all seeds "linearly" within the alignment window. Figure 8.10 illustrates the stitching process. All black regions are the hits of anchor seeds in the alignment window. By simple dynamic programming, two adjacent seeds are stitched together allowing at most one gap and several mismatches. For each read R, among all alignments of R in all windows, we report alignments whose scores are better than the user-defined threshold. If the best scoring (main) alignment does not cover the entire read, STAR tries to identify other alignment windows and check if they can form chimeric alignments. (Those chimeric alignments are potential fusion genes.) Figure 8.11 gives the pseudocode of STAR.

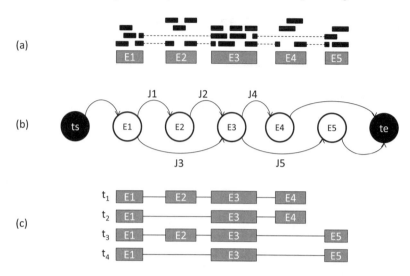

FIGURE 8.12: The steps to construct isoforms. (a) From the mapping reads, we identify all exons and splice junctions. (b) Form a splice graph. (c) From the splice graph, reconstruct all possible transcripts.

8.6 Construction of isoforms

The previous section described methods to map RNA-seq reads onto a reference genome. After the mapping, the next step is to identify the transcripts. A number of methods can recover the isoforms. They include G-Mo.R-Se [63], MapPER [108], Scripture [93] and Cufflinks [297]. Below, we will detail G-Mo.R-Se [63].

Given a set of aligned RNA-seq reads, G-Mo.R-Sc recovers isoforms using three steps.

The first step of G-Mo.R-Se maps the RNA-seq reads by SOAP on the reference genome allowing at most 2 mismatches. Then, regions, called islands, are identified that are covered by some minimum number of reads. Those islands are candidate exons. For unmapped reads, G-Mo.R-Sc searches putative splice sites within 100 bp around each island. If some unmapped reads exist that link two islands, a splice junction is identified (see Figure 8.12(a)).

The second step forms the splice graph by combining the candidate exons and candidate splice junctions. In the splice graph, every candidate exon is a node while every candidate splice junction generates an edge between two nodes (see Figure 8.12(b)).

The last step of G-Mo.R-Se extracts all possible transcripts from the splice graph (see Figure 8.12(c)). Then, the coding region of each transcript is identified and G-Mo.R-Se will filter those candidates that are unlikely to be real.

8.7 Estimating expression level of each transcript

The previous section described methods for reconstructing transcripts in a transcriptome. This section is devoted to estimate their expression levels (or their abundances). Precisely, this section aims to solve the following problem. Let $\mathcal{T} = \{t_1, \ldots, t_K\}$ be the set of transcripts in the transcriptome. Suppose each transcript $t_i \in \mathcal{T}$ has c_{t_i} copies in the transcriptome. Then, we define $\rho_{t_i} = \frac{c_{t_i}}{\sum_{t_j \in \mathcal{T}} c_{t_j}}$ be the relative expression level of the transcript t_i. Note that $\sum_{t_i \in \mathcal{T}} \rho_{t_i} = 1$. This section aims to estimate $\rho = (\rho_{t_1}, \ldots, \rho_{t_K})$.

We can estimate the expression levels of transcripts by RNA-seq. First, RNA-seq reads are aligned on the transcripts in $\mathcal{T}$. Then, by counting the number of reads aligned on each transcript, we can estimate its expression level. Below subsections detail these ideas.

The organization of this section is as follows. Section 8.7.1 covers this basic approach for estimating ρ assuming every RNA-seq read aligns on exactly one transcript. Then, Section 8.7.2 relaxes this assumption and gives methods that estimate the expression levels of transcripts assuming reads may map to multiple isoforms of a gene. Finally, Section 8.7.3 describes methods for estimating the expression level of each gene, which has multiple isoforms.

8.7.1 Estimating transcript abundances when every read maps to exactly one transcript

Consider the transcriptome of a sample. Let $\mathcal{T} = \{t_1, \ldots, t_K\}$ be the set of transcripts in the transcriptome. Let $\mathcal{R} = \{r_1, \ldots, r_N\}$ be the set of N RNA-seq fragments of the transcriptome. Suppose every fragment $r_j \in \mathcal{R}$ aligns to exactly one transcript $t_i \in \mathcal{T}$. This section describes a method to estimate the expression level ρ_{t_i} for every transcript $t_i \in \mathcal{T}$.

We first define the effective length of a transcript. For every transcript $t \in \mathcal{T}$, let ℓ_t be its length. The effective length $\widetilde{\ell}_t$ of t is defined as the number of possible RNA-seq fragments that can be extracted from t. When every fragment is represented by a RNA-seq read, every fragment is of some fixed length m. The number of possible fragments that can be extracted from t is $\ell_t - m + 1$. Hence, we define $\widetilde{\ell}_t = \ell_t - m + 1$. When every fragment is represented by a RNA-seq paired-end reads, the fragments are of different lengths. Let $F(m)$ be the fragment length distribution (i.e., $F(m)$ is the probability that a fragment is of length m). Then, the effective length of the transcript t is $\widetilde{\ell}_t = \sum_{m=1}^{\ell_t} F(\ell)(\ell_t - m + 1)$.

For example, suppose all fragments are RNA-seq reads of length 100. Let the lengths of three transcripts t_1, t_2 and t_3 be 1099, 3099 and 2099, respectively. The effective lengths of the three transcripts are $\widetilde{\ell}_{t_1} = 1000$, $\widetilde{\ell}_{t_2} = 3000$ and $\widetilde{\ell}_{t_3} = 2000$, respectively.

We can determine the expression level ρ_{t_i} of each transcript t_i by counting the number of RNA-seq fragments r_j aligned on t_i. Let $\mathcal{R}_t \subseteq \mathcal{R}$ be the set of RNA-seq fragments that are aligned on the transcript t, for each $t \in \mathcal{T}$. Note that $\mathcal{R} = \bigcup_{t \in \mathcal{T}} \mathcal{R}_t$ and $N = \sum_{t \in \mathcal{T}} |\mathcal{R}_t|$.

Naively, we expect the expression level ρ_t of each transcript t in $\mathcal{T}$ is proportional to $|\mathcal{R}_t|$ and inversely proportional to $\tilde{\ell}_t$. Hence, we expect $\rho_t \propto \frac{|\mathcal{R}_t|}{\tilde{\ell}_t}$.

For example, consider a set of $N = 1000$ RNA-seq fragments, each of length $m = 100$bp. Assume $|\mathcal{R}_1| = 300$, $|\mathcal{R}_2| = 300$ and $|\mathcal{R}_3| = 400$. We have $\rho_{t_1} : \rho_{t_2} : \rho_{t_3} = \frac{300}{1000} : \frac{300}{3000} : \frac{400}{2000}$. Hence, we have $\rho = (\frac{1}{2}, \frac{1}{6}, \frac{1}{3})$.

Although this expression level estimation method is simple, we can mathematically show that this approach is correct under the assumption that every read is sampled from the transcriptome uniformly and randomly.

Let α_t be the chance that a RNA-seq fragment is extracted from t, i.e., $\alpha_t = Pr(t|\rho)$. By definition, α_t is just the proportional to both ρ_t and the effectively length of t (i.e., $\alpha_t \propto \rho_t \tilde{\ell}_t$). Hence,

$$\alpha_t = Pr(t|\rho) = \frac{\rho_t \tilde{\ell}_t}{\sum_{s \in \mathcal{T}} \rho_s \tilde{\ell}_s}. \tag{8.1}$$

Note that $\sum_{t \in \mathcal{T}} \alpha_t = 1$. Since every RNA-seq fragment is independently sampled from $\mathcal{T}$, the likelihood function $L(\rho|\mathcal{R})$ is as follows.

$$\begin{aligned} L(\rho|\mathcal{R}) = Pr(\mathcal{R}|\rho) &= \prod_{t \in \mathcal{T}} \prod_{r \in \mathcal{R}_t} Pr(r|\rho) \\ &= \prod_{t \in \mathcal{T}} \prod_{r \in \mathcal{R}_t} Pr(r|t)Pr(t|\rho) \\ &= \prod_{t \in \mathcal{T}} \prod_{r \in \mathcal{R}_t} \left(\alpha_t \frac{1}{\tilde{\ell}_t} \right) \\ &= \prod_{t \in \mathcal{T}} \left(\frac{\alpha_t}{\tilde{\ell}_t} \right)^{|\mathcal{R}_t|} \\ &\propto \prod_{t \in \mathcal{T}} (\alpha_t)^{|\mathcal{R}_t|} \end{aligned}$$

We aim to find ρ that maximizes the likelihood $L(\rho|\mathcal{R})$. The following lemma gives the result.

Lemma 8.2 *To maximize $L(\rho|\mathcal{R})$, we have:*

$$\rho_t = \frac{\frac{|\mathcal{R}_t|}{\tilde{\ell}_t}}{\sum_{s \in \mathcal{T}} \frac{|\mathcal{R}_s|}{\tilde{\ell}_s}} \propto \frac{|\mathcal{R}_t|}{\tilde{\ell}_t}. \tag{8.2}$$

Proof Since $\log L(\rho|\mathcal{R})$ is a linear equation, it admits a closed form solution. By partial differentiation, the likelihood $L(\rho|\mathcal{R})$ is maximized when $\alpha_t = \frac{|\mathcal{R}_t|}{N}$ for all $t \in \mathcal{T}$. By substituting $\alpha_t = \frac{|\mathcal{R}_t|}{N}$ into Equation 8.1 for all $t \in \mathcal{T}$, we obtain a system of simultaneous equations with K equations. By solving the K equations (see Exercise 5), ρ can be uniquely determined and the lemma follows. ∎

By spike-in transcripts of different lengths with known expression levels, Mortazavi et al. [207] experimentally validated that $\rho_t \propto \frac{|\mathcal{R}_t|}{\tilde{\ell}_t}$ for every transcript t.

Although $\frac{|\mathcal{R}_t|}{\tilde{\ell}_t}$ can measure transcript expression level, its value is affected by the total number of RNA-seq fragments N. To normalize N, RPKM (stands for Reads Per Kilobase of exon model per Million mapped reads) and FPKM (stands for Fragments Per Kilobase of exon model per Million mapped reads) are proposed. The FPKM of the transcript t is defined to be

$$\frac{10^9 |\mathcal{R}_t|}{\tilde{\ell}_t N}.$$

When each fragment is a RNA-seq read, the FPKM measurement is also known as RPKM.

For the above example where $\tilde{\ell}_{t_1} = 1000$, $\tilde{\ell}_{t_2} = 3000$, $\tilde{\ell}_{t_3} = 2000$, $|\mathcal{R}_1| = 300$, $|\mathcal{R}_2| = 300$ and $|\mathcal{R}_3| = 400$. The RPKMs (or FPKMs) of the three transcripts are 3×10^5, 1×10^5 and 2×10^5, respectively.

RPKM and FPKM are suitable for comparing the expression levels of different transcripts within the same sample. However, it may not be suitable for comparing expression levels of transcripts between different samples. To solve the problem, another measurement TPM (Transcripts Per Million) is proposed. The TPM of a transcript t equals

$$\frac{|\mathcal{R}_t|}{\tilde{\ell}_t} \cdot \frac{10^6}{\sum_{s \in \mathcal{T}} \frac{|\mathcal{R}_s|}{\tilde{\ell}_s}}.$$

Note that, for different samples, the sum of the FPKM (or RPKM) measurements of all transcripts can be different. So, when a transcript has the same RPKM (or FPKM) measurement in two different samples, we do not sure if the same proportion of RNA-seq fragments are mapped to this transcript in the two samples.

On the other hand, for any sample, observe that the sum of the TPM measurements of all transcripts is always equal to 10^6. So, when a transcript has the same TPM measurement in two samples, we know that the same proportion of RNA-seq fragments are mapped to this transcript in the two samples. Hence, TPM is more suitable for comparing transcript expression levels across samples.

8.7.2   Estimating transcript abundances when a read maps to multiple isoforms

Section 8.7.1 assumes every read maps to exactly one transcript. When some reads map to multiple isoforms of the same gene, it cannot work.

Consider N reads $\mathcal{R} = \{r_1, \ldots, r_N\}$ and K transcripts $\mathcal{T} = \{t_1, \ldots, t_K\}$. For each transcript $t \in \mathcal{T}$, recall that $\widetilde{\ell}_t$ is the effective length of t, ρ_t is the abundance of the transcript t, and $\alpha_t = Pr(t|\rho)$ is the chance that a read is extracted from the transcript t (α_t can be computed by Equation 8.1).

Let Y be an $N \times K$ matrix where $y_{r_i, t_k} = 1$ if the read r_i aligns to transcript t_k; and 0 otherwise. The matrix Y is called the compatible matrix. We aim to find ρ that maximizes the likelihood $L(\rho|\mathcal{R})$, which equals

$$Pr(\mathcal{R}|\rho) = \prod_{r \in \mathcal{R}} Pr(r|\rho) \tag{8.3}$$

$$= \prod_{r \in \mathcal{R}} \left(\sum_{t \in \mathcal{T}} Pr(r|t) Pr(t|\rho) \right) \tag{8.4}$$

$$= \prod_{r \in \mathcal{R}} \left(\sum_{t \in \mathcal{T}} \frac{y_{r,t}}{\widetilde{\ell}_t} \cdot \alpha_t \right) \tag{8.5}$$

$$= \prod_{r \in \mathcal{R}} \left(\frac{\sum_{t \in \mathcal{T}} y_{r,t} \rho_t}{\sum_{s \in \mathcal{T}} \rho_s \widetilde{\ell}_s} \right). \tag{8.6}$$

There is no closed form for this problem. Moreover, when the matrix Y is of full rank, $L(\rho|\mathcal{R})$ is guaranteed to have a unique global maximum.

As an example, consider three transcript models $\mathcal{T} = \{t_1, t_2, t_3\}$ in Figure 8.13(a). Assume all three transcripts are of the same length. Suppose there are 6 reads $\mathcal{R} = \{r_1, \ldots, r_6\}$ mapping on them. Then, we can construct a compatible matrix Y as shown in Figure 8.13(b). Note that one read may map on more than one transcript. One example is read r_3 which maps on both transcripts t_1 and t_3. Hence, we set $y_{r_3, t_1} = 1$ and $y_{r_3, r_3} = 1$. We aim to estimate the abundances $\rho_{t_1}, \rho_{t_2}, \rho_{t_3}$. Since all transcripts have the same length L and $\rho_{t_1} + \rho_{t_2} + \rho_{t_3} = 1$, we have $\sum_{s \in \mathcal{T}} \rho_s \widetilde{\ell}_s = L$. Thus, by Equation 8.6, we have

$$L(\rho|\mathcal{R}) = \frac{1}{L^6} \rho_{t_1} (\rho_{t_1} + \rho_{t_2})(\rho_{t_1} + \rho_{t_3})(\rho_{t_2} + \rho_{t_3})^2.$$

Since Y is of rank 3, a unique global maximum exists for $L(\rho|\mathcal{R})$. We can show that the optimal solution is $(\rho_{t_1}, \rho_{t_2}, \rho_{t_3}) = (\frac{1}{\sqrt{5}}, \frac{5-\sqrt{5}}{10}, \frac{5-\sqrt{5}}{10}) = (0.447, 0.276, 0.276)$.

Although a closed form is not available for this problem, we can still estimate ρ using an EM algorithm. The algorithm will iteratively improve the estimation of ρ. Denote $\rho^{(c)}$ as the estimation of ρ at the c-th iteration. Initially, we assume all transcripts in $\mathcal{T}$ have the same abundances, that is, $\rho^{(0)} = (\rho_{t_1}^{(0)}, \ldots, \rho_{t_K}^{(0)})$ where $\rho_{t_k}^{(0)} = \frac{1}{K}$. Then, the EM algorithm iteratively

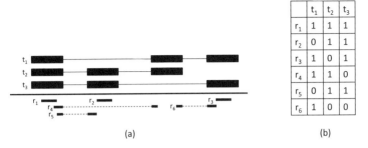

	t_1	t_2	t_3
r_1	1	1	1
r_2	0	1	1
r_3	1	0	1
r_4	1	1	0
r_5	0	1	1
r_6	1	0	0

(a)   (b)

FIGURE 8.13: (a) shows three isoforms $\{t_1, t_2, t_3\}$ of a gene (all isoforms are of the same length) and 6 RNA-seq reads $\{r_1, \ldots, r_6\}$ that map on the three isoforms. (b) is the compatibility matrix Y. The entry y_{r_i, t_k} equals 1 if and only if the read r_i maps on the transcript t_k.

computes $\rho^{(c)} = (\rho_{t_1}^{(c)}, \ldots, \rho_{t_K}^{(c)})$ by performing two steps: the E-step and the M-step.

- **E-step**: The E-step computes the expected read count X_t for each transcript t. It has two steps. Let $p_{r,t} = Pr(r|t)$ be the probability that r is extracted from the transcript t. The first step estimates $p_{r,t} = Pr(r|t)$ by $\frac{y_{r,t}\rho_t^{(c)}}{\sum_{s \in \mathcal{T}} y_{r,s}\rho_s^{(c)}}$, for every read $r \in \mathcal{R}$ and every transcript $t \in \mathcal{T}$. Then, the expected read count X_t can be computed by setting $X_t = \sum_{r \in \mathcal{R}} p_{r,t}$.

- **M-step**: The M-step finds $\rho^{(c+1)}$ that maximizes $L(\rho^{(c+1)}|X_{t_1}, \ldots, X_{t_K})$. The maximum likelihood can be solved using a similar idea as Lemma 8.2 (see Exercise 6). We have:

$$\rho_t^{(c+1)} = \frac{\frac{X_t}{\ell_t}}{\sum_{s \in \mathcal{T}} \frac{X_s}{\ell_t}}. \tag{8.7}$$

Figure 8.14 states the detail of the EM algorithm.

We demonstrate the execution of the EM algorithm using the example in Figure 8.13. Initially, we set $\rho^{(0)} = (\rho_{t_1}^{(0)}, \rho_{t_2}^{(0)}, \rho_{t_3}^{(0)}) = (1/3, 1/3, 1/3)$. Figure 8.15 illustrates three iterations of the EM algorithm. The values of $\rho^{(1)}, \rho^{(2)}$ and $\rho^{(3)}$ are $(0.39, 0.31, 0.31)$, $(0.42, 0.29, 0.29)$ and $(0.43, 0.28, 0.28)$, respectively. We can see that the values are converging to the true optimal values, that is, $(\frac{1}{\sqrt{5}}, \frac{5-\sqrt{5}}{10}, \frac{5-\sqrt{5}}{10})$.

The above discussion assumes the RNA-seq reads are uniformly randomly extracted from every transcript in the transcriptome. However, it has been shown that the RNA-reads are not uniformly distributed along the transcript due to site-specific bias, position bias and mapping bias. These biases can be adjusted by giving different weight to each RNA-seq read depending on its sequence and mapping position [110].

Algorithm EstimateAbundance(Y)

Require: The $N \times K$ matrix Y where $y_{r_i,t_k} = 1$ if the read r_i aligns to the transcript t_k

Ensure: The relative transcript abundances $\rho = \{\rho_{t_1}, \ldots, \rho_{t_K}\}$ of the K transcripts;

1: Initialize ρ, say, uniform distribution (i.e., $\rho_{t_k}^{(0)} = \frac{1}{K}$ for $k = 1, \ldots, K$);
2: $c = 0$;
3: **while** $c <$ the maximum number of iteration or the improvement on $L(\rho|R)$ is small **do**
4:     For every read $r \in R$ and $t \in T$, $p_{r,t} = \frac{y_{r,t}\rho_t^{(c)}}{\sum_{s \in T} y_{r,s}\rho_s^{(c)}}$;
5:     For every $t \in T$, set $X_t = \sum_{r \in R} p_{r,t}$;
6:     For every $t \in T$, set $\rho_t^{(c+1)} = \frac{\frac{X_t}{\ell_t}}{\sum_{s \in T} \frac{X_s}{\ell_t}}$;
7:     $c = c + 1$;
8: **end while**

FIGURE 8.14: The EM algorithm for estimating the relative transcript abundances ρ of all transcripts in T.

8.7.3   Estimating gene abundance

Sections 8.7.1–8.7.2 describe methods to estimate the abundance of each transcript. This section describes methods to estimate the abundance of each gene. When a gene only has one isoform, the gene abundance is the same as the transcript abundance. When a gene has multiple isoforms, there are three possible ways to estimate the gene abundance [307, 297, 295].

The first approach is called the exon-union model. We just count the total number of reads that fall on all exons of the gene. Then, the RPKM value is computed.

For example, consider the gene in Figure 8.16. It has 3 isoforms and consists of 4 exons. All exons have the same length L. Assume the RNA-seq library has $N = 10^9$ reads. Using the exon-union model, we compute the RPKM using the read counts of all 4 exons. Hence, the RPKM measurement of the gene is $\frac{10^9(4+3+3+2)}{(4L)N} = \frac{3}{L}$. This approach is simple. However, it will under-estimate the abundance.

Another way is called the exon-intersection model. It only counts the number of reads that fall on exons which appear in all isoforms of the gene. Then, the RPKM value is computed. For the example in Figure 8.16, all three transcripts of the gene share exon E1. We compute the RPKM using the read count of exon E1 only. Hence, the RPKM measurement of the gene is $\frac{10^9(4)}{LN} = \frac{4}{L}$.

This method can estimate the abundance correctly. However, it does not utilize the full information.

The last approach is called the isoform-based model. It first estimates

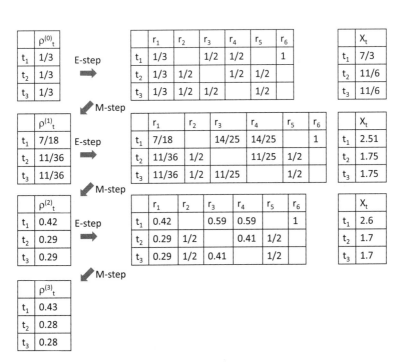

FIGURE 8.15: Consider the example in Figure 8.13, which illustrates three iterations of the EM algorithm that estimate the transcript abundances of $\rho = (\rho_{t_1}, \rho_{t_2}, \rho_{t_3})$. In each iteration, we iterate the E-step and M-step: Given $\rho^{(c)}$, the E-step computes the table for $p_{r,t}$ for every $r \in \mathcal{R}$ and every $t \in \mathcal{T}$; then, it computes the table for X_t for every $t \in \mathcal{T}$. The M-step computes $\rho^{(c+1)}$ given X_t. Note that the optimal value of $\rho = (\rho_{t_1}, \rho_{t_2}, \rho_{t_3}) = (0.447, 0.276, 0.276)$.

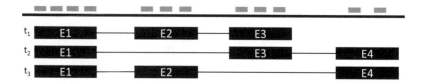

FIGURE 8.16: For this example gene, it has 3 isoforms and consists of 4 exons: E1, E2, E3 and E4. The read counts of the four exons are 4, 3, 3 and 2 respectively.

the abundance of each isoform of the gene. Then, the abundance of the gene equals the sum of the abundance of all its isoforms. For the example in Figure 8.16, we first estimate the relative abundance of each isoform. We can show that $(\rho_{t_1}, \rho_{t_2}, \rho_{t_3}) = (1/2, 1/4, 1/4)$ (see Exercise 7). Since the total number of reads is 12, the estimated read counts for the three isoforms t_1, t_2, t_3 are $6, 3, 3$, respectively. Hence, the RPKM measurements of the 3 isoforms are $\left(\frac{10^9(6)}{(3L)N}, \frac{10^9(3)}{(3L)N}, \frac{10^9(3)}{(3L)N}\right) = (\frac{2}{L}, \frac{1}{L}, \frac{1}{L})$. The RPKM measurement of the gene is their sum, which is $\frac{2}{L} + \frac{1}{L} + \frac{1}{L} = \frac{4}{L}$.

8.8   Summary and further reading

For a review of the RNA-seq methods, please read [308, 86, 41]. Reference [220] describes the complete pipeline to analyze RNA-seq data.

Please read [281] for a comparison of methods for estimating differential expressions of genes.

Previously, people think that protein and mRNA expression levels are not correlated. It is shown that protein and mRNA expression levels are in fact correlated (once the non-linear bias in the mass spectrometry data is removed) [3, 173].

8.9   Exercises

1. The simplest model for the donor site of an intron is the frequency of the first two bases at the 5' end of the intron. Suppose you are given a set of introns, can you describe a method to compute the model for the donor site? Similarly, can you propose a method to compute the model for the acceptor site?

2. Suppose you obtain a set of RNA-seq reads from a human sample. Assume the length of the reads is 150 bp. Assume the RNA-seq reads are uniformly distributed along each transcript. Can you compute the expected percentages of reads that cover at least one splice junction? (Hint: You need to use the statistics of the human transcriptome stated in Section 8.1.)

3. Consider a reference genome $T[1..n]$ and a read $R[1..m]$. Given the suffix array of T, show that $MMP(R, i, T)$ can be computed in $O(\ell \log n)$ time, where ℓ is the length of $MMP(R, i, T)$.

4. Consider a set of paired-end reads whose length distribution is $F(301) = 0.2, F(302) = 0.3, F(303) = 0.3$ and $F(304) = 0.2$. For a transcript t of length $\ell = 600$, what is its effective length $\tilde{\ell}$?

5. Suppose $\alpha_t = \frac{\rho_t \tilde{\ell}_t}{\sum_{s \in \mathcal{T}} \rho_s \tilde{\ell}_s}$, for all $t \in \mathcal{T}$. Please show that

$$\rho_t = \frac{\frac{\alpha_t}{\ell_t}}{\sum_{s \in \mathcal{T}} \frac{\alpha_s}{\ell_s}}.$$

6. This question refers to Section 8.7.2. Can you give the likelihood function $L(\rho^{(c+1)}|X_{t_1}, \ldots, X_{t_K})$? To maximize the likelihood, please show that Equation 8.7 is correct, that is,

$$\rho_t^{(c+1)} = \frac{\frac{X_t}{\ell_t}}{\sum_{s \in \mathcal{T}} \frac{X_s}{\ell_t}}.$$

7. For the example in Figure 8.16, there are 12 reads and we denote them as $R = \{r_1, \ldots, r_{12}\}$. Let $\rho = (\rho_{t_1}, \rho_{t_2}, \rho_{t_3})$ be the relative transcript abundances of the 3 transcripts.

   (a) Can you give the likelihood function formula for $L(\rho|R)$?
   (b) Can you compute ρ that maximizes $L(\rho|R)$?
   (c) We can compute ρ using the algorithm in Figure 8.14. Can you illustrate three iterations of the EM algorithm?

8. Section 8.7.3 mentioned that gene abundance estimated based on the exon-union model is only correct for a gene with a single isoform. When there is more than one isoform, the abundance is under-estimated. Can you show that this is true?

Chapter 9

Peak calling methods

9.1 Introduction

Previous chapters describe methods for identifying genomic variations. This chapter studies another type of method known as peak callers. After sequencing reads are aligned on the reference genome, peak callers identify genomic regions that have a high density of reads. These regions are called peaks.

Peak callers find applications in studying protein-DNA interactions and epigenetics based on ChIP-seq (see Sections 9.2.1 and 9.2.2). They can also be used in some specialized assays like DNase-seq, MNase-seq and ATAC-seq.

This chapter is devoted to discussing methods for peak calling. We first briefly describe techniques like ChIP-seq that generate the sequencing datasets. Then, we detail the peak calling methods.

9.2 Techniques that generate density-based datasets

Peak calling finds applications in many different next-generation sequencing datasets. Below, we briefly describe them.

9.2.1 Protein DNA interaction

Within a cell, some proteins called transcription factors (TFs) or DNA-binding proteins can interact with the genome. These interactions finally regulate the expression of the genes. For example, Oct4 and Klf4 are master TFs which regulate the expression of stem cell related genes and they can reprogram human neural stem cells into human ES cells [141]. Another example is Estrogen Receptor (ER), which is the key TF related to breast cancer [35]. Hence, it is important to understand the TF−DNA interaction.

To understand the TF−DNA interaction, we need is to find the loci on our genome where the TFs interact. These loci are called transcription factor

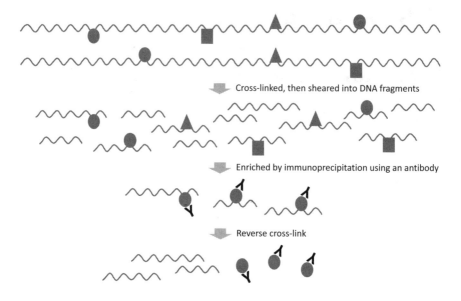

FIGURE 9.1: The workflow of the ChIP experiment. The curly line is the genomic DNA. The triangle, circle and square are the transcription factors. The Y symbol represents the antibody for the circle transcription factor.

binding sites (TFBSs). ChIP-seq is the technology that enables us to identify TF binding sites on a genome-wide scale.

ChIP-seq involves two steps. The first step performs Chromatin Immuno-precipitation (ChIP). ChIP is a type of immunoprecipitation experimental technique. It requires a specific antibody that has high binding affinity with the selected transcription factor. By enriching the selected TF by the anti-body, we can capture DNAs that interact with the selected transcription factor in the cell. The workflow of the ChIP experiment is shown in Figure 9.1. First, TFs are cross-linked to their genomic DNA targets in vivo. Second, the protein-DNA complexes are extracted from cells and the bounded DNA is further sheared by sonication into DNA fragments. Third, using an antibody for the selected transcription factors, the cross-linked DNA fragments with the selected transcription factor are enriched by immunoprecipitation (IP). Finally, by reverse cross-linking, the ChIP fragments (i.e., IP-enriched DNA fragments) are obtained. These ChIP fragments will be sequenced by next-generation sequencing in the second step. These sequencing reads are expected to be extracted from regions close to the transcription factor binding sites.

There are many variations of the ChIP-seq protocol. Table 9.1 summarizes the current available ChIP-based techniques. ChIP-seq sequences one end of the ChIP fragments. There is a paired-end version that sequences both ends

TABLE 9.1: Different ChIP-based techniques.

Technique	detection method	Resolution
ChIP-PET [312]	paired-end read sequencing	100 bp
ChIP-seq [124]	short read sequencing	100 bp
ChIP-exo [243]	short read sequencing	10 bp

TABLE 9.2: Different histone modifications and their genomic functions

Histone modification	Genomic function
H3K27me3	marker for inactive promoter
H3K4me3	marker for promoter
H3K4me1	marker for enhancer
H3K36me3	marker for active gene body (hallmark of elongation)
H3K27ac	marker for active promoter or enhancer
H3K9me2/3	marker for heterochromatin (a repressive mark)
H4K20me3	marker for heterochromatin
H3K9ac	marker for active promoter
H3K14ac	marker for active promoter

of the ChIP fragments, which is known as ChIP-PET [312]. ChIP-exo [243] is an enhanced version of ChIP-seq which applies a lambda exonuclease to further cut the unbound parts of the ChIP fragments. ChIP-exo improves the resolution of the binding sites.

9.2.2 Epigenetics of our genome

Within a cell, our genome wraps around histones. Histones are subjected to post-translational modifications. These modifications include methylation, acetylation, phosphorylation and ubiquitination. Similar to our genome, histone modifications are inherited when a cell divides [11, 239]. Hence, cells of the same tissue types have the same histone modification profile.

Biologists discovered that histone modifications mark genomic functions of the DNA that wipe around them. For example, DNA regions that wipe around histones with H3K27me3 correspond to inactive promoters of some genes. DNA regions that wipe around histones with H3K4me3 correspond to active promoters. Table 9.2 gives the genomic functions of some well-known histone modifications. Understanding the histone modification profiles for different tissue types can help us to decode their functions. It also helps us understand the cause of diseases.

A histone modification profile can be identified on a genome-wide scale by applying ChIP-seq. The idea is the same as identifying transcription factor−DNA interaction. For each histone modification, there is some specific antibody that can interact with the histone modification. The workflow consists of two steps. The first step performs ChIP with the antibody specific for the histone modification. Then, by next-generation sequencing, the second step obtains reads near the histone modifications.

Unlike binding sites, the peaks for histone modifications are usually broader.

9.2.3   Open chromatin

Open chromatin is defined as the chromatin regions that are accessible (by TFs and other molecules). Most of the DNA-related events are expected to happen around open chromatin regions. For instance, it is known that most TFs bind to open chromatin regions [282]. Researchers found that different cell types have different open chromatin profiles [216]. These studies imply that open chromatin is important for understanding the gene regulation mechanism.

Open chromatins can be identified by DNase-seq [30, 101], FAIRE-seq [280] and ATAC-seq [32] on a genome-wide scale.

9.3   Peak calling methods

The wet-lab techniques in Section 9.2, like ChIP-seq, generate a set of reads. After these reads are aligned on the reference genome, the computational problem is to find regions in the reference genome enriched with aligned reads. These regions are called peaks and they are expected to have some useful genomic features. This section focuses on methods for calling peaks.

We first present a simple peak caller. The input is a set of aligned reads (in SAM or BAM format). The simple peak caller runs two steps. First, it generates a signal profile by piling up the aligned reads; then, from the signal profile, the second step reports peaks that have a signal higher than a selected cutoff threshold θ. The algorithm is stated in Figure 9.2. Figure 9.3 gives an example to illustrate this process. Figure 9.3(a) gives a set of aligned reads. Figure 9.3(b) gives the signal profile formed by piling up all aligned reads. (The pileup of the aligned reads can be stored in bigWig [135] or cWig [112] (see Section 2.6).) In this illustration, our cutoff threshold is $\theta = 3$. Each peak is defined to be a maximal consecutive region where the signal is higher than θ. In our example, we obtain 5 peaks.

Although the simple peak caller can work, its accuracy is low. Below, we refine the simple peak caller. We will discuss a few issues:

(1) **ChIP fragment length**: In the ChIP-seq protocol, our genome is sheared into fragments. The fragment size (also called the insert size)

Algorithm SimplePeakCaller($\mathcal{R}, \theta$)

Require: $\mathcal{R}$ is a set of aligned ChIP-seq reads and a cutoff threshold θ

Ensure: A list of peaks

1: By piling up the reads in $\mathcal{R}$, compute the signal profile S such that $S[i]$ is the number of reads covering position i;

2: Report all maximal regions $r_1..r_2$ such that $S[i] \geq \theta$ for $r_1 \leq i \leq r_2$;

FIGURE 9.2: A simple peak caller.

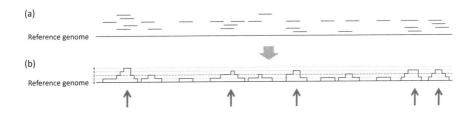

FIGURE 9.3: (a) is the visualization of a set of aligned reads on a reference genome. (b) is the signal profile formed by piling up the aligned reads in (a). The arrows indicate the 5 peaks whose signals are at least 3.

will affect the performance of peak callers. We will discuss the effect of insert size and how to determine insert size.

(2) **Model noise**: Although the ChIP protocol enriches DNA fragments with TF binding, noise DNA fragments exist. However, noise DNA fragments are not uniformly distributed on the genome due to various biases. False peaks are created due to noise. We will discuss the use of a control library to reduce noise.

(3) **Noise rate**: Let α be the proportion of ChIP fragments that do not have a TF binding. Depending on the quality of the antibody and various protocol-specific factors, α has different values for different ChIP-seq experiments. To improve peak calling, we will discuss how to determine α.

(4) **Cutoff threshold θ**: The cutoff threshold θ is manually determined in the simple peak caller. This threshold is difficult to set. We will discuss how to set θ depending on the p-value or the false discovery rate.

(5) **Issue on the reference genome**: The genome of the sample is different from our reference genome. In particular, some genomic regions have different copy numbers. This will create noise in peak calling. We will discuss this issue and study a solution based on a sponge database.

(6) **Kernel model**: The simple peak caller assumes every loci covered by a ChIP fragment has an equal chance to be a TF binding site. It may not be exactly correct. Often, the center of the ChIP fragment is more likely to be the actual binding site. To incorporate this idea, we will discuss a method based on kernel function.

9.3.1   Model fragment length

By performing ChIP on a sample, ChIP fragments are obtained. Then, either single-end reads or paired-end reads can be sequenced from the ChIP fragments. When we perform paired-end sequencing (known as ChIP-PET), the two endpoints of each ChIP fragment can be determined by the alignment of the paired-end read. Then, we can pile up the aligned ChIP fragments and obtain a signal profile like Figure 9.3.

When we perform single-end sequencing (known as ChIP-seq), only one end of each ChIP fragment will be sequenced and we can only determine one endpoint of each ChIP fragment. Figure 9.4(a) illustrates this process. There are 8 ChIP fragments in this example. One short read is sequenced for each ChIP fragment. After piling up the alignments of these short reads, we obtain the signal profile in Figure 9.4(b). For this example, since the aligned short reads do not overlap, the simple peak caller will fail to call a peak in this region if the cutoff threshold $\theta = 2$.

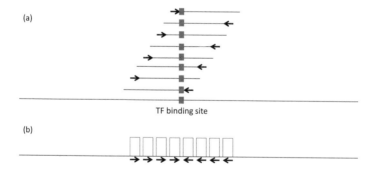

FIGURE 9.4: (a) is the visualization of a set of aligned reads on a reference genome. (b) is the signal profile formed by piling up the aligned reads in (a).

This problem arises since the ChIP-seq reads do not appear exactly on the binding sites. If we know the insert size (i.e., the average ChIP fragment length), this problem can be solved as follows. Observe that every read has an alignment orientation. From the alignment strand of each read, we know whether the binding site is on the left or right of the read. There are two possible solutions: (1) shifting of the reads and (2) extension of the reads.

For (1), we shift the read by $L/2$ in a strand-specific fashion. After shifting, the reads are expected to overlap with the binding site. This approach is used by MACS [331], QuEST [299], SiSSRs [127] and spp [137]. For (2), each read is extended by L in a strand-specific fashion. This approach is used by XSET [247].

As an illustration, consider the set of aligned reads in Figure 9.5(a). The pile-up signal will contain two peaks, where neither peak overlaps the binding site. Approach (1) will shift the read by $L/2$. Then, we obtain Figure 9.5(b). The pile-up signal will contain one signal peak that overlaps the binding site. Note that the peak intensity is roughly double when it is compared with the two peaks in Figure 9.5(a). This approach generates a very sharp peak. However, if L is estimated incorrectly, this approach may fail to detect the correct peak.

For approach (2), every read is extended by L. Then, we obtain Figure 9.5(c). Similar to approach (1), the pile-up signal will contain one signal peak that overlaps the binding site. The peak intensity is also roughly double when it is compared with the two peaks in Figure 9.5(a). As long as L is estimated to be long enough, this approach always reports the correct peak. However, the peak predicted by this approach may be very broad.

Figure 9.6 details the algorithm to generate the signal profile.

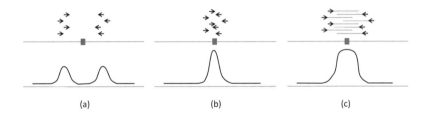

(a)   (b)   (c)

FIGURE 9.5: (a) A set of aligned ChIP-seq reads and the corresponding signal profile. (b) The signal profile formed after the reads are shifted by $L/2$ in a strand-specific fashion. (c) The signal profile formed after the reads are extended by L in a strand-specific manner.

As discussed above, we need an accurate estimation of the insert size L, in particular for approach (1). L can be estimated from wet-lab. However, the accuracy of the estimation depends on the laboratory and the experimenter. Another choice is to estimate L computationally. This can be done by studying the cross-correlation between positive and negative strand reads.

Here, we discuss the method used by spp [137]. Suppose a set of ChIP-seq reads are aligned on a reference genome $T[1..n]$, where the read length is ℓ. For every loci i in the genome T, let $f(i)$ be 1 if some read is aligned on the +ve strand of $T[i..i + \ell - 1]$; and 0 otherwise. Let $g(i)$ be 1 if some read is aligned on the −ve strand of $T[i - \ell + 1..i]$; and 0 otherwise. If the average fragment length is L, we will expect that if $f(i) = 1$, then the chance that $g(i + L) = 1$

Algorithm CreateSignalProfile($\mathcal{R}$)

Require: $\mathcal{R}$ is a set of aligned ChIP-seq reads
Ensure: A signal profile S
  1: Estimate the average fragment length L;
  2: Shift the read by $L/2$bp (for approach 1) or extend the read by Lbp (for approach 2);
  3: By piling-up the reads in $\mathcal{R}$, compute the signal profile S such that $S[i]$ is the number of reads covering position i;

FIGURE 9.6: Algorithm to generate the signal profile from a set of aligned ChIP-seq reads.

pos	1	2	3	4	5	6	7	8	9	10
$f(i)$	1	1	0	1	0	0	1	0	1	1
$map(i)$	1	1	0	1	0	0	1	1	1	1
$g(i+L)$	1	0	0	0	0	1	1	0	1	0
$map(i+L-\ell+1)$	1	0	1	1	0	1	1	1	1	0

FIGURE 9.7: An example illustrates the computation of cross-correlation. For some fixed L, row 1 shows the genomic position while rows 2−5 show $f(i)$, $map(i)$, $g(i+L)$ and $map(i+L-\ell+1)$.

is higher. To check if this is true, we can compute the cross-correlation of the pairs in $\{(f(i), g(i+L)) \mid i = 1, \ldots, n-L\}$, which equals

$$\rho_L = \frac{1}{n-L} \sum_{i=1}^{n-L} \frac{[f(i) - \mu_f][g(i+L) - \mu_g]}{\sigma_f \sigma_g}$$

where μ_f and σ_f are the mean and standard derivation, respectively, of $\{f(1), \ldots, f(n)\}$ and μ_g and σ_g are the mean and standard derivation, respectively, of $\{g(1), \ldots, g(n)\}$. Precisely, $\mu_f = \frac{1}{n}\sum_{i=1}^{n} f(i)$, $\mu_g = \frac{1}{n}\sum_{i=1}^{n} g(i)$, $\sigma_f^2 = \frac{1}{n}\sum_{i=1}^{n}[f(i) - \mu_f]^2$ and $\sigma_g^2 = \frac{1}{n}\sum_{i=1}^{n}[g(i) - \mu_g]^2$. Note that $\sigma_f^2 = \mu_f(1 - \mu_f)$ and $\sigma_g^2 = \mu_g(1 - \mu_g)$ (Exercise 3).

For the example in Figure 9.7, $\mu_f = \frac{6}{10} = 0.6$, $\mu_g = \frac{4}{10} = 0.4$, $\sigma_f^2 = 0.6 * 0.4 = 0.24$ and $\sigma_g^2 = 0.4 * 0.6 = 0.24$. By the above equation, the cross-correlation score $\rho_L = 0.25$.

The average fragment length is estimated to be the length L that maximizes ρ_L.

The above method is sometimes inaccurate. The reason is that different parts of the genome have different mappability. Mappability measures whether a read originated from some position i can be mapped back to position i unambiguously. Mappability introduces bias, which affects the estimation of the fragment length.

Precisely, mappability is defined as follows. For any position i, the position

i is mappable if $T[i..i + \ell - 1]$ occurs exactly once in the genome. We denote $map(i) = 1$ if position i is mappable; and 0 otherwise. Note that when a position i is not mappable, no read can map to position i on the +ve strand or position $i + \ell - 1$ on the −ve strand.

Ramachandran et al. [240] suggested a method to remove the bias contributed by mappability. The idea is to include $(f(i), g(i + L))$ in the cross-correlation computation only when both $map(i) = 1$ and $map(i+L-\ell+1) = 1$. For the example in Figure 9.7, we only include $(f(i), g(i + L))$ for $i = 1, 4, 7, 8, 9$. Then, we have $\mu_f = 0.8$, $\mu_g = 0.6$, $\sigma_f^2 = 0.8 * 0.2 = 0.16$ and $\sigma_g^2 = 0.6 * 0.4 = 0.24$. The cross-correlation score $\rho_L = 0.612$.

9.3.2 Modeling noise using a control library

For the simple peak calling method in Figure 9.2, a cutoff threshold θ is required to determine if a peak is enriched with ChIP-seq reads. Suppose the ChIP-seq library contains N reads, each of length ℓ. Then, the expected coverage is $\lambda = \frac{N\ell}{n}$, where n is the genome size. To get a set of peaks with fold enrichment f, we can set the cutoff threshold $\theta = f\lambda$.

The above solution assumes noisy reads uniformly distribute on the reference genome. However, the uniform noise model is shown to be too ideal. Due to the existence of sequencing and mapping biases, chromatin structure and genomic copy number variations, many false positive regions are enriched with ChIP-seq reads [300].

To filter the false peaks, Johnson et al. [124] proposed to sequence a control library in addition to the sample library. A control library is the same as the sample library except that we do not enrich DNA fragments that have TF binding. The control library provides a more accurate background, which reduces the number of false positive peaks.

Figure 9.8 illustrates the usage of the control library. The figure has two regions in the sample library where the number of overlapping reads is 3. By the simple peak caller, both regions are predicted as peaks. Moreover, when we check the control library, the left peak also has three overlapping reads in the control library. This implies that the left peak is noise.

Motivated by this idea, we obtain an algorithm in Figure 9.9. Instead of comparing the number of overlapping reads of each region with a fixed cutoff threshold, Step 2 identifies peak regions $r_1..r_2$ such that its signal in the sample is significantly higher than that in the control. Suppose the control library has M reads while the sample library has N reads. Let S_s and S_c be the signal profiles of the sample and control, respectively (i.e., $S_s[i]$ and $S_c[i]$ are the numbers of reads covering the position i in the sample and control, respectively). Let $\lambda_s(r_1..r_2)$ and $\lambda_c(r_1..r_2)$ be the normalized average signal in the region $r_1..r_2$ for both sample and control, respectively. That is, $\lambda_s(r_1..r_2) = \frac{1}{N} \sum_{i=r_1}^{r_2} S_s[i]$ and $\lambda_c(r_1..r_2) = \frac{1}{M} \sum_{i=r_1}^{r_2} S_c[i]$. When $\lambda_s(r_1..r_2)$ is significantly bigger than $\lambda_c(r_1..r_2)$, the region $r_1..r_2$ is considered to be a peak.

In real life, $\lambda_c(r_1..r_2)$ may be zero. To avoid this problem, MACS [331]

Our sample library:

Control library:

FIGURE 9.8: An example of a ChIP-seq library for the sample (top) and the control (bottom).

Algorithm CallPeakWithControl($\mathcal{R}_s, \mathcal{R}_c$)

Require: $\mathcal{R}_s$ and $\mathcal{R}_c$ are the sets of mapped ChIP-seq reads for both sample and control, respectively.

Ensure: A list of peaks

1: Generate the signal profiles S_s = CreateSignalProfile($\mathcal{R}_s$) and S_c = CreateSignalProfile($\mathcal{R}_c$);

2: Set $\mathcal{S}$ be an empty set;

3: **for** every fixed size region $r_1..r_2$ **do**

4:     Include the region $r_1..r_2$ into $\mathcal{S}$ if $\lambda_s(r_1..r_2) >> \lambda_c(r_1..r_2)$;

5: **end for**

6: Overlapping regions in $\mathcal{S}$ are merged; We denote these regions as peaks;

7: Score and rank every peak in $\mathcal{S}$ and report them;

FIGURE 9.9: Algorithm to generate a list of peaks gives two sets of aligned ChIP-seq reads from the sample and control.

suggested that the signal in the control can be determined by $\lambda_c(r_1..r_2) = (r_2 - r_1 + 1)\max\{\lambda_{bg}, \lambda(h, 1000), \lambda(h, 5000), \lambda(h, 10000)\}$, where λ_{bg} is the genome-wide average read coverage per loci, $\lambda(h, \ell) = \frac{1}{\ell}\sum_{i=h-\ell/2}^{h+\ell/2} S_c[i]$, and h is the summit of the peak region $r_1..r_2$, i.e., $h = argmax_{r_1 \leq i \leq r_2} S_s[i]$.

9.3.3 Noise in the sample library

The above solution assumes the control library follows the noise model while the sample library follows the signal model. However, this is not exactly correct. The sample library is in fact a mixture of the signal and noise models. Figure 9.10 illustrates the idea.

Let α ($0 \leq \alpha \leq 1$) be the proportion of reads coming from the noise model, that is, the noise rate. If the ChIP experiment is efficient, we will expect α is close to 1. Otherwise, if the antibody is not good, we expect α is close to 0. We assume the noise model in the sample library is the same as the noise model

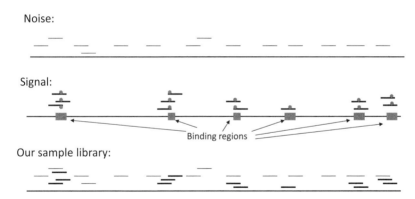

FIGURE 9.10: The linear signal-noise model. Our sample library is the mixture of the signal model and the noise model. The sample ChIP fragments are represented by thick edges while the noise fragments are represented by thin edges.

in the control library. A line of evidence supporting the above assumption was given by Rozowsky et al. [253]. They generated a scatter plot of ChIP and control read counts on a list of genomic regions with no detectable ChIP enrichment. A linear and nearly symmetric scatter pattern was observed.

We aim to estimate α. Suppose E is the set of regions with a real ChIP signal while $\bar{E}$ is the set of the remaining regions (i.e., background). By definition, for any region $r_1..r_2$ in $\bar{E}$, we have $\frac{\lambda_s(r_1..r_2)}{\lambda_c(r_1..r_2)} = \frac{\frac{1}{N}\sum_{i=r_1}^{r_2} S_s[i]}{\frac{1}{M}\sum_{i=r_1}^{r_2} S_c[i]} \approx \alpha$.

By this observation, α can be estimated iteratively as follows. We partition the genome into a set of bins. Initially, S_s is assumed to be a noise profile, that is, $\alpha = 1$. Then, two steps are iterated. The first step finds $\bar{E}$. Basically, a bin $r_1..r_2$ is considered to be a background region if $\frac{\lambda_s(r_1..r_2)}{\lambda_c(r_1..r_2)} \leq \alpha$. Then, we obtain a set the background regions $\bar{E}$. The second step computes the noise rate $\alpha = \frac{\sum_{r_1..r_2 \in \bar{E}} \lambda_s(r_1..r_2)}{\sum_{r_1..r_2 \in \bar{E}} \lambda_c(r_1..r_2)}$. The algorithm is shown in Figure 9.11. This algorithm is used in CCAT [322]. It is shown that the convergence is fast. In the spike-in simulation, the noise rate is converged in about 5 iterations. The noise rate estimation is accurate. The relative error is less than 5%.

9.3.4 Determination if a peak is significant

In Sections 9.3.2 and 9.3.3, a region $r_1..r_2$ is determined to be a peak if $\lambda_s(r_1..r_2) \gg \lambda_c(r_1..r_2)$. However, we do not state the exact method to determine if a peak is significant or not. This section aims to discuss this issue.

Recall that α is the noise ratio. To determine if the region $r_1..r_2$ is a signif-

---

Algorithm EstimateNoiseRate(S_s, S_c)

Require: S_s and S_c are the signal profiles of sample and control, respectively.

Ensure: The noise rate α

1: Set $\alpha = 1$;

2: Partition the whole genome into a set of equal-size bins.

3: **while** α is not stabilized **do**

4:     Identify a set $\bar{E}$ of bins $r_1..r_2$ such that $\dfrac{\frac{1}{N}\sum_{i=r_1}^{r_2} S_s[i]}{\frac{1}{M}\sum_{i=r_1}^{r_2} S_c[i]} \leq \alpha$;

5:     Set $\alpha = \dfrac{\frac{1}{N}\sum_{r_1..r_2 \in \bar{E}}\sum_{i=r_1}^{r_2} S_s[i]}{\frac{1}{M}\sum_{r_1..r_2 \in \bar{E}}\sum_{i=r_1}^{r_2} S_c[i]}$;

6: **end while**

7: Report α;

FIGURE 9.11: Algorithm to estimate the noise rate α.

icant peak, the simplest solution is to compute the fold enrichment $\frac{\lambda_s(r_1..r_2)}{\alpha\lambda_c(r_1..r_2)}$. If the fold enrichment is bigger than some user-defined cutoff (say, 5), the region is determined to be a significant peak.

Another way is to model the background distribution of the reads in the sample library. Assume reads are uniformly distributed. Then, the number of reads X that fall in the region $r_1..r_2$ can be modeled as a Poisson distribution $Poisson(\alpha\lambda_c(r_1..r_2))$. The region $r_1..r_2$ is called a significant peak if $Pr(X > \lambda_s(r_1..r_2))$ is smaller than some user-defined p-value cutoff (say 10^{-5}). This approach is used in MACS [331]. Apart from the Poisson distribution, other distributions are also used in the literature, including binomial distribution [253] and negative binomial distribution [119] (see Exercise 4).

In real life, the assumption that reads are uniformly distributed does not work properly. The reason is that the wet-lab noise is not uniformly distributed in the genome. Hence, the observed standard derivation is usually higher in real life. Another choice is to learn an empirical distribution. One solution is to use library swapping [331, 322]. The idea is as follows. Let $\mathcal{R}_s$ and $\mathcal{R}_c$ be the set of reads in the sample library and the control library, respectively. We run the algorithm in Figure 9.9 twice. First, by calling the algorithm CallPeakWithControl($\mathcal{R}_s, \mathcal{R}_c$), we compare the sample with respect to the control and obtain a set of peaks called signal peaks $\mathcal{P}_s$. Second, we swap the two libraries. By calling the algorithm CallPeakWithControl($\mathcal{R}_c, \mathcal{R}_s$), we compare the control with respect to the sample and obtain a set of peaks called control peaks $\mathcal{P}_c$. For every peak $r_1..r_2 \in \mathcal{P}_s \cup \mathcal{P}_c$, let $score(r_1..r_2)$ be its score (for example, the fold enrichment score $\frac{\lambda_s(r_1..r_2)}{\alpha\lambda_c(r_1..r_2)}$).

Suppose we want to control the false discovery rate (FDR) to be γ (i.e., the proportion of false peaks is γ). The cutoff score threshold is set to be θ such that we will accept at most γ percent of the control peaks. More precisely, θ

	Fold enrichment scores of peaks
Sample peaks	24.8, 23.2, 20.1, 19.5, 18.4, 7.5, 5.4, 4.2, 3.3, 3.1
Control peaks	4.3, 4.2, 4.1, 3.9, 3.9, 3.8, 3.7, 3.7, 3.5, 3.2

FIGURE 9.12: An example consists of 10 sample peaks and 10 control peaks. There are 7 sample peaks and 1 control peak whose fold enrichment scores are at least 4.3. With the cutoff threshold 4.3, the FDR is $1/8 = 0.125$.

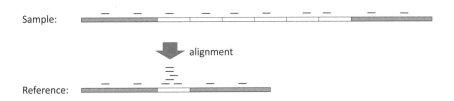

FIGURE 9.13: An example of a sample genome whose white region has 8 copies. Due to incompleteness of the reference, the white region appears exactly once in the reference genome. Assume one background ChIP-seq read occurs in each copy of the white region. All these 8 ChIP-seq reads will be aligned to the white region in the reference genome. This generates a false positive peak in the reference.

is set to be the cutoff such that $\frac{|\{score(r_1..r_2)>\theta|r_1..r_2\in\mathcal{P}_c\}|}{|\{score(r_1..r_2)>\theta|r_1..r_2\in\mathcal{P}_c\cup\mathcal{P}_s\}|} \leq \gamma$. Figure 9.12 gives an example.

9.3.5 Unannotated high copy number regions

Due to the limitation of current technologies, our reference genome is still incomplete. Some sequences are missed or under-represented in our reference genome. Here, we show that the incomplete genome affects the peak calling. (By analogy, Section 6.11 shows that the incomplete genome affects the variation calling.)

Figure 9.13 illustrates the problem. In the example, a region appears exactly once in the reference genome. However, such region actually has eight copies in the actual genome. When we align the reads generated from this repeat region, all of them are aligned to the same region and form a false peak [232].

To address this problem, we can align ChIP-seq reads on both the primary reference genome and the decoy sequences. (For example, the reference hs37d5 is formed by the primary reference genome (hg19) and the decoy sequences (see Section 6.11).) This approach filters some mis-aligned ChIP-seq reads, which reduces the number of false positive peaks.

However, the decoy sequences do not cover all missing and/or under-represented regions. Also, they cannot capture the copy numbers of the sequences. Miga et al. [201] proposed to build a sponge database. The sponge database is formed by all WGS reads that cannot be mapped to the primary reference genome. The database consists of 246M bp unmapped reads (which is roughly 8.2% of the human genome). Most of the reads in the sponge database are tandem repeats which include ribosomal repeats and satellite DNAs. Many of these sequences are enriched in centromeric regions and acrocentric short arms, which are not well assembled in our reference genome.

Given the sponge database, ChIP-seq reads are aligned to both the primary reference and the sponge database. When a read has multiple alignments, it is randomly assigned to one of the sites. This approach assigns the correct number of reads to each region and reduces the alignment bias. Miga et al. [201] showed that this approach reduces the number of false positive peaks while it will not affect the sensitivity of peak calling.

9.3.6 Constructing a signal profile by Kernel methods

The algorithm CreateSignalProfile in Figure 9.6 computes the signal profile by counting the number of ChIP-seq reads covering each loci. This means that each read assigns equal weight to every loci covered by it. However, we expect the binding site should be close to the center of the read (or the ChIP-seq fragment). Hence, it is good if we give higher weight to the center of each ChIP-seq fragment. Such a weighting scheme can be modeled by a smoothing kernel such as a Gaussian kernel. This idea is used by Fseq [29] and PeaKDEck [196]. Below, we describe the signal profile constructed by Fseq.

Let $\{x_1, \ldots, x_N\}$ be the centers of N ChIP-seq fragments (or reads). Each ChIP-seq fragment centered at x_i is represented by a Gaussian kernel with mean x_i and variance b, where b is a bandwidth that controls the smoothness. In Fseq, the default value of b is 600. By summing up all Gaussian kernels, we obtain the signal profile. Precisely, for position x, the signal equals $\hat{\rho}(x) = \frac{1}{bN} \sum_{i=1}^{N} K(\frac{x-x_i}{b})$, where $K(v) = \frac{1}{\sqrt{2\pi}} e^{-\frac{1}{2}v^2}$ is the Gaussian kernel with mean 0 and variance 1. Figure 9.14 illustrates the pileup of 10 Gaussian kernels.

To compute the signals $\hat{\rho}(x)$ for all positions x, we can directly apply the above formula. However, the computation takes $O(nN)$ time, where n is the size of the genome, which is slow. Fseq proposes a heuristic to speed up the computation. Precisely, Fseq ignores the values of the kernel functions that are smaller than $10^{-\gamma}$ for some user-defined γ. Denote w as the smallest integer such that $K(\frac{w}{b}) > 10^{-\gamma}$. Typically, w is in the order of thousands. For instance, when $b = 600$ and $\gamma = 6$, w equals 3047. Then, we can approximate $\hat{\rho}(x)$ by $\frac{1}{bN} \sum \{K(\frac{x-x_i}{b}) \mid |x - x_i| \le w\}$. (Note that when there are no points x_i where $|x - x_i| \le w$, we estimate $\hat{\rho}(x) = 0$.) With this heuristic, the signal profile can be computed in $O(wN)$ time using the algorithm in Figure 9.15 (see Exercise 7).

By kernel methods, we can improve the signal profile to the base pair resolution. However, this signal profile may contain many local noisy peaks.

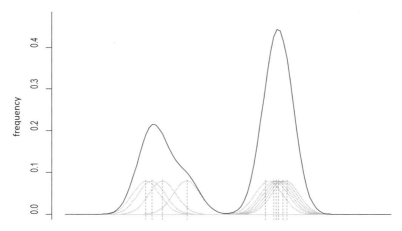

FIGURE 9.14: This is an example with 10 points $\{x_1, \ldots, x_{10}\}$ where each point is represented by a vertical line. Each point is represented as a Gaussian kernel. By piling up the 10 Gaussian kernels, we obtain the signal profile (the curve above the 10 Gaussian kernels).

9.4 Sequencing depth of the ChIP-seq libraries

To call peaks, we should sequence enough reads from the ChIP-seq library. This section discusses the issue of determining the sequencing depth.

Given a set $\mathcal{R}$ of N ChIP-seq reads, we can run the peak caller A to predict a list of peaks P. One important question is as follows: Can we identify more peaks by increasing the ChIP-seq sequencing depth? If the answer is yes, the biologist may want to do more ChIP-seq sequencing.

This question can be answered using saturation analysis. Given a random subset $\mathcal{R}'$ of ChIP-seq reads of $\mathcal{R}$, we can run the peak caller A and identify another set of peaks $\mathcal{P}'$. If $\mathcal{P}'$ and $\mathcal{P}$ are highly overlapped, we say the library is saturated. In other words, we will not discover a lot more peaks even if we increase the sequencing depth.

In particular, Kharchenko et al. [137] suggested the following criteria to check if a ChIP-seq library is saturated. Consider a set $\mathcal{R}$ of N ChIP-seq reads. For any subset $\mathcal{R}'$ of $\mathcal{R}$, let $\mathcal{P}_{\mathcal{R}'}$ be the set of peaks found by the peak caller A. Consider k size-x subsets $\mathcal{R}_1, \ldots, \mathcal{R}_k$ of $\mathcal{R}$. Denote $F_A(N, x)$ as the average proportion of peaks in $\mathcal{P}_{\mathcal{R}_i}$ that appear in $\mathcal{P}$, that is, $\frac{1}{k} \sum_{i=1}^{k} \frac{|\mathcal{P}_{\mathcal{R}_i} \cap \mathcal{P}|}{|\mathcal{P}|}$. If $F_A(N, N-10^5) > 0.99$, the ChIP-seq library $\mathcal{R}$ is said to be saturated. When a library is saturated, this means that we do not expect to discover more peaks even if we increase the sequencing depth.

Note that we can run the peak caller A under a different cutoff thresh-

---

Algorithm ComputeKernelSignalProfile(X, b, w)

Require: $X = \{x_1, \ldots, x_N\}$ is the centers of N ChIP fragments, b is the
  bandwidth and w is the window size.

Ensure: $\hat{\rho}(x) = \frac{1}{bN} \sum_{|x-x_i| \leq w} K(\frac{x-x_i}{b})$

1: **for** $j = -w$ to w **do**
2:     Compute $K(\frac{j}{b}) = \frac{1}{\sqrt{2\pi}} e^{-\frac{1}{2}(\frac{j}{b})^2}$;
3: **end for**
4: **for** each x_i **do**
5:     **for** $x = x_i - w$ to $x_i + w$ **do**
6:       Set $\hat{\rho}(x) = 0$;
7:     **end for**
8: **end for**
9: **for** each x_i **do**
10:     **for** $x = x_i - w$ to $x_i + w$ **do**
11:       $\hat{\rho}(x) = \hat{\rho}(x) + \frac{1}{bN} K(\frac{x-x_i}{b})$;
12:     **end for**
13: **end for**
14: Report $\hat{\rho}(x)$ which are non-zero;

FIGURE 9.15: Algorithm to compute the kernel signal profile.

old θ. Denote A_θ as the peak caller with cutoff threshold θ. Another way to ask the question is as follows: Under the sequencing depth of N reads, what is the minimum cutoff θ so that the peak caller A_θ can discover all peaks? This question can be answered by computing the minimal saturated enrichment ratio (MSER) [137]. Precisely, MSER is the minimum cutoff θ such that $F_{A_\theta}(N, N - 10^5) > 0.99$.

9.5 Further reading

This chapter discusses different computational techniques for calling ChIP-seq peaks given the sample ChIP-seq library and control ChIP-seq library. For a recent review of the ChIP-seq technologies, please read [224]. For a review of computational methods used in ChIP-seq, please read [118, 271].

When a control library is not available, we can still call peaks by assuming the noise is uniformly distributed. However, this approach will give many false peaks. To reduce the number of false peaks, we need to correct the background distribution of the reads based on the GC content, mappability, and copy number variation. This idea is used by PeakSeq [253].

This chapter gives methods that call peaks of a sample ChIP-seq library with respect to a control ChIP-seq library. These methods assume the two libraries are performed under the same experimental conditions like the same cell type. We may sometimes want to call peaks by comparing two ChIP-seq libraries under different experimental conditions. For instance, we may want to fix the TF and find its differential binding sites under two different tissues.

One solution is to conduct peak calling between the two ChIP-seq conditions treating one as the control. This solution might erroneously detect a region that is weak in both conditions but significantly enriched in one condition over the other.

Another method is to run peak calling in separate conditions followed by intersection analysis to identify unique peaks for each condition, but it might miscall a region when it is barely above and below the peak calling cutoff in the respective conditions. Some better methods are proposed. They include ChIPDiff [323], MAnorm [269] and dPCR [120].

Apart from DNA−protein interaction, there is a method called ChiRP-seq, which allows us to discover regions in our genome which are bound by a specific RNA.

9.6 Exercises

1. You are provided with the ChIP-seq datasets for Sox2, H3K4me1 and H3K4me3. Can you suggest a way to check if Sox2 is a promoter TF or an enhancer TF?

2. Suppose you are given a list of reads $\{R_1, \ldots, R_N\}$. Let $pos(R_i)$ be the leftmost position of each read R_i. Assume every read is of length m. Assume the genome is of length n. Let $S[1..n]$ be the signal profile, that is, $S[i]$ equals the number of reads covering position i. Can you give an $O(N + n)$ time algorithm to compute $S[1..n]$? (You can assume $pos(R_1) \leq pos(R_2) \leq \ldots \leq pos(R_N)$.)

3. Suppose $f(i)$ is a binary function (i.e. $f(i) \in \{0, 1\}$). Let $\mu_f = \frac{1}{n} \sum_{i=1}^{n} f(i)$ and $\sigma_f^2 = \frac{1}{n} \sum_{i=1}^{n} [f(i) - \mu_f]^2$. Please show that $\sigma_f^2 = \mu_f(1 - \mu_f)$. (Hint: Observe that $f(i)^2 = f(i)$.)

4. Suppose we only sequence a sample library. Assume there is no sequencing and mapping bias due to CG content and mappability. Suppose we partition the genome into a set of bins. Let a_i be the number of reads in the ith bin. We aim to determine if the ith bin is a peak. Can you describe a way to determine if the ith bin is a peak assuming the background distribution of the reads follows (a) a Poisson distribution, (b) a binomial distribution and (c) a negative binomial distribution?

5. We can model the background distribution of the reads using binomial distribution, Poisson distribution and negative binomial distribution. Can you comment which background distribution is a better fit for the ChIP-seq data?

6. Let S_s and S_c be the set of reads in a sample library and a control library, respectively. By calling the algorithm CallPeakWithControl(S_s, S_c), we obtain a list of sample peaks $(p_1, \ldots, p_{15})$ whose fold change scores are 18.2, 17.8, 16.5, 15.2, 14.1, 13.7, 13.2, 12.1, 10.8, 9.2, 8.5, 7.4, 6.1, 5.7. By calling the algorithm CallPeakWithControl(S_c, S_s), we obtain a list of control peaks $(q_1, \ldots, q_{10})$ whose fold change scores are 11.2, 10.3, 9.9, 8.7, 8.5, 7.3, 6.8, 5.4, 4.8, 4.1. Suppose we want the FDR to be 10%. Can you estimate the fold change score cutoff?

7. Please show that the algorithm in Figure 9.15 correctly computes $\hat{\rho}(x) = \frac{1}{bN} \sum_{|x-x_i| \le w} K(\frac{x-x_i}{b})$ for all positions x. Please also show that the time complexity of the algorithm is $O(wN)$.

Chapter 10

Data compression techniques used in NGS files

10.1 Introduction

NGS technologies generate hundreds of billions of DNA bases per day per machine. The throughput is expected to further increase in the future. At the same time, the demand of NGS service is increasing. The NGS service is no longer for research use only. Personalized genome sequencing is now available as a commercial service. Furthermore, the sequencing cost of NGS is reasonably low now. The cost of sequencing a human genome is approximately US$1000. The cost is expected to drop further.

Hence, the amount of NGS data is expected to increase dramatically. Even today, the amount of NGS data is already too much and the data management cost is already surpassed the sequencing cost. An important aspect of NGS data management is in raw data storage. Since the amount of raw data is huge, the current practice is to compress the data using standard and general-purpose data compression methods (e.g., gzip or bzip2) and archive them on tape. However, this approach makes data access inconvenient.

Another issue is data communication. We need to transfer the NGS data from one site to another site (say, transfer the data from the sequencing center to the research lab). If the dataset is huge, the network communication bandwidth may be too slow for transferring NGS data.

To resolve these problems, one solution is to reduce the size of the NGS files. This chapter covers approaches for NGS data compression. It is organized as follows. Sections 10.2−10.5 describe the strategies for compressing fasta/fastq files. Section 10.6 briefly covers the compression methods for other NGS data formats. (Before reading this chapter, it is recommended to read Section 3.6 first.)

10.2 Strategies for compressing fasta/fastq files

The NGS raw reads are represented in fasta or fastq format (see Section 2.2). They can be compressed using general-purpose data compression methods like gzip or bzip2. For example, fasta and fastq are usually compressed into fasta.gz and fastq.gz, respectively, by gzip. This approach can reduce the size to 30% of the original size. One important question is whether we can further compress them.

Observe that the fasta/fastq file can be separated into different fields. For instance, for a fastq file, the 3 fields are (a) identifiers, (b) DNA bases and (c) quality scores. Since different fields use different datatypes, it may be possible to improve the compression ratio by compressing different fields separately. In fact, we can improve the compression ratio by the following simple two-step method: (1) partition the fastq file into 3 files, each storing a particular field, (2) compress the 3 files separately using gzip. As a bonus, this method runs faster. The improvement is because similar datatypes share similar features, which make the data more compressible. For instance, each DNA base can be encoded using at most 2 bits. However, if we mix the DNA bases with other data, the compression method does not have prior knowledge and it may use more than 2 bits to store each DNA base.

This principle has been applied by many methods to compress fasta and fastq files, including fqzcomp [26], DSRC1 [64], DSRC2 [249] and LFQC [225]. G-SQZ [293] also uses this approach. Moreover, it treated each DNA base and its associated quality score together and recoded them using Huffman code. (Note that Huffman code has better compression efficiency when the alphabet size is big.)

Sections 10.3−10.5 are devoted to discussing specific compression methods for (a) identifiers, (b) DNA bases and (c) quality scores.

10.3 Techniques to compress identifiers

In fasta/fastq files, each read has an identifier. This section discusses methods used to compress identifiers. In particular, we discuss the technique used in fqzcomp [26] and LFQC [225].

Figure 10.1 lists some examples of identifiers. Observe that identifiers have uniform format. If we tokenize the identifiers and align them vertically, many columns will look similar. We can improve the compression by compressing each column using a different compression scheme. Precisely, for alphabetical columns, we can use run-length encoding (see Section 3.6.7) to encode them. For numerical columns, if the difference between the current value and the

```
@M01853:169:000000000-AE4HA:1:1101:14385:1130 1:N:0:1
@M01853:169:000000000-AE4HA:1:1101:16874:1132 1:N:0:1
@M01853:169:000000000-AE4HA:1:1101:14623:1133 1:N:0:1
@M01853:169:000000000-AE4HA:1:1101:10934:1134 1:N:0:1
@M01853:169:000000000-AE4HA:1:1101:12674:1134 1:N:0:1
```

FIGURE 10.1: An example of the read names extracted from a fastq file (from Illumina Mi-seq sequencer). We can observe that many columns look alike.

previous value is smaller than 255, we store their differences; otherwise, we store the current value itself. This method works well for identifiers that can be tokenized. For a free text identifier, gzip will be good enough.

10.4 Techniques to compress DNA bases

The alphabet for DNA bases is $\{A, C, G, T\}$. Hence, each nucleotide can be represented using $\log_2 4 = 2$ bits. Here, we aim to compress DNA bases using less than 2 bits per base. As shown in the literature, the entropy of a DNA sequence is between $1.8 - 2.0$ bits per base. This means that DNA raw reads are highly incompressible. If we compress DNA raw reads by general compressors (like gzip and bzip2), it is unlikely to give a good compression ratio. In this section, we discuss various techniques for compressing DNA bases. There are 4 techniques:

1. Statistical-based approach: e.g., Quip [126] and fqzcomp [26].

2. Reference-based approach: e.g., GenCompress [59], SlimGene [148], Quip [126] and Cramtool [107].

3. BWT approach (without reference): e.g., BEETL [56].

4. Assembly-based approach: e.g., Quip [126].

10.4.1 Statistical-based approach

The statistical-based approach tries to compress the DNA reads based on statistical properties. Note that reads are not pure random sequences. They are extracted from the genome of some species. When we have a long enough context (say, length-k substring), we can predict the next base. Based on this observation, we can use a k-th order Markov chain to model the reads (see Section 3.6.6).

Below, we detail the method. Consider a set of reads $\mathcal{R}$. First, we learn the k-th order Markov chain model. That is, we learn the probability $Pr(b|b_1 \ldots b_k)$ that b is the nucleotide following the k-mer $b_1 \ldots b_k$ in $\mathcal{R}$. As stated in Section 3.6.6, $Pr(b|b_1 \ldots b_k)$ can be learned by checking all $(k+1)$-mers in $\mathcal{R}$. Next, we encode every read $R \in \mathcal{R}$. We first store $R[1..k]$ using 2 bits per base. For $i > k$, the i-th base $R[i]$ is encoded by the Huffman code (or the arithmetic code) based on the frequency $Pr(R[i]|R[i-k..i-1])$. Figure 10.3 details the algorithm. Figure 10.2 gives an example for encoding three length-7 reads R_1, R_2, R_3 using the 3rd order Markov chain. For each read, the first three bases are encoded by the table in Figure 10.2(a); then, the remaining bases are encoded by the table in Figure 10.2(b). The lengths of the encodings of R_1, R_2, and R_3 are 7, 9 and 8, respectively. (As a comparison, if each length-7 read is encoded by the 2-bit code in Figure 10.2(a), each read requires 14 bits.)

The k-th order Markov chain approach is used by a few methods, including Quip [126] (with $k = 12$), fqzcomp [26] (with $k = 14$), DSRC1 [64] and DSRC2 [249] (with $k = 9$).

10.4.2 BWT-based approach

Let $\mathcal{R} = \{R_1, \ldots, R_m\}$ be a set of reads in a fasta file. $\mathcal{R}$ can be compressed using bzip2 as follows. First, all reads in $\mathcal{R}$ are concatenated and separated by end markers \$, i.e., $S = R_1 \$ R_2 \$ \ldots R_m \$$, where \$ is a non-nucleotide symbol that represents the end marker. Then, bzip2 first applies the Burrow–Wheeler transform (BWT, see Section 3.5.3) to convert S to a string $B[1..|S|]$. B has a few good properties: (1) it can convert back to S and (2) it has relatively more consecutive runs of the same symbols. Hence, B is more compressible. Bzip2 compresses B by applying different encoding techniques, like run-length encoding (see Section 3.6.7) and Huffman code (see Section 3.6.4).

For example, consider a set of 4 reads $S = \{$TCGCAT, AAGCAT, CTTCAT, GAGCAT$\}$. After adding end markers, we have

$$S = \text{TCGCAT\$AAGCAT\$CTTCAT\$GAGCAT\$}.$$

The second column of Figure 10.4 gives the string B, which is the BWT of S. We can see that B has many runs of the same nucleotides. Hence, with the help of run-length encoding, bzip2 can compress B better than S.

Although bzip2 can compress B better than S, BEETL [56] showed that we can further reduce the size by reordering B. Cox et al. [56] observed that the characters in B can be reordered while B still represents all reads in $\mathcal{R}$. Precisely, for elements with the same suffix (excluding the end marker), we reorder the characters in lexicographical order. The resulting string is called B_{SAP} and we can show that B_{SAP} also represents the same set of reads $\mathcal{R}$ (see Exercise 6). Furthermore, B_{SAP} has more consecutive runs of the same nucleotide than B. Hence, B_{SAP} is more compressible. BEETL compresses B_{SAP} using 7-zip.

base	code
A	00
C	01
G	10
T	11

(a)

$b_1b_2b_3b_4$	freq	$Pr(b_4|b_1b_2b_3)$	code
ACGT	1	1	
CGTC	2	2/4=0.5	0
CGTG	1	1/4=0.25	10
CGTT	1	1/4=0.25	11
GTCG	2	1	
GTTC	1	1	
TCGT	3	1	
TTCC	1	1	

(b)

	read	1	2	3	4	5	6	7	code
R_1	ACGTCGT	00	01	10		0			0001100
R_2	CGTCGTG	01	10	11	0			10	011011010
R_3	TCGTTCC	11	01	10		11			11011011

(c)

FIGURE 10.2: Consider three reads R_1 = ACGTCGT, R_2 = CGTCGTG and R_3 = TCGTTCC. Encode them using a 3rd order Markov chain. (a) The encoding table for each individual base. (b) The table of all 4-mers $b_1b_2b_3b_4$ in the R_1, R_2, R_3. The second column shows the frequency of the 4-mers. The third column shows $Pr(b_4|b_1b_2b_3)$. The fourth column gives the Huffman code. (c) shows the encoding for each base in R_1, R_2, R_3.

Algorithm k-Markov($\mathcal{R}$)

Require: $\mathcal{R}$ is a set of reads

Ensure: The encoding of every read in $\mathcal{R}$

1: For every length-$(k+1)$ substring $b_1 \ldots b_{k+1}$, set $c(b_1 \ldots b_{k+1})$ be its number of occurrences in $\mathcal{R}$;

2: **for** every length-k substring $b_1 \ldots b_k$ **do**

3:     **for** every base $b \in \{A, C, G, T\}$ **do**

4:         Set $Pr(b|b_1 \ldots b_k) = \frac{c(b_1 \ldots b_k b)}{c(b_1 \ldots b_k A) + c(b_1 \ldots b_k C) + c(b_1 \ldots b_k G) + c(b_1 \ldots b_k T)}$;

5:     **end for**

6:     Given the frequency $Pr(A|b_1 \ldots b_k), Pr(C|b_1 \ldots b_k), Pr(G|b_1 \ldots b_k)$ and $Pr(T|b_1 \ldots b_k)$, we compute Huffman code $H(b|b_1 \ldots b_k)$ for every base $b \in \{A, C, G, T\}$;

7: **end for**

8: **for** every read $R \in \mathcal{R}$ **do**

9:     We encode every base in $R[1..k]$ using 2 bits;

10:     **for** $i = k+1$ to $|R|$ **do**

11:         We encode $R[i]$ using Huffman code $H(R[i]|R[i-k..i-1])$;

12:     **end for**

13:     Report the encoding of R;

14: **end for**

FIGURE 10.3: The algorithm k-Markov($\mathcal{R}$) computes the encoding of all reads in $\mathcal{R}$.

As an illustration, for the string B in the second column of Figure 10.4, there are 5 intervals (highlighted by the curly brackets) whose suffixes are the same (excluding the end markers). We can reorder the characters in these intervals in lexicographical order to generate more consecutive runs of the same nucleotide. The resulting string is B_{SAP}, which is the last column of Figure 10.4. B_{SAP} is in fact the BWT of $S' = $ TCGCAT\$CTTCAT\$GAGCAT\$AAGCAT\$. Note that B_{SAP} still represents all reads in $\mathcal{R}$, though the ordering of the reads in S' is different from that in S.

10.4.3   Reference-based approach

Given the reference genome T (say human), NGS reads are expected to be perfectly or near perfectly aligned to T. By this observation, each NGS read can be represented by (1) its aligned interval in T and (2) the variations between the read and T. Storing the read as an interval requires less space than storing the actual DNA sequence itself. Empirical study showed that it can achieve 40−fold compression [148]. A number of methods use this approach to reduce the storage space. They include GenCompress [59], SlimGene [148], Cramtool [107] and Quip [126].

Below, we present the reference-based scheme used by Cramtool. In Cramtool, each read R is described by three features: the mapping position, the strand flag (1 bit, indicates if the read is mapped on the +ve or the −ve strand) and the match flag (1 bit, indicates if the read matches the reference T perfectly). If the match flag equals 0, we also store the list of variations. Each variation is stored as (1) its position on the read, (2) the variation type (substitution, insertion and deletion) and (3) the corresponding base change or length of deletion. For example, Figure 10.5(a) gives the alignments of three reads on the reference genome. Figure 10.5(b) represents the three reads by their features.

To store the features for each read space efficiently, we use different encoding schemes. The mapping positions are in increasing order. Instead of storing the absolute positions, we store the relatively positions of the reads. We observe that most of the relative positions are small. Hence, they can be encoded space-efficiently by Golomb code (see Section 3.6.3). (Note that Golomb code requires a parameter M, which is set to be the average of the relative distances between adjacent reads.) For the strand flag and the match flag, each is stored explicitly in 1 bit. Each variation has three features. For (1), the position is stored by Golomb code. (The parameter M for Golomb code is set to be the average distance between successive variations.) For (2), the three variation types are encoded in two bits. Precisely, substitution, insertion and deletion are encoded by 0, 10 and 11, respectively. Feature (3) describes the corresponding base change or length of deletion. For deletion, the length of deletion is encoded by Gamma code (see Section 3.6.2). For substitution, if base i is substituted with base j, we store $((C(j) - C(i)) \mod 5) - 1$ in 2 bits, where $C(\text{A}) = 0$, $C(\text{C}) = 1$, $C(\text{G}) = 2$, $C(\text{T}) = 3$, and $C(\text{N}) = 4$. For example,

sorted suffixes	B[]	$B_{SAP}[]$
$\$_1$	T	T
$\$_2$	T	T
$\$_3$	T	T
$\$_4$	T	T
AAGCAT$\$_2$	\$	\$
AGCAT$\$_2$	A	A
AGCAT$\$_4$	G	G
AT$\$_1$	C	C
AT$\$_2$	C	C
AT$\$_3$	C	C
AT$\$_4$	C	C
CAT$\$_1$	G	G
CAT$\$_2$	G	G
CAT$\$_3$	T	G
CAT$\$_4$	G	T
CGCAT$\$_1$	T	T
CTTCAT$\$_3$	\$	\$
GAGCAT$\$_4$	\$	\$
GCAT$\$_1$	C	A
GCAT$\$_2$	A	A
GCAT$\$_4$	A	C
T$\$_1$	A	A
T$\$_2$	A	A
T$\$_3$	A	A
T$\$_4$	A	A
TCAT$\$_3$	T	T
TCGCAT$\$_1$	\$	\$
TTCAT$\$_3$	C	C

FIGURE 10.4: Consider S = TCGCAT\$AAGCAT\$CTTCAT\$GAGCAT\$. The first column of the figure is the list of suffixes of S sorted in alphabetical order. As shorthand, we denote $\$_1$, $\$_2$, $\$_3$ and $\$_4$ as \$AAGCAT\$CTTCAT\$GAGCAT\$, \$CTTCAT\$GAGCAT\$, \$GAGCAT\$ and \$, respectively. We also highlight elements with the same suffix using the curly brackets. The second column is the BWT $B[]$ of S. The last column is $B_{SAP}[]$, which is formed by the reordering of the characters in B whose corresponding suffixes (excluding the end markers) are the same.

```
                1           2          3
        12345678901234567890123456789012  345678901234567
Ref: CTACAAGACATGCCAGACATGA--CACGTACCAGAGGAT
r1                              TGAGTCACGT
r2        CAAGTCA---CAG
r3                                  CGTACCAGAG
```
(a)

Name	Pos	Strand	Match	Variation 1	Variation 2
r2	4	+ (1)	0	(5,S,A→T)	(8,D,3)
r1	20	+ (1)	0	(4,I,GT)	
r3	25	+ (1)	1		

(b)

Golomb$_8$(4), 1, 0, [Golomb$_4$(5), 0, 10], [Golomb$_4$(3), 11, Gamma(3)]

Golomb$_8$(16), 1, 0, [Golomb$_4$(4), 10, 100, 101, 111]

Golomb$_8$(5), 1, 1

(c)

FIGURE 10.5: (a) An example of alignments of three reads to the reference. (b) The reads are sorted in increasing order of their mapping locations. The table shows the strand flag, match flag, and the list of variations. (c) The binary encoding of the three reads. (*Gamma(v)*) stands for Gamma encoding of v and $Golomb_M(v)$ stands for Golomb encoding of v with parameter M.)

for the substitution from A to G, since $((C(G) - C(A)) \mod 5) - 1 = 1$, such substitution is represented by 01. For insertion, we encode A, C, G, T and N by 00, 01, 100, 101 and 110, respectively. We also use 111 to denote the end of the sequence. Figure 10.5(c) gives the actual binary encodings of the three reads.

Paired-end reads can be compressed similarly. In addition, we need to store paired-end read information, which includes a bit indicating whether a read is paired, whether the read is the first of the pair, and the relative offset between the positions of the pair of reads (using Golomb code). An example is shown in Figure 10.6.

10.4.4 Assembly-based approach

Reads can be compressed effectively if the reference genome is known (Section 10.4.3). If the reference genome is not available, we can apply genome assembly to reconstruct the genome (see Chapter 5). Then, by treating the reconstructed assembly as a reference, we can store the reads using the reference-based approach. This approach has a good compression ratio. However, computing a full assembly is slow and memory intensive, which makes the approach impractical.

```
              1                 2              3              4
   1234567890123456789012  3456789012345678901234
Ref: ATCGAACTGACTGGACTACAAC**CGTGACTGGACTGATCTAGACT
R1        AAGTG--TCG................................GGTGTGAT
R2           TGACTGGA.........................GACGGGAC
R3                   CAGCGTCG
```

(a)

Name	Pos	Strand	Match	Variation 1	Variation 2	Variation 3	Paired	First	Offset
R1	5	+ (1)	0	(3,S,C→G)	(6,D,2)	(7,S,G→C)	1	1	4
R2	8	+ (1)	1				1	1	2
R3	19	+ (1)	0	(3,S,A→G)	(5,I,GC)		0		
R2	26	- (0)	0	(4,S,T→G)			1		
R1	30	- (0)	0	(3,S,A→T)	(4,S,C→G)		1		

(b)

Golomb$_5$(5), 1, 0, [Golomb$_2$(3), 0, 00], [Golomb$_2$(3), 11, Gamma(2)], [Golomb$_2$(1), 0, 11], 1, 1, Golomb$_3$(4)

Golomb$_5$(3), 1, 1, 1, 1, Golomb$_3$(2)

Golomb$_5$(11), 1, 0, [Golomb$_2$(3), 0, 01], [Golomb$_2$(2), 10, 01, 100, 111], 0

Golomb$_5$(7), 0, 0, [Golomb$_2$(4), 0, 10], 1

Golomb$_5$(4), 0, 0, [Golomb$_2$(3), 0, 10], [Golomb$_2$(1), 0, 00], 1

(c)

FIGURE 10.6: Consider two paired-end reads $R_1 = (\text{AAGTGTCG}, \text{ATCACACC})$ and $R_2 = (\text{TGACTGGA}, \text{GTCCCGTC})$ and one single-end read $R_3 = \text{CAGCGTCG}$. (a) An example of alignments of R_1, R_2 and R_3 on the reference. (b) The reads are sorted in increasing order of their mapping locations. The table shows the strand flag, match flag, the list of variations, and the paired-end read information. (c) The binary encoding of the three reads. (*Gamma*(v) stands for Gamma encoding of v and *Golomb$_M$*(v) stands for Golomb encoding of v with parameter M.)

Instead of reconstructing an accurate genome, computing an approximate genome may be good enough for the purpose of compression. This approach is used by Quip [126]. Quip proposed a simplified greedy genome assembler. Although this assembler is not accurate, it is fast and generates a good enough reference for compression. The algorithm of the assembler is as follows. First, k-mers (by default, k=25) are extracted from the first 2.5 million reads (by default) in the fastq file. These k-mers are stored in a d-left counting Bloom filter (Section 3.4.3). A read sequence is used as a seed and a contig is obtained by extending the seed on both ends base-by-base. The extension is done greedily by extending the most frequent k-mer (the most frequent k-mer is determined by querying the counting Bloom filter) one by one. These contigs are then used as the reference database to encode the reads.

For example, consider the following set of reads: CAAGAT, AAGATC, GATCGA and ATCGAT. Assume $k = 4$. The method first stores all 4-mers using a counting Bloom Filter. Then, a read, say, GATCGA is chosen as a seed. The seed is extended in both directions. To extend the seed to the left base-by-base, there are two possible extensions: A and C with AGAT occurs twice and CGAT occurs once. By the greedy approach, we extend to the left by A and obtain AGATCGA. By applying this greedy approach iteratively to the left, the seed will be extended to CAAGATCGA. Similarly, we can extend the seed to the right and, finally, we obtain the contig CAAGATCGAT.

With this contig as the reference, the four reads CAAGAT, AAGATC, GATCGA and ATCGAT can be represented by positions 1, 2, 4 and 5, respectively.

10.5 Quality score compression methods

Apart from the DNA sequences, the quality scores are another important part of the NGS raw sequence data. The quality score of each base typically has $41-46$ distinct values. Hence, $6(= \lceil \log_2 41 \rceil)$ bits are required to store each quality score. This section discusses techniques that reduce the storage for quality scores. Roughly speaking, we can classify the techniques into lossless compression and lossy compression.

- Lossless compression

  - Storing quality score differences between adjacent bases (e.g., Slim-Gene [148] and Wan et al. [302])

  - Storing quality scores using a Markov chain (e.g., SlimGene [148] and Quip)

- Lossy compression

  - Quality budgeting (e.g., Cramtool [107])

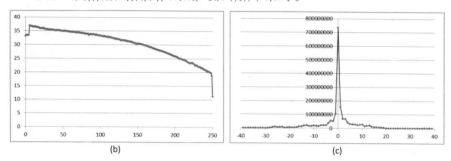

(a) @M01730:133:000000000-ADYHE:1:1101:14991:1655 1:N:0:1
    TGTACCTCTCTATCGCTCTCTCTCTCTTTCTCTTCTTCCTCTTCTTCTCTTGTCGACAACAGCATCACAACACCAACGCTGGCACGCCGTCAGC
    GTCAGCTGAGCAGCCTGCTTTTAGTTTGCTTTCGCGCTATCCAGTCGCCTTTTTCCGTTCGTGTTGAACACACGCGTTTTTCAGCTCCACGCCG
    TTACTCGCCTTTCACATCTCTTTTCCTTTTGTATTTGTTTTCCGTCGTCGTTCACTCCCCATC
    +
    68A8CEEFFFF9FEF@EEFCEFFEFFEFFFFFFFGFFFEEEFFFEEECFFGG,CF,,@,+C,,;,;C,C,,C8@@,,B,7B,+89,8@@++@,,<
    +,B,,BE,,,B,,4,,,:9FF,,,:CE,<EFDE+8++8++9=,,:C:++8C=E3:+8>A+@+,8,,,8,8,@+6+8>B@+4,,,,,,,,*44*
    *6*469***4;<;,,,5514=<,,;4,4=,,*++4++4;**3</+*))1))*/*+;+**1*0*

FIGURE 10.7: (a) shows one read in fastq format, where the read is of length 251. (b) shows the average quality scores at positions 1−251. The quality scores decrease as the positions increase. (c) shows the distribution of the differences between successive quality scores. Most of the differences are zero.

 – Binning (e.g., SlimGene [148] and Wan et al. [302])

Below, we discuss these techniques in detail.

10.5.1   Lossless compression

Before we discuss methods for compressing quality scores, we first look to see if there are some useful properties. For illustration, we use an in-house fastq file generated by Illumina Mi-seq where every read is of length 251 bp. Figure 10.7(a) shows one of the reads in the file. From the quality scores of the read, we observe that the quality scores decrease as the positions increase and there are many runs of the same quality scores. In fact, this observation is true in general. Figure 10.7(b) shows the average quality scores at different positions on the reads. Although the quality score has a big range, the quality scores gradually decrease as the positions increase. Another observation is that the quality scores at adjacent positions are usually similar. This is illustrated in Figure 10.7(c). It shows that the differences between successive quality scores are mostly zero. Furthermore, SlimGene also observed that the entropy is low at the beginning (all values are high), and at the end (all values are low), but increases in the middle, with an average entropy of 3.85. By these observations, Fqzcomp [26], DSRC1 [64], DSRC2 [249], Wan et al. [302], SlimGene [148] and Quip [126] suggest that the quality scores can be compressed. Below, we briefly describe their ideas.

Suppose the quality scores for a length-n read are $(q_1, q_2, \ldots, q_n)$. Wan et al. [302] first convert it to $(q_1 - 33, q_2 - q_1, \ldots, q_n - q_{n-1})$, where 33 is the minimum quality score. By Figure 10.7(c), we expect most of these successive differences are small integers. These numbers can be positive or negative. They are converted to positive values by bisection, i.e., $c(v) = 2|v| - 1$ if v is positive and $2|v|$, otherwise. (Precisely, $0 \to 0, 1 \to 1, -1 \to 2, 2 \to 3, -2 \to 4, \ldots$) After the conversion, we obtain $(c(q_1 - 33), c(q_2 - q_1), \ldots, c(q_n - q_{n-1}))$. The transformed values are called T-gaps. Using Golomb code, Wan et al. showed that each quality score can be encoded using approximately 3 bits (which is smaller than 6 bits).

Another compression scheme is proposed by SlimGene [148]. It is also based on the fact that two successive quality scores q_i and q_{i+1} have a small difference. In other words, the quality score q_{i+1} is in the range $[q_i - 1, q_i + 1]$ with high probability. Precisely, we have

$$Pr(q_i \to q_{i+1}) = \frac{\text{number of pairs of successive quality scores } (q_i, q_{i+1})}{\text{number of quality score } q_i}.$$

Given this probability, SlimGene [148] models the quality scores by a 1-order Markov model (see Section 3.6.6). Every transition from q_i to q_{i+1} is encoded using a Huffman code of $-\log(Pr(q_i \to q_{i+1}))$ bits. The entropy of the Markov model is approximately 3.3 bits.

Using a similar principle as [148], Quip [126] uses a 3-order Markov model instead. Furthermore, Quip observed that some reads may have highly variable quality scores. A read is considered as having highly variable quality scores if it has many large jumps (i.e., the positions where $|q_i - q_{i+1}| > 1$). Quip modified the Markov model so that it not only depends on the quality scores of the preceding 3 positions, it also depends on the running count of the number of large jumps. For DSRC1 [64] and DSRC2 [249], the quality values are stored in one of the two ways: (1) an order-1 Huffman coder or (2) arithmetic code with context lengths up to 6.

Fqzcomp [26] observed that the initial quality value for each read tended to be a common maximum value; then, the quality values decline along the length of the sequence (see Figure 10.7(b)). Fqzcomp predicts q_i using the following contexts: (1) q_{i-1}, (2) $\max\{q_{i-2}, q_{i-3}\}$, (3) a bit indicates if $q_{i-2} = q_{i-3}$, (4) $\min\{7, \lfloor \frac{1}{8} \sum_{j=1}^{i-1} \max\{0, q_{j-1} - q_j\} \rfloor\}$, and (5) $\min\{7, \lfloor i/8 \rfloor\}$.

10.5.2 Lossy compression

Although lossless compression improves the compression of quality scores, we still need approximately 3 bits per base. It is resource demanding to store quality scores.

Unlike the base information, it is not necessary to store the exact quality scores. Hence, a number of lossy compression methods have been suggested. They include quality budgeting and various different binning methods.

Quality budgeting is used in Cramtool [107]. It is assumed that quality scores are mostly useful for positions with variations. Instead of storing the quality score for every base, we only store quality scores for positions with variations. People can store additional portions of quality scores based on their preference. The list of selected quality scores is then further compressed using a Huffman-based code.

Another solution reduces the space by binning. Note that the PHRED quality score converts probability $0..1$ into 41 quality values in the range $33..73$. It is in fact a form of binning. For lossy transformation, we further reduce the number of bins. SlimGene [148] is the first group that tried to use binning approach. It stores the quality score using b bits where b is a user-defined value. Precisely, for any quality score Q, we denote $LQscore_b(Q) = rrand(\frac{(Q-Q_{min})2^b}{(Q_{max}-Q_{min})})$, where Q_{max} is the upper bound of the quality score (73, default), Q_{min} is the lower bound of the quality score (33, default) and $rrand(x) = \lceil x \rceil$ with probability $(x - \lfloor x \rfloor)$; and $\lfloor x \rfloor$ otherwise. Then,

$$Q_{approx} = Q_{min} + LQscore_b(Q).$$

Given the approximate quality score Q_{approx} for every base, SlimGene applies a 1-order Markov chain to further reduce the storage space (see Section 10.5.1).

Wan et al. [302] tried to systematically study different binning strategies. They suggested three ways to perform binning:

- UniBinning: The probability $0..1$ is uniformly partitioned into k bins. For example, if we partition the probability into 5 bins, the probability ranges of the 5 bins are $1.0-0.8$, $0.8-0.6$, $0.6-0.4$, $0.4-0.2$ and $0.2-0.0$. If the probability is p, the corresponding quality score is $rrand(10 \log_{10} p)$. Hence, the quality scores for the 5 bins are $0-33$, $34-34$, $35-35$, $36-37$ and $38-40$, respectively.

- Truncating: Transforming $33..126$ to $33..(33 + k - 1)$. In other words, if the quality score is in the range $(33 + k - 1)..126$, we replace it by $(33 + k - 1)$.

- LogBinning: Transforming $33..126$ into k equal bins. For example, if we would like to obtain $k = 19$ bins, then the first 18 bins each have 5 scores while the last bin has 4 scores, i.e., $123..126$.

Wan et al. [302] suggested that logBinning is the best. Note that the binning approach used by SlimGene [148] is logBinning.

10.6   Compression of other NGS data

Previous sections focused on methods for compressing fasta/fastq files. They are useful for archiving data.

In NGS analysis, we have other file types that are for genome annotation and for data analysis. These file types include SAM, BED, VCF, Wiggle, bed-Graph, etc (see Chapter 2). They are not only for archiving. Users will query these files to perform analysis. Hence, we not only need to reduce their sizes, but also need to enable efficient query. The standard solution is as follows. The features are sorted by genomic locations. Then, they are partitioned into blocks $B_1, \ldots, B_z$, where each block B_i represents features in a particular interval of the genome. Each block is compressed using gzip-like method. To enable random access, an R-tree data structure is built on top of all blocks. The R-tree indexes the starting positions of all blocks. To identify information appearing at position p in the genome, we just query the R-tree to find the block containing position p; then, information can be extracted from that block.

The above standard approach is widely used for compressing various NGS file types. For SAM, the compressed format is BAM. For BED, the compressed format is bigBed [135]. For VCF, the compressed format is BCF. For Wiggle and bedGraph formats, the compressed format is bigWig [135].

For bigWig, in addition to enabling fast access to data, it also supports the following query types for any selected interval: max, min, average, standard derivation (as shown in Section 2.6). These queries facilitate efficient downstream analysis and enable fast visualization of the data.

For example, bigWig supports the mean intensity query, which reports the mean intensity $\frac{1}{(q-p+1)} \sum_{i=p}^{q} r_i$ for every genomic region $p..q$, where r_i is the intensity for the genome position i.

To support efficient mean intensity query, bigWig file stores more information. Every block (starts at position s) in the bigWig file stores the intensity sum $\sum_{i=1}^{s-1} r_i$. To compute the mean intensity of the region $p..q$, we can query the R-tree to find the block B_x whose starting position p' is just bigger than p and the block B_y whose starting position q' is just smaller than q. Note that $\frac{1}{q-p+1} \sum_{i=p}^{q} r_i$ equals

$$\frac{1}{q-p+1} \left(\sum_{i=p}^{p'-1} r_i + (\sum_{i=1}^{q'-1} r_i - \sum_{i=1}^{p'-1} r_i) + \sum_{i=q'}^{q} r_i \right).$$

The second and third summations in the equation can be extracted from the blocks B_x and B_y. The first and the last summations can be computed by extracting the information from blocks B_{x-1} and B_y, respectively.

Utilizing bigWig, the UCSC genome browser can support interactive browsing of density data. BigWig is one of the most popular track types. In the hg19 browser, about 4400 tracks (10% of all hg19 tracks) are bigWig tracks; and they use 1.6 TB (it is equivalent to 31% of the total space used to store all UCSC hg19 tracks). In the future, the number of bigWig tracks is expected to increase. Hence, a new file format cWig [112] is proposed. It reduces the file size of bigWig threefold while it improves the query time by 10- to 100- folds.

10.7 Exercises

1. Consider the sequence AACGTAACGTAACGTACGTC. What is the 0-order entropy, the 1-order entropy, and the 2-order entropy?

2. Consider a sequence $R = $ ACTCGAAGCA and its corresponding sequence of quality scores, $Q = (40, 30, 20, 30, 10, 20, 40, 20, 30, 20)$.

    (a) What is the space required to encode R using Huffman code?

    (b) What is the space required to encode Q using Huffman code?

    (c) What is the space required to encode $\{(R[i], Q[i]) \mid i = 1, \ldots, 10\}$ using Huffman code? Is it better than storing R and Q separately?

3. Consider two DNA reads $R_1 = $ CGCGCT and $R_2 = $ TCGCGA. Can you encode them using a second-order Markov chain? What are the lengths of the encoded reads?

4. Consider two DNA reads R_1 and R_2 of length 6. This table shows the frequencies of all 3-mers in R_1 and R_2. Using a second-order Markov chain, R_1 and R_2 are encoded as 010100011 and 01011010, respectively. Suppose $R_2 = $ CCTCCT. Can you guess what R_1 is?

$b_1 b_2 b_3$	CCA	CCC	CCT	CTC	CTC
freq	1	3	2	1	1

5. Can you build the BWT $B[]$ and $B_{SAP}[]$ for the string $S = $ GTCAGA\$CCTCG\$ACCAGA\$GGTCG\$?

6. Consider a set of reads $\mathcal{S}$. Let S be the concatenation of all the reads in $\mathcal{S}$, separated by the end marker \$. Let B be the BWT of S. Let B_{SAP} be the sequence formed by the reordering of the characters in B whose corresponding suffixes (excluding the end markers) are the same. Can you show that both B and B_{SAP} represent the same set of reads $\mathcal{S}$?

7. Consider three reads $R_1 = $ (CTAGACAGAT, GCGGCATCT), $R_2 = $ TAAACTCTCT, $R_3 = $ (GTCTCGTG, CACAGTAG). The following figure shows the alignment of the three reads. Please encode these three reads by the encoding scheme of Cramtools.

```
              1              2              3            4
     12345  67890123456789012345  678901234567890123456789
  T: CACTA-ACAGATTTAGCCACAGTAGA-GCCGCATTGCCACGATACAGCCCA
  R1    CTAGACAGAT..........................AGATGCCGC
  R2      AGAGAG-TTTA
  R3                   CACAGTAG...................CACGAGAC
```

8. Consider a read of length 10. Suppose the corresponding quality scores for the 10 bases are $(60, 59, 59, 40, 58, 59, 57, 56, 58, 57)$. Can you transform it to T-gaps?

9. Suppose we aim to partition the quality scores $33-73$ into 4 bins. For UniBinning, Trancating and LogBinning, can you state the mapping from the quality scores to the 4 bins for (a) UniBinning, (b) Trancating and (c) LogBinning?

10. Section 10.6 discussed the data structure of bigWig for computing mean intensity of any genomic regions. Can you propose a data structure for computing the standard derivation of intensities in any genomic region?

References

[1] 1000 Genomes Project Consortium, G. R. Abecasis, A. Auton, L. D. Brooks, M. A. DePristo, R. M. Durbin, R. E. Handsaker, H. M. Kang, G. T. Marth, and G. A. McVean. An integrated map of genetic variation from 1,092 human genomes. *Nature*, 491(7422):56–65, Nov 2012.

[2] A. Abyzov, A. E. Urban, M. Snyder, and M. Gerstein. CNVnator: An approach to discover, genotype, and characterize typical and atypical CNVs from family and population genome sequencing. *Genome Research*, 21(6):974–984, Jun 2011.

[3] E. Ahrné, L. Molzahn, T. Glatter, and A. Schmidt. Critical assessment of proteome-wide label-free absolute abundance estimation strategies. *Proteomics*, 13(17):2567–2578, Sep 2013.

[4] C. A. Albers, G. Lunter, D. G. MacArthur, G. McVean, W. H. Ouwehand, and R. Durbin. Dindel: Accurate indel calls from short-read data. *Genome Research*, 21(6):961–973, Jun 2011.

[5] S. F. Altschul, T. L. Madden, A. A. Schäffer, J. Zhang, Z. Zhang, W. Miller, and D. J. Lipman. Gapped BLAST and PSI-BLAST: A new generation of protein database search programs. *Nucleic Acids Res*, 25(17):3389–3402, Sep 1997.

[6] A. Ameur, A. Wetterbom, L. Feuk, and U. Gyllensten. Global and unbiased detection of splice junctions from RNA-seq data. *Genome Biol*, 11(3):R34, 2010.

[7] A. Amir, M. Lewenstein, and E. Porat. Faster algorithms for string matching with k mismatches. *Journal of Algorithms*, 50(2):257–275, Feb 2004.

[8] E. L. Anson and E. W. Myers. ReAligner: A program for refining DNA sequence multi-alignments. *Journal of Computational Biology*, 4(3):369–383, 1997.

[9] P. N. Ariyaratne and W.-K. Sung. PE-Assembler: De novo assembler using short paired-end reads. *Bioinformatics*, 27(2):167–174, 2011.

[10] K. F. Au, H. Jiang, L. Lin, Y. Xing, and W. H. Wong. Detection of splice junctions from paired-end RNA-seq data by SpliceMap. *Nucleic Acids Res*, 38(14):4570–4578, Aug 2010.

[11] P. N. C. B. Audergon, S. Catania, A. Kagansky, P. Tong, M. Shukla, A. L. Pidoux, and R. C. Allshire. Restricted epigenetic inheritance of H3K9 methylation. *Science*, 348(6230):132–135, Apr 2015.

[12] O. T. Avery, C. M. Macleod, and M. McCarty. Studies on the chemical nature of the substance inducing transformation of pneumococcal types: Induction of transformation by a desoxyribonucleic acid fraction isolated from pneumococcus type III. *J Exp Med*, 79(2):137–158, Feb 1944.

[13] A. Bankevich, S. Nurk, D. Antipov, A. A. Gurevich, M. Dvorkin, A. S. Kulikov, V. M. Lesin, S. I. Nikolenko, S. Pham, A. D. Prjibelski, A. V. Pyshkin, A. V. Sirotkin, N. Vyahhi, G. Tesler, M. A. Alekseyev, and P. A. Pevzner. SPAdes: A new genome assembly algorithm and its applications to single-cell sequencing. *Journal of Computational Biology*, 19(5):455–477, May 2012.

[14] V. Bansal, O. Harismendy, R. Tewhey, S. S. Murray, N. J. Schork, E. J. Topol, and K. A. Frazer. Accurate detection and genotyping of SNPs utilizing population sequencing data. *Genome Research*, 20(4):537–545, Apr 2010.

[15] H. Bao, Y. Xiong, H. Guo, R. Zhou, X. Lu, Z. Yang, Y. Zhong, and S. Shi. MapNext: A software tool for spliced and unspliced alignments and SNP detection of short sequence reads. *BMC Genomics*, 10 Suppl 3:S13, 2009.

[16] D. W. Barnett, E. K. Garrison, A. R. Quinlan, M. P. Strömberg, and G. T. Marth. BamTools: A C++ API and toolkit for analyzing and managing BAM files. *Bioinformatics*, 27(12):1691–1692, Jun 2011.

[17] C. Bartenhagen and M. Dugas. Robust and exact structural variation detection with paired-end and soft-clipped alignments: SoftSV compared with eight algorithms. *Brief Bioinform*, 17(1):51–62, Jan 2016.

[18] Y. Benjamini and T. P. Speed. Summarizing and correcting the GC content bias in high-throughput sequencing. *Nucleic Acids Res*, 40(10):e72, May 2012.

[19] D. R. Bentley, S. Balasubramanian, H. P. Swerdlow, G. P. Smith, J. Milton, C. G. Brown, K. P. Hall, D. J. Evers, C. L. Barnes, H. R. Bignell, J. M. Boutell, J. Bryant, R. J. Carter, R. Keira Cheetham, A. J. Cox, D. J. Ellis, M. R. Flatbush, N. A. Gormley, S. J. Humphray, L. J. Irving, M. S. Karbelashvili, S. M. Kirk, H. Li, X. Liu, K. S. Maisinger, L. J. Murray, B. Obradovic, T. Ost, M. L. Parkinson, M. R. Pratt, I. M. J. Rasolonjatovo, M. T. Reed, R. Rigatti, C. Rodighiero, M. T. Ross, A. Sabot, S. V. Sankar, A. Scally, G. P. Schroth, M. E. Smith, V. P. Smith, A. Spiridou, P. E. Torrance, S. S. Tzonev, E. H. Vermaas, K. Walter, X. Wu, L. Zhang, M. D. Alam, C. Anastasi, I. C. Aniebo,

D. M. D. Bailey, I. R. Bancarz, S. Banerjee, S. G. Barbour, P. A. Baybayan, V. A. Benoit, K. F. Benson, C. Bevis, P. J. Black, A. Boodhun, J. S. Brennan, J. A. Bridgham, R. C. Brown, A. A. Brown, D. H. Buermann, A. A. Bundu, J. C. Burrows, N. P. Carter, N. Castillo, M. Chiara E Catenazzi, S. Chang, R. Neil Cooley, N. R. Crake, O. O. Dada, K. D. Diakoumakos, B. Dominguez-Fernandez, D. J. Earnshaw, U. C. Egbujor, D. W. Elmore, S. S. Etchin, M. R. Ewan, M. Fedurco, L. J. Fraser, K. V. Fuentes Fajardo, W. Scott Furey, D. George, K. J. Gietzen, C. P. Goddard, G. S. Golda, P. A. Granieri, D. E. Green, D. L. Gustafson, N. F. Hansen, K. Harnish, C. D. Haudenschild, N. I. Heyer, M. M. Hims, J. T. Ho, A. M. Horgan, K. Hoschler, S. Hurwitz, D. V. Ivanov, M. Q. Johnson, T. James, T. A. Huw Jones, G.-D. Kang, T. H. Kerelska, A. D. Kersey, I. Khrebtukova, A. P. Kindwall, Z. Kingsbury, P. I. Kokko-Gonzales, A. Kumar, M. A. Laurent, C. T. Lawley, S. E. Lee, X. Lee, A. K. Liao, J. A. Loch, M. Lok, S. Luo, R. M. Mammen, J. W. Martin, P. G. McCauley, P. McNitt, P. Mehta, K. W. Moon, J. W. Mullens, T. Newington, Z. Ning, B. Ling Ng, S. M. Novo, M. J. O'Neill, M. A. Osborne, A. Osnowski, O. Ostadan, L. L. Paraschos, L. Pickering, A. C. Pike, A. C. Pike, D. Chris Pinkard, D. P. Pliskin, J. Podhasky, V. J. Quijano, C. Raczy, V. H. Rae, S. R. Rawlings, A. Chiva Rodriguez, P. M. Roe, J. Rogers, M. C. Rogert Bacigalupo, N. Romanov, A. Romieu, R. K. Roth, N. J. Rourke, S. T. Ruediger, E. Rusman, R. M. Sanches-Kuiper, M. R. Schenker, J. M. Seoane, R. J. Shaw, M. K. Shiver, S. W. Short, N. L. Sizto, J. P. Sluis, M. A. Smith, J. Ernest Sohna Sohna, E. J. Spence, K. Stevens, N. Sutton, L. Szajkowski, C. L. Tregidgo, G. Turcatti, S. Vandevondele, Y. Verhovsky, S. M. Virk, S. Wakelin, G. C. Walcott, J. Wang, G. J. Worsley, J. Yan, L. Yau, M. Zuerlein, J. Rogers, J. C. Mullikin, M. E. Hurles, N. J. McCooke, J. S. West, F. L. Oaks, P. L. Lundberg, D. Klenerman, R. Durbin, and A. J. Smith. Accurate whole human genome sequencing using reversible terminator chemistry. *Nature*, 456(7218):53–59, Nov 2008.

[20] K. Berlin, S. Koren, C.-S. Chin, J. P. Drake, J. M. Landolin, and A. M. Phillippy. Assembling large genomes with single-molecule sequencing and locality-sensitive hashing. *Nat Biotechnol*, 33(6):623–630, Jun 2015.

[21] T. R. Bhangale, M. J. Rieder, R. J. Livingston, and D. A. Nickerson. Comprehensive identification and characterization of diallelic insertion-deletion polymorphisms in 330 human candidate genes. *Hum Mol Genet*, 14(1):59–69, Jan 2005.

[22] J. Blom, T. Jakobi, D. Doppmeier, S. Jaenicke, J. Kalinowski, J. Stoye, and A. Goesmann. Exact and complete short-read alignment to microbial genomes using graphics processing unit programming. *Bioinformatics*, 27(10):1351–1358, May 2011.

[23] B. H. Bloom. Space/time trade-offs in hash coding with allowable errors. *Commun. ACM*, 13(7):422–426, July 1970.

[24] M. Boetzer, C. V. Henkel, H. J. Jansen, D. Butler, and W. Pirovano. Scaffolding pre-assembled contigs using SSPACE. *Bioinformatics*, 27(4):578–579, Feb 2011.

[25] V. Boeva, A. Zinovyev, K. Bleakley, J.-P. Vert, I. Janoueix-Lerosey, O. Delattre, and E. Barillot. Control-free calling of copy number alterations in deep-sequencing data using GC-content normalization. *Bioinformatics*, 27(2):268–269, Jan 2011.

[26] J. K. Bonfield and M. V. Mahoney. Compression of FASTQ and SAM format sequencing data. *PLoS One*, 8(3):e59190, 2013.

[27] F. Bonomi, M. Mitzenmacher, R. Panigrahy, S. Singh, and G. Varghese. An improved construction for counting bloom filters. In *European Symp on Algorithms (ESA)*, pages 684–695, 2006.

[28] A. Bowe, T. Onodera, K. Sadakane, and T. Shibuya. Succinct de Bruijn graphs. In *Workshop on Algorithms in Bioinformatics (WABI)*, pages 225–235, 2012.

[29] A. P. Boyle, J. Guinney, G. E. Crawford, and T. S. Furey. F-Seq: A feature density estimator for high-throughput sequence tags. *Bioinformatics*, 24(21):2537–2538, Nov 2008.

[30] A. P. Boyle, L. Song, B.-K. Lee, D. London, D. Keefe, E. Birney, V. R. Iyer, G. E. Crawford, and T. S. Furey. High-resolution genome-wide in vivo footprinting of diverse transcription factors in human cells. *Genome Research*, 21(3):456–464, 2011.

[31] S. Brenner, M. Johnson, J. Bridgham, G. Golda, D. H. Lloyd, D. Johnson, S. Luo, S. McCurdy, M. Foy, M. Ewan, R. Roth, D. George, S. Eletr, G. Albrecht, E. Vermaas, S. R. Williams, K. Moon, T. Burcham, M. Pallas, R. B. DuBridge, J. Kirchner, K. Fearon, J. Mao, and K. Corcoran. Gene expression analysis by massively parallel signature sequencing (MPSS) on microbead arrays. *Nat Biotechnol*, 18(6):630–634, Jun 2000.

[32] J. D. Buenrostro, P. G. Giresi, L. C. Zaba, H. Y. Chang, and W. J. Greenleaf. Transposition of native chromatin for fast and sensitive epigenomic profiling of open chromatin, DNA-binding proteins and nucleosome position. *Nature Methods*, 10(12):1213–1218, 2013.

[33] S. Burkhardt and J. Karkkainen. Better filtering with gapped q-grams. In *Combinatorial Pattern Matching (CPM)*, pages 73–85, 2001.

[34] J. Butler, I. MacCallum, M. Kleber, I. A. Shlyakhter, M. K. Belmonte, E. S. Lander, C. Nusbaum, and D. B. Jaffe. ALLPATHS: De novo assembly of whole-genome shotgun microreads. *Genome Research*, 18(5):810–820, May 2008.

[35] J. S. Carroll, C. A. Meyer, J. Song, W. Li, T. R. Geistlinger, J. Eeckhoute, A. S. Brodsky, E. K. Keeton, K. C. Fertuck, G. F. Hall, Q. Wang, S. Bekiranov, V. Sementchenko, E. A. Fox, P. A. Silver, T. R. Gingeras, X. S. Liu, and M. Brown. Genome-wide analysis of estrogen receptor binding sites. *Nat Genet*, 38(11):1289–1297, Nov 2006.

[36] N. P. Carter. Methods and strategies for analyzing copy number variation using DNA microarrays. *Nat Genet*, 39(7 Suppl):S16–S21, Jul 2007.

[37] M. Chaisson, P. Pevzner, and H. Tang. Fragment assembly with short reads. *Bioinformatics*, 20(13):2067–2074, Sep 2004.

[38] M. J. Chaisson and P. A. Pevzner. Short read fragment assembly of bacterial genomes. *Genome Research*, 18(2):324–330, Feb 2008.

[39] M. J. Chaisson and G. Tesler. Mapping single molecule sequencing reads using basic local alignment with successive refinement (BLASR): Application and theory. *BMC Bioinformatics*, 13:238, 2012.

[40] M. J. P. Chaisson, J. Huddleston, M. Y. Dennis, P. H. Sudmant, M. Malig, F. Hormozdiari, F. Antonacci, U. Surti, R. Sandstrom, M. Boitano, J. M. Landolin, J. A. Stamatoyannopoulos, M. W. Hunkapiller, J. Korlach, and E. E. Eichler. Resolving the complexity of the human genome using single-molecule sequencing. *Nature*, 517(7536):608–611, Jan 2015.

[41] G. Chen, C. Wang, and T. Shi. Overview of available methods for diverse RNA-seq data analyses. *Sci China Life Sci*, 54(12):1121–1128, Dec 2011.

[42] K. Chen, L. Chen, X. Fan, J. Wallis, L. Ding, and G. Weinstock. TIGRA: A targeted iterative graph routing assembler for breakpoint assembly. *Genome Research*, 24(2):310–317, Feb 2014.

[43] K. Chen, J. W. Wallis, M. D. McLellan, D. E. Larson, J. M. Kalicki, C. S. Pohl, S. D. McGrath, M. C. Wendl, Q. Zhang, D. P. Locke, X. Shi, R. S. Fulton, T. J. Ley, R. K. Wilson, L. Ding, and E. R. Mardis. BreakDancer: An algorithm for high-resolution mapping of genomic structural variation. *Nat Methods*, 6(9):677–681, Sep 2009.

[44] D. Y. Chiang, G. Getz, D. B. Jaffe, M. J. T. O'Kelly, X. Zhao, S. L. Carter, C. Russ, C. Nusbaum, M. Meyerson, and E. S. Lander. High-resolution mapping of copy-number alterations with massively parallel sequencing. *Nat Methods*, 6(1):99–103, Jan 2009.

[45] C.-S. Chin, D. H. Alexander, P. Marks, A. A. Klammer, J. Drake, C. Heiner, A. Clum, A. Copeland, J. Huddleston, E. E. Eichler, S. W. Turner, and J. Korlach. Nonhybrid, finished microbial genome assemblies from long-read SMRT sequencing data. *Nat Methods*, 10(6):563–569, Jun 2013.

[46] C.-S. Chin, J. Sorenson, J. B. Harris, W. P. Robins, R. C. Charles, R. R. Jean-Charles, J. Bullard, D. R. Webster, A. Kasarskis, P. Peluso, E. E. Paxinos, Y. Yamaichi, S. B. Calderwood, J. J. Mekalanos, E. E. Schadt, and M. K. Waldor. The origin of the Haitian cholera outbreak strain. *N Engl J Med*, 364(1):33–42, Jan 2011.

[47] H. Chitsaz, J. L. Yee-Greenbaum, G. Tesler, M.-J. Lombardo, C. L. Dupont, J. H. Badger, M. Novotny, D. B. Rusch, L. J. Fraser, N. A. Gormley, O. Schulz-Trieglaff, G. P. Smith, D. J. Evers, P. A. Pevzner, and R. S. Lasken. Efficient de novo assembly of single-cell bacterial genomes from short-read data sets. *Nat Biotechnol*, 29(10):915–921, Oct 2011.

[48] M. Choi, U. I. Scholl, W. Ji, T. Liu, I. R. Tikhonova, P. Zumbo, A. Nayir, A. Bakkaloğlu, S. Ozen, S. Sanjad, C. Nelson-Williams, A. Farhi, S. Mane, and R. P. Lifton. Genetic diagnosis by whole exome capture and massively parallel DNA sequencing. *Proc Natl Acad Sci USA*, 106(45):19096–19101, Nov 2009.

[49] H.-T. Chu, W. W. L. Hsiao, J.-C. Chen, T.-J. Yeh, M.-H. Tsai, H. Lin, Y.-W. Liu, S.-A. Lee, C.-C. Chen, T. T. H. Tsao, and C.-Y. Kao. EBAR-Denovo: Highly accurate de novo assembly of RNA-seq with efficient chimera-detection. *Bioinformatics*, 29(8):1004–1010, Apr 2013.

[50] K. Cibulskis, M. S. Lawrence, S. L. Carter, A. Sivachenko, D. Jaffe, C. Sougnez, S. Gabriel, M. Meyerson, E. S. Lander, and G. Getz. Sensitive detection of somatic point mutations in impure and heterogeneous cancer samples. *Nat Biotechnol*, 31(3):213–219, Mar 2013.

[51] K. Cibulskis, A. McKenna, T. Fennell, E. Banks, M. DePristo, and G. Getz. ContEst: Estimating cross-contamination of human samples in next-generation sequencing data. *Bioinformatics*, 27(18):2601–2602, Sep 2011.

[52] N. Cloonan, A. R. R. Forrest, G. Kolle, B. B. A. Gardiner, G. J. Faulkner, M. K. Brown, D. F. Taylor, A. L. Steptoe, S. Wani, G. Bethel, A. J. Robertson, A. C. Perkins, S. J. Bruce, C. C. Lee, S. S. Ranade, H. E. Peckham, J. M. Manning, K. J. McKernan, and S. M. Grimmond. Stem cell transcriptome profiling via massive-scale mRNA sequencing. *Nat Methods*, 5(7):613–619, Jul 2008.

[53] N. Cloonan, Q. Xu, G. J. Faulkner, D. F. Taylor, D. T. P. Tang, G. Kolle, and S. M. Grimmond. RNA-MATE: A recursive mapping strategy for high-throughput RNA-sequencing data. *Bioinformatics*, 25(19):2615–2616, Oct 2009.

[54] D. F. Conrad, D. Pinto, R. Redon, L. Feuk, O. Gokcumen, Y. Zhang, J. Aerts, T. D. Andrews, C. Barnes, P. Campbell, T. Fitzgerald, M. Hu, C. H. Ihm, K. Kristiansson, D. G. Macarthur, J. R. Macdonald, I. Onyiah, A. W. C. Pang, S. Robson, K. Stirrups, A. Valsesia, K. Walter, J. Wei, W. T. C. C. C., C. Tyler-Smith, N. P. Carter, C. Lee, S. W. Scherer, and M. E. Hurles. Origins and functional impact of copy number variation in the human genome. *Nature*, 464(7289):704–712, Apr 2010.

[55] T. A. Cooper, L. Wan, and G. Dreyfuss. RNA and disease. *Cell*, 136(4):777–793, Feb 2009.

[56] A. J. Cox, M. J. Bauer, T. Jakobi, and G. Rosone. Large-scale compression of genomic sequence databases with the Burrows-Wheeler transform. *Bioinformatics*, 28(11):1415–1419, Jun 2012.

[57] M. P. Cox, D. A. Peterson, and P. J. Biggs. SolexaQA: At-a-glance quality assessment of Illumina second-generation sequencing data. *BMC Bioinformatics*, 11:485, 2010.

[58] R. Dahm. Discovering DNA: Friedrich Miescher and the early years of nucleic acid research. *Hum Genet*, 122(6):565–581, Jan 2008.

[59] K. Daily, P. Rigor, S. Christley, X. Xie, and P. Baldi. Data structures and compression algorithms for high-throughput sequencing technologies. *BMC Bioinformatics*, 11:514, 2010.

[60] P. Danecek, A. Auton, G. Abecasis, C. A. Albers, E. Banks, M. A. DePristo, R. E. Handsaker, G. Lunter, G. T. Marth, S. T. Sherry, G. McVean, R. Durbin, and 1000 Genomes Project Analysis Group. The variant call format and VCFtools. *Bioinformatics*, 27(15):2156–2158, Aug 2011.

[61] A. Dayarian, T. P. Michael, and A. M. Sengupta. SOPRA: Scaffolding algorithm for paired reads via statistical optimization. *BMC Bioinformatics*, 11:345, 2010.

[62] F. De Bona, S. Ossowski, K. Schneeberger, and G. Rätsch. Optimal spliced alignments of short sequence reads. *Bioinformatics*, 24(16):i174–i180, Aug 2008.

[63] F. Denoeud, J.-M. Aury, C. Da Silva, B. Noel, O. Rogier, M. Delledonne, M. Morgante, G. Valle, P. Wincker, C. Scarpelli, O. Jaillon, and F. Artiguenave. Annotating genomes with massive-scale RNA sequencing. *Genome Biol*, 9(12):R175, 2008.

[64] S. Deorowicz and S. Grabowski. Compression of DNA sequence reads in FASTQ format. *Bioinformatics*, 27(6):860–862, Mar 2011.

[65] M. A. DePristo, E. Banks, R. Poplin, K. V. Garimella, J. R. Maguire, C. Hartl, A. A. Philippakis, G. del Angel, M. A. Rivas, M. Hanna, A. McKenna, T. J. Fennell, A. M. Kernytsky, A. Y. Sivachenko, K. Cibulskis, S. B. Gabriel, D. Altshuler, and M. J. Daly. A framework for variation discovery and genotyping using next-generation DNA sequencing data. *Nat Genet*, 43(5):491–498, May 2011.

[66] M. T. Dimon, K. Sorber, and J. L. DeRisi. HMMSplicer: A tool for efficient and sensitive discovery of known and novel splice junctions in RNA-seq data. *PLoS One*, 5(11):e13875, 2010.

[67] H. Do and W. Sung. Compressed directed acyclic word graph with application in local alignment. *Algorithmica*, 67(2):125–141, 2013.

[68] A. Dobin, C. A. Davis, F. Schlesinger, J. Drenkow, C. Zaleski, S. Jha, P. Batut, M. Chaisson, and T. R. Gingeras. STAR: Ultrafast universal RNA-seq aligner. *Bioinformatics*, 29(1):15–21, Jan 2013.

[69] J. C. Dohm, C. Lottaz, T. Borodina, and H. Himmelbauer. SHARCGS, a fast and highly accurate short-read assembly algorithm for de novo genomic sequencing. *Genome Research*, 17(11):1697–1706, Nov 2007.

[70] N. Donmez and M. Brudno. SCARPA: Scaffolding reads with practical algorithms. *Bioinformatics*, 29(4):428–434, Feb 2013.

[71] J. Eid, A. Fehr, J. Gray, K. Luong, J. Lyle, G. Otto, P. Peluso, D. Rank, P. Baybayan, B. Bettman, A. Bibillo, K. Bjornson, B. Chaudhuri, F. Christians, R. Cicero, S. Clark, R. Dalal, A. Dewinter, J. Dixon, M. Foquet, A. Gaertner, P. Hardenbol, C. Heiner, K. Hester, D. Holden, G. Kearns, X. Kong, R. Kuse, Y. Lacroix, S. Lin, P. Lundquist, C. Ma, P. Marks, M. Maxham, D. Murphy, I. Park, T. Pham, M. Phillips, J. Roy, R. Sebra, G. Shen, J. Sorenson, A. Tomaney, K. Travers, M. Trulson, J. Vieceli, J. Wegener, D. Wu, A. Yang, D. Zaccarin, P. Zhao, F. Zhong, J. Korlach, and S. Turner. Real-time DNA sequencing from single polymerase molecules. *Science*, 323(5910):133–138, Jan 2009.

[72] A. C. English, S. Richards, Y. Han, M. Wang, V. Vee, J. Qu, X. Qin, D. M. Muzny, J. G. Reid, K. C. Worley, and R. A. Gibbs. Mind the gap: Upgrading genomes with Pacific Biosciences RS long-read sequencing technology. *PLoS One*, 7(11):e47768, 2012.

[73] B. Ewing, L. Hillier, M. C. Wendl, and P. Green. Base-calling of automated sequencer traces using phred. I. Accuracy assessment. *Genome Research*, 8(3):175–185, Mar 1998.

[74] L. Fan, P. Cao, J. Almeida, and A. Z. Broder. Summary cache: A scalable Wide-area Web cache sharing protocol. *IEEE/ACM Transactions on Networking*, 8(3):281–293, June 2000.

[75] H. Fang, Y. Wu, G. Narzisi, J. A. O'Rawe, L. T. J. Barrón, J. Rosenbaum, M. Ronemus, I. Iossifov, M. C. Schatz, and G. J. Lyon. Reducing INDEL calling errors in whole genome and exome sequencing data. *Genome Med*, 6(10):89, 2014.

[76] M. Farach. Optimal suffix tree construction with large alphabets. In *IEEE Symposium on Foundations of Computer Science (FOCS)*, pages 137–143, 1997.

[77] J. Fernandez-Banet, N. P. Lee, K. T. Chan, H. Gao, X. Liu, W.-K. Sung, W. Tan, S. T. Fan, R. T. Poon, S. Li, K. Ching, P. A. Rejto, M. Mao, and Z. Kan. Decoding complex patterns of genomic rearrangement in hepatocellular carcinoma. *Genomics*, 103(2-3):189–203, 2014.

[78] P. Ferragine and G. Manzini. Opportunistic data structures with applications. In *IEEE Symposium on Foundations of Computer Science (FOCS)*, pages 390–398, 2000.

[79] M. Ferrarini, M. Moretto, J. A. Ward, N. Šurbanovski, V. Stevanović, L. Giongo, R. Viola, D. Cavalieri, R. Velasco, A. Cestaro, and D. J. Sargent. An evaluation of the PacBio RS platform for sequencing and de novo assembly of a chloroplast genome. *BMC Genomics*, 14:670, 2013.

[80] L. Feuk, A. R. Carson, and S. W. Scherer. Structural variation in the human genome. *Nat Rev Genet*, 7(2):85–97, Feb 2006.

[81] N. A. Fonseca, J. Rung, A. Brazma, and J. C. Marioni. Tools for mapping high-throughput sequencing data. *Bioinformatics*, 28(24):3169–3177, Dec 2012.

[82] X. Fu, N. Fu, S. Guo, Z. Yan, Y. Xu, H. Hu, C. Menzel, W. Chen, Y. Li, R. Zeng, and P. Khaitovich. Estimating accuracy of RNA-seq and microarrays with proteomics. *BMC Genomics*, 10:161, 2009.

[83] E. R. Gamazon, R. S. Huang, M. E. Dolan, and N. J. Cox. Copy number polymorphisms and anticancer pharmacogenomics. *Genome Biol*, 12(5):R46, 2011.

[84] S. Gao, D. Bertrand, B. K. H. Chia, and N. Nagarajan. OPERA-LG: Efficient and exact scaffolding of large, repeat-rich eukaryotic genomes with performance guarantees. *Genome Biol*, 17:102, 2016.

[85] S. Gao, W.-K. Sung, and N. Nagarajan. Opera: Reconstructing optimal genomic scaffolds with high-throughput paired-end sequences. *Journal of Computational Biology*, 18(11):1681–1691, Nov 2011.

[86] M. Garber, M. G. Grabherr, M. Guttman, and C. Trapnell. Computational methods for transcriptome annotation and quantification using RNA-seq. *Nat Methods*, 8(6):469–477, Jun 2011.

[87] A. Gillet-Markowska, H. Richard, G. Fischer, and I. Lafontaine. Ulysses: Accurate detection of low-frequency structural variations in large insert-size sequencing libraries. *Bioinformatics*, 31(6):801–808, Mar 2015.

[88] G. Gonnella and S. Kurtz. Readjoiner: A fast and memory efficient string graph-based sequence assembler. *BMC Bioinformatics*, 13:82, 2012.

[89] M. G. Grabherr, B. J. Haas, M. Yassour, J. Z. Levin, D. A. Thompson, I. Amit, X. Adiconis, L. Fan, R. Raychowdhury, Q. Zeng, Z. Chen, E. Mauceli, N. Hacohen, A. Gnirke, N. Rhind, F. di Palma, B. W. Birren, C. Nusbaum, K. Lindblad-Toh, N. Friedman, and A. Regev. Full-length transcriptome assembly from RNA-seq data without a reference genome. *Nat Biotechnol*, 29(7):644–652, Jul 2011.

[90] G. R. Grant, M. H. Farkas, A. D. Pizarro, N. F. Lahens, J. Schug, B. P. Brunk, C. J. Stoeckert, J. B. Hogenesch, and E. A. Pierce. Comparative analysis of RNA-seq alignment algorithms and the RNA-seq unified mapper (RUM). *Bioinformatics*, 27(18):2518–2528, Sep 2011.

[91] M. Griffith, O. L. Griffith, J. Mwenifumbo, R. Goya, A. S. Morrissy, R. D. Morin, R. Corbett, M. J. Tang, Y.-C. Hou, T. J. Pugh, G. Robertson, S. Chittaranjan, A. Ally, J. K. Asano, S. Y. Chan, H. I. Li, H. McDonald, K. Teague, Y. Zhao, T. Zeng, A. Delaney, M. Hirst, G. B. Morin, S. J. M. Jones, I. T. Tai, and M. A. Marra. Alternative expression analysis by RNA sequencing. *Nat Methods*, 7(10):843–847, Oct 2010.

[92] W. Gu, F. Zhang, and J. R. Lupski. Mechanisms for human genomic rearrangements. *Pathogenetics*, 1(1):4, 2008.

[93] M. Guttman, M. Garber, J. Z. Levin, J. Donaghey, J. Robinson, X. Adiconis, L. Fan, M. J. Koziol, A. Gnirke, C. Nusbaum, J. L. Rinn, E. S. Lander, and A. Regev. Ab initio reconstruction of cell type-specific transcriptomes in mouse reveals the conserved multi-exonic structure of lincRNAs. *Nat Biotechnol*, 28(5):503–510, May 2010.

[94] I. Hajirasouliha, F. Hormozdiari, C. Alkan, J. M. Kidd, I. Birol, E. E. Eichler, and S. C. Sahinalp. Detection and characterization of novel sequence insertions using paired-end next-generation sequencing. *Bioinformatics*, 26(10):1277–1283, May 2010.

[95] J. Hardy and A. Singleton. Genomewide association studies and human disease. *N Engl J Med*, 360(17):1759–1768, Apr 2009.

[96] P. J. Hastings, G. Ira, and J. R. Lupski. A microhomology-mediated break-induced replication model for the origin of human copy number variation. *PLoS Genet*, 5(1):e1000327, Jan 2009.

[97] P. J. Hastings, J. R. Lupski, S. M. Rosenberg, and G. Ira. Mechanisms of change in gene copy number. *Nat Rev Genet*, 10(8):551–564, Aug 2009.

[98] A. Hatem, D. Bozdağ, A. E. Toland, and U. V. Catalyürek. Benchmarking short sequence mapping tools. *BMC Bioinformatics*, 14:184, 2013.

[99] M. Hayes, Y. S. Pyon, and J. Li. A model-based clustering method for genomic structural variant prediction and genotyping using paired-end sequencing data. *PLoS One*, 7(12):e52881, 2012.

[100] D. Hernandez, P. François, L. Farinelli, M. Osterås, and J. Schrenzel. De novo bacterial genome sequencing: Millions of very short reads assembled on a desktop computer. *Genome Research*, 18(5):802–809, May 2008.

[101] J. R. Hesselberth, X. Chen, Z. Zhang, P. J. Sabo, R. Sandstrom, A. P. Reynolds, R. E. Thurman, S. Neph, M. S. Kuehn, W. S. Noble, et al. Global mapping of protein-DNA interactions in vivo by digital genomic footprinting. *Nature Methods*, 6(4):283–289, 2009.

[102] M. Holtgrewe, A.-K. Emde, D. Weese, and K. Reinert. A novel and well-defined benchmarking method for second generation read mapping. *BMC Bioinformatics*, 12:210, 2011.

[103] N. Homer, B. Merriman, and S. F. Nelson. BFAST: An alignment tool for large scale genome resequencing. *PLoS One*, 4(11):e7767, 2009.

[104] W.-K. Hon, K. Sadakane, and W.-K. Sung. Breaking a time-and-space barrier in constructing full-text indices. *SIAM Journal on Computing*, 38(6):2162–2178, 2009.

[105] F. Hormozdiari, C. Alkan, E. E. Eichler, and S. C. Sahinalp. Combinatorial algorithms for structural variation detection in high-throughput sequenced genomes. *Genome Research*, 19(7):1270–1278, Jul 2009.

[106] F. Hormozdiari, I. Hajirasouliha, P. Dao, F. Hach, D. Yorukoglu, C. Alkan, E. E. Eichler, and S. C. Sahinalp. Next-generation Variation-Hunter: Combinatorial algorithms for transposon insertion discovery. *Bioinformatics*, 26(12):i350–i357, Jun 2010.

[107] M. Hsi-Yang Fritz, R. Leinonen, G. Cochrane, and E. Birney. Efficient storage of high throughput DNA sequencing data using reference-based compression. *Genome Research*, 21(5):734–740, May 2011.

[108] Y. Hu, K. Wang, X. He, D. Y. Chiang, J. F. Prins, and J. Liu. A probabilistic framework for aligning paired-end RNA-seq data. *Bioinformatics*, 26(16):1950–1957, Aug 2010.

[109] S. Huang, J. Zhang, R. Li, W. Zhang, Z. He, T.-W. Lam, Z. Peng, and S.-M. Yiu. SOAPsplice: Genome-wide ab initio detection of splice junctions from RNA-seq data. *Front Genet*, 2:46, 2011.

[110] Y. Huang, Y. Hu, C. D. Jones, J. N. MacLeod, D. Y. Chiang, Y. Liu, J. F. Prins, and J. Liu. A robust method for transcript quantification with RNA-seq data. *Journal of Computational Biology*, 20(3):167–187, Mar 2013.

[111] D. H. Huson, K. Reinert, and E. W. Myers. The greedy path-merging algorithm for contig scaffolding. *J. ACM*, 49(5):603–615, Sept. 2002.

[112] D. Huy Hoang and W.-K. Sung. CWig: Compressed representation of Wiggle/BedGraph format. *Bioinformatics*, 30(18):2543–2550, Sep 2014.

[113] R. M. Idury and M. S. Waterman. A new algorithm for DNA sequence assembly. *Journal of Computational Biology*, 2(2):291–306, 1995.

[114] S. Ivakhno, T. Royce, A. J. Cox, D. J. Evers, R. K. Cheetham, and S. Tavaré. CNAseg: A novel framework for identification of copy number changes in cancer from second-generation sequencing data. *Bioinformatics*, 26(24):3051–3058, Dec 2010.

[115] M. Jain, I. T. Fiddes, K. H. Miga, H. E. Olsen, B. Paten, and M. Akeson. Improved data analysis for the MinION nanopore sequencer. *Nat Methods*, 12(4):351–356, Apr 2015.

[116] G. Jean, A. Kahles, V. T. Sreedharan, F. De Bona, and G. R?tsch. RNA-seq read alignments with PALMapper. *Curr Protoc Bioinformatics*, Chapter 11:Unit 11.6, Dec 2010.

[117] W. R. Jeck, J. A. Reinhardt, D. A. Baltrus, M. T. Hickenbotham, V. Magrini, E. R. Mardis, J. L. Dangl, and C. D. Jones. Extending assembly of short DNA sequences to handle error. *Bioinformatics*, 23(21):2942–2944, Nov 2007.

[118] H. Ji. Computational analysis of ChIP-seq data. *Methods Mol Biol*, 674:143–159, 2010.

[119] H. Ji, H. Jiang, W. Ma, D. S. Johnson, R. M. Myers, and W. H. Wong. An integrated software system for analyzing ChIP-chip and ChIP-seq data. *Nat Biotechnol*, 26(11):1293–1300, Nov 2008.

[120] H. Ji, X. Li, Q.-f. Wang, and Y. Ning. Differential principal component analysis of ChIP-seq. *Proc Natl Acad Sci USA*, 110(17):6789–6794, Apr 2013.

[121] H. Jiang and W. H. Wong. SeqMap: Mapping massive amount of oligonucleotides to the genome. *Bioinformatics*, 24(20):2395–2396, Oct 2008.

[122] Y. Jiang, A. L. Turinsky, and M. Brudno. The missing indels: An estimate of indel variation in a human genome and analysis of factors that impede detection. *Nucleic Acids Res*, 43(15):7217–7228, Sep 2015.

[123] Y. Jiang, Y. Wang, and M. Brudno. PRISM: Pair-read informed split-read mapping for base-pair level detection of insertion, deletion and structural variants. *Bioinformatics*, 28(20):2576–2583, Oct 2012.

[124] D. S. Johnson, A. Mortazavi, R. M. Myers, and B. Wold. Genome-wide mapping of in vivo protein-DNA interactions. *Science*, 316(5830):1497–1502, Jun 2007.

[125] J. M. Johnson, J. Castle, P. Garrett-Engele, Z. Kan, P. M. Loerch, C. D. Armour, R. Santos, E. E. Schadt, R. Stoughton, and D. D. Shoemaker. Genome-wide survey of human alternative pre-mRNA splicing with exon junction microarrays. *Science*, 302(5653):2141–2144, Dec 2003.

[126] D. C. Jones, W. L. Ruzzo, X. Peng, and M. G. Katze. Compression of next-generation sequencing reads aided by highly efficient de novo assembly. *Nucleic Acids Res*, 40(22):e171, Aug 2012.

[127] R. Jothi, S. Cuddapah, A. Barski, K. Cui, and K. Zhao. Genome-wide identification of in vivo protein-DNA binding sites from ChIP-seq data. *Nucleic Acids Res*, 36(16):5221–5231, Sep 2008.

[128] A. Kalsotra and T. A. Cooper. Functional consequences of developmentally regulated alternative splicing. *Nat Rev Genet*, 12(10):715–729, Oct 2011.

[129] Z. Kan, H. Zheng, X. Liu, S. Li, T. Barber, Z. Gong, H. Gao, K. Hao, M. D. Willard, J. Xu, R. Hauptschein, P. A. Rejto, J. Fernandez, G. Wang, Q. Zhang, B. Wang, R. Chen, J. Wang, N. P. Lee, W. Zhou, Z. Lin, Z. Peng, K. Yi, S. Chen, L. Li, X. Fan, J. Yang, R. Ye, J. Ju, K. Wang, H. Estrella, S. Deng, P. Wei, M. Qiu, I. H. Wulur, J. Liu, M. E. Ehsani, C. Zhang, A. Loboda, W. K. Sung, A. Aggarwal, R. T. Poon, S. T. Fan, J. Hardwick, J. Wang, C. Reinhard, H. Dai, Y. Li, J. M. Luk, and M. Mao. Whole genome sequencing identifies recurrent mutations in hepatocellular carcinoma. *Genome Research*, 23(9):1422–1433, Jun 2013.

[130] C. Kanduri, L. Ukkola-Vuoti, J. Oikkonen, G. Buck, C. Blancher, P. Raijas, K. Karma, H. Lähdesmäki, and I. Järvelä. The genome-wide landscape of copy number variations in the MUSGEN study provides evidence for a founder effect in the isolated Finnish population. *Eur J Hum Genet*, 21(12):1411–1416, Dec 2013.

[131] E. Karakoc, C. Alkan, B. J. O'Roak, M. Y. Dennis, L. Vives, K. Mark, M. J. Rieder, D. A. Nickerson, and E. E. Eichler. Detection of structural variants and indels within exome data. *Nat Methods*, 9(2):176–178, Feb 2012.

[132] T. M. Keane, K. Wong, and D. J. Adams. RetroSeq: Transposable element discovery from next-generation sequencing data. *Bioinformatics*, 29(3):389–390, Feb 2013.

[133] J. D. Kececioglu and E. W. Myers. Combinatorial algorithms for DNA sequence assembly. *Algorithmica*, 13(1/2):7–51, 1995.

[134] W. J. Kent. BLAT — the BLAST-like alignment tool. *Genome Research*, 12(4):656–664, Apr 2002.

[135] W. J. Kent, A. S. Zweig, G. Barber, A. S. Hinrichs, and D. Karolchik. BigWig and BigBed: Enabling browsing of large distributed datasets. *Bioinformatics*, 26(17):2204–2207, Sep 2010.

[136] H. Keren, G. Lev-Maor, and G. Ast. Alternative splicing and evolution: Diversification, exon definition and function. *Nat Rev Genet*, 11(5):345–355, May 2010.

[137] P. V. Kharchenko, M. Y. Tolstorukov, and P. J. Park. Design and analysis of ChIP-seq experiments for DNA-binding proteins. *Nat Biotechnol*, 26(12):1351–1359, Dec 2008.

[138] J. M. Kidd, T. Graves, T. L. Newman, R. Fulton, H. S. Hayden, M. Malig, J. Kallicki, R. Kaul, R. K. Wilson, and E. E. Eichler. A human genome structural variation sequencing resource reveals insights into mutational mechanisms. *Cell*, 143(5):837–847, Nov 2010.

[139] S. M. Kielbasa, R. Wan, K. Sato, P. Horton, and M. C. Frith. Adaptive seeds tame genomic sequence comparison. *Genome Research*, 21(3):487–493, jan 2011.

[140] D. Kim, G. Pertea, C. Trapnell, H. Pimentel, R. Kelley, and S. L. Salzberg. Tophat2: Accurate alignment of transcriptomes in the presence of insertions, deletions and gene fusions. *Genome Biol*, 14(4):R36, Apr 2013.

[141] J. B. Kim, H. Zaehres, G. Wu, L. Gentile, K. Ko, V. Sebastiano, M. J. Araúzo-Bravo, D. Ruau, D. W. Han, M. Zenke, and H. R. Schöler. Pluripotent stem cells induced from adult neural stem cells by reprogramming with two factors. *Nature*, 454(7204):646–650, Jul 2008.

[142] T.-M. Kim, L. J. Luquette, R. Xi, and P. J. Park. rSW-seq: Algorithm for detection of copy number alterations in deep sequencing data. *BMC Bioinformatics*, 11:432, 2010.

[143] D. C. Koboldt, K. Chen, T. Wylie, D. E. Larson, M. D. McLellan, E. R. Mardis, G. M. Weinstock, R. K. Wilson, and L. Ding. VarScan: Variant detection in massively parallel sequencing of individual and pooled samples. *Bioinformatics*, 25(17):2283–2285, Sep 2009.

[144] D. C. Koboldt, Q. Zhang, D. E. Larson, D. Shen, M. D. McLellan, L. Lin, C. A. Miller, E. R. Mardis, L. Ding, and R. K. Wilson. VarScan 2: Somatic mutation and copy number alteration discovery in cancer by exome sequencing. *Genome Research*, 22(3):568–576, Mar 2012.

[145] J. O. Korbel, A. Abyzov, X. J. Mu, N. Carriero, P. Cayting, Z. Zhang, M. Snyder, and M. B. Gerstein. PEMer: A computational framework with simulation-based error models for inferring genomic structural variants from massive paired-end sequencing data. *Genome Biol*, 10(2):R23, 2009.

[146] S. Koren, M. C. Schatz, B. P. Walenz, J. Martin, J. T. Howard, G. Ganapathy, Z. Wang, D. A. Rasko, W. R. McCombie, E. D. Jarvis, and Adam M Phillippy. Hybrid error correction and de novo assembly of single-molecule sequencing reads. *Nat Biotechnol*, 30(7):693–700, Jul 2012.

[147] S. Koren, M. C. Schatz, B. P. Walenz, J. Martin, J. T. Howard, G. Ganapathy, Z. Wang, D. A. Rasko, W. R. McCombie, E. D. Jarvis, and Adam M Phillippy. Hybrid error correction and de novo assembly of single-molecule sequencing reads. *Nat Biotechnol*, 30(7):693–700, Jul 2012.

[148] C. Kozanitis, C. Saunders, S. Kruglyak, V. Bafna, and G. Varghese. Compressing genomic sequence fragments using SlimGene. *Journal of Computational Biology*, 18(3):401–413, Mar 2011.

[149] P. Krawitz, C. Rödelsperger, M. Jäger, L. Jostins, S. Bauer, and P. N. Robinson. Microindel detection in short-read sequence data. *Bioinformatics*, 26(6):722–729, Mar 2010.

[150] H. Y. K. Lam, X. J. Mu, A. M. Stütz, A. Tanzer, P. D. Cayting, M. Snyder, P. M. Kim, J. O. Korbel, and M. B. Gerstein. Nucleotide-resolution analysis of structural variants using BreakSeq and a breakpoint library. *Nat Biotechnol*, 28(1):47–55, Jan 2010.

[151] T. W. Lam, W.-K. Sung, S.-L. Tam, C.-K. Wong, and S.-M. Yiu. Compressed indexing and local alignment of DNA. *Bioinformatics*, 24(6):791–797, 2008.

[152] T. W. Lam, A. Tam, E. Wu, R. Li, S. Wong, and S. M. Yiu. High throughput short read alignment via bi-directional BWT. In *IEEE International Conference on Bioinformatics and Biomedicine (BIBM)*, pages 31–36, 2009.

[153] G. M. Landau and U. Vishkin. Introducing efficient parallelism into approximate string matching and a new serial algorithm. In *Annual ACM Symposium on Theory of Computing (STOC)*, pages 220–230, 1986.

[154] B. Langmead and S. L. Salzberg. Fast gapped-read alignment with bowtie 2. *Nat Methods*, 9(4):357–359, Apr 2012.

[155] B. Langmead, C. Trapnell, M. Pop, and S. L. Salzberg. Ultrafast and memory-efficient alignment of short DNA sequences to the human genome. *Genome Biol*, 10(3):R25, 2009.

[156] J. Laserson, V. Jojic, and D. Koller. Genovo: De novo assembly for metagenomes. *Journal of Computational Biology*, 18(3):429–443, Mar 2011.

[157] R. M. Layer, C. Chiang, A. R. Quinlan, and I. M. Hall. LUMPY: A probabilistic framework for structural variant discovery. *Genome Biol*, 15(6):R84, 2014.

[158] J. A. Lee, C. M. B. Carvalho, and J. R. Lupski. A DNA replication mechanism for generating nonrecurrent rearrangements associated with genomic disorders. *Cell*, 131(7):1235–1247, Dec 2007.

[159] S. Lee, F. Hormozdiari, C. Alkan, and M. Brudno. MoDIL: Detecting small indels from clone-end sequencing with mixtures of distributions. *Nat Methods*, 6(7):473–474, Jul 2009.

[160] S. Lee, E. Xing, and M. Brudno. MoGUL: Detecting dommon insertions and deletions in a population. In *Annual Intl Conf on Comp Molecular Biology (RECOMB)*, pages 357–368, 2010.

[161] W.-P. Lee, M. P. Stromberg, A. Ward, C. Stewart, E. P. Garrison, and G. T. Marth. MOSAIK: A hash-based algorithm for accurate next-generation sequencing short-read mapping. *PLoS One*, 9(3):e90581, 2014.

[162] J. Z. Levin, M. Yassour, X. Adiconis, C. Nusbaum, D. A. Thompson, N. Friedman, A. Gnirke, and A. Regev. Comprehensive comparative analysis of strand-specific RNA sequencing methods. *Nat Methods*, 7(9):709–715, Sep 2010.

[163] H. Li. A statistical framework for SNP calling, mutation discovery, association mapping and population genetical parameter estimation from sequencing data. *Bioinformatics*, 27(21):2987–2993, Nov 2011.

[164] H. Li. Exploring single-sample SNP and INDEL calling with whole-genome de novo assembly. *Bioinformatics*, 28(14):1838–1844, Jul 2012.

[165] H. Li. Toward better understanding of artifacts in variant calling from high-coverage samples. *Bioinformatics*, 30(20):2843–2851, Oct 2014.

[166] H. Li. Minimap and miniasm: Fast mapping and de novo assembly for noisy long sequences. *Bioinformatics*, 32(14):2103–2110, Jul 2016.

[167] H. Li and R. Durbin. Fast and accurate short read alignment with Burrows-Wheeler transform. *Bioinformatics*, 25(14):1754–1760, Jul 2009.

[168] H. Li and R. Durbin. Fast and accurate long-read alignment with Burrows-Wheeler transform. *Bioinformatics*, 26(5):589–595, Mar 2010.

[169] H. Li, B. Handsaker, A. Wysoker, T. Fennell, J. Ruan, N. Homer, G. Marth, G. Abecasis, R. Durbin, and 1000 Genome Project Data Processing Subgroup. The Sequence Alignment/Map format and SAMtools. *Bioinformatics*, 25(16):2078–2079, Aug 2009.

[170] H. Li, B. Handsaker, A. Wysoker, T. Fennell, J. Ruan, N. Homer, G. Marth, G. Abecasis, R. Durbin, and 1000 Genome Project Data Processing Subgroup. The Sequence Alignment/Map format and SAMtools. *Bioinformatics*, 25(16):2078–2079, Aug 2009.

[171] H. Li and N. Homer. A survey of sequence alignment algorithms for next-generation sequencing. *Brief Bioinform*, 11(5):473–483, Sep 2010.

[172] H. Li, J. Ruan, and R. Durbin. Mapping short DNA sequencing reads and calling variants using mapping quality scores. *Genome Research*, 18(11):1851–1858, Nov 2008.

[173] J. J. Li, P. J. Bickel, and M. D. Biggin. System wide analyses have underestimated protein abundances and the importance of transcription in mammals. *PeerJ*, 2:e270, feb 2014.

[174] R. Li, W. Fan, G. Tian, H. Zhu, L. He, J. Cai, Q. Huang, Q. Cai, B. Li, Y. Bai, Z. Zhang, Y. Zhang, W. Wang, J. Li, F. Wei, H. Li, M. Jian, J. Li, Z. Zhang, R. Nielsen, D. Li, W. Gu, Z. Yang, Z. Xuan, O. A. Ryder, F. C.-C. Leung, Y. Zhou, J. Cao, X. Sun, Y. Fu, X. Fang, X. Guo, B. Wang, R. Hou, F. Shen, B. Mu, P. Ni, R. Lin, W. Qian, G. Wang, C. Yu, W. Nie, J. Wang, Z. Wu, H. Liang, J. Min, Q. Wu, S. Cheng, J. Ruan, M. Wang, Z. Shi, M. Wen, B. Liu, X. Ren, H. Zheng, D. Dong, K. Cook, G. Shan, H. Zhang, C. Kosiol, X. Xie, Z. Lu, H. Zheng, Y. Li, C. C. Steiner, T. T.-Y. Lam, S. Lin, Q. Zhang, G. Li, J. Tian, T. Gong, H. Liu, D. Zhang, L. Fang, C. Ye, J. Zhang, W. Hu, A. Xu, Y. Ren, G. Zhang, M. W. Bruford, Q. Li, L. Ma, Y. Guo, N. An, Y. Hu, Y. Zheng, Y. Shi, Z. Li, Q. Liu, Y. Chen, J. Zhao, N. Qu, S. Zhao, F. Tian, X. Wang, H. Wang, L. Xu, X. Liu, T. Vinar, Y. Wang, T.-W. Lam, S.-M. Yiu, S. Liu, H. Zhang, D. Li, Y. Huang, X. Wang, G. Yang, Z. Jiang, J. Wang, N. Qin, L. Li, J. Li, L. Bolund, K. Kristiansen, G. K.-S. Wong, M. Olson, X. Zhang, S. Li, H. Yang, J. Wang, and J. Wang. The sequence and de novo assembly of the giant panda genome. *Nature*, 463(7279):311–317, Jan 2010.

[175] R. Li, Y. Li, K. Kristiansen, and J. Wang. SOAP: Short oligonucleotide alignment program. *Bioinformatics*, 24(5):713–714, Mar 2008.

[176] R. Li, H. Zhu, J. Ruan, W. Qian, X. Fang, Z. Shi, Y. Li, S. Li, G. Shan, K. Kristiansen, S. Li, H. Yang, J. Wang, and J. Wang. De novo assembly of human genomes with massively parallel short read sequencing. *Genome Research*, 20(2):265–272, Feb 2010.

[177] S. Li, R. Li, H. Li, J. Lu, Y. Li, L. Bolund, M. H. Schierup, and J. Wang. SOAPindel: Efficient identification of indels from short paired reads. *Genome Research*, 23(1):195–200, Jan 2013.

[178] J.-Q. Lim, C. Tennakoon, P. Guan, and W.-K. Sung. BatAlign: An incremental method for accurate alignment of sequencing reads. *Nucleic Acids Res*, 43(16):e107, Jul 2015.

[179] H. Lin, Z. Zhang, M. Q. Zhang, B. Ma, and M. Li. ZOOM! zillions of oligos mapped. *Bioinformatics*, 24(21):2431–2437, Nov 2008.

[180] M. R. Lindberg, I. M. Hall, and A. R. Quinlan. Population-based structural variation discovery with Hydra-Multi. *Bioinformatics*, 31(8):1286–1289, Apr 2015.

[181] C.-M. Liu, T. Wong, E. Wu, R. Luo, S.-M. Yiu, Y. Li, B. Wang, C. Yu, X. Chu, K. Zhao, R. Li, and T.-W. Lam. SOAP3: Ultra-fast GPU-based parallel alignment tool for short reads. *Bioinformatics*, 28(6):878–879, Mar 2012.

[182] Y. Liu and B. Schmidt. Long read alignment based on maximal exact match seeds. *Bioinformatics*, 28(18):i318–i324, Sep 2012.

[183] Y. Liu, B. Schmidt, and D. L. Maskell. CUSHAW: A CUDA compatible short read aligner to large genomes based on the Burrows-Wheeler transform. *Bioinformatics*, 28(14):1830–1837, Jul 2012.

[184] J. R. Lupski. Genomic rearrangements and sporadic disease. *Nat Genet*, 39(7 Suppl):S43–S47, Jul 2007.

[185] A. Magi, M. Benelli, S. Yoon, F. Roviello, and F. Torricelli. Detecting common copy number variants in high-throughput sequencing data by using JointSLM algorithm. *Nucleic Acids Res*, 39(10):e65, May 2011.

[186] U. Manber and G. Myers. Suffix arrays: A new method for on-line string searches. *SIAM Journal on Computing*, 22(5):935–948, 1993.

[187] G. Manzini. An analysis of the Burrows-Wheeler transform. *J. ACM*, 48(3):407–430, 2001.

[188] G. Marçais and C. Kingsford. A fast, lock-free approach for efficient parallel counting of occurrences of k-mers. *Bioinformatics*, 27(6):764–770, Mar 2011.

[189] S. Marco-Sola, M. Sammeth, R. Guigó, and P. Ribeca. The GEM mapper: Fast, accurate and versatile alignment by filtration. *Nat Methods*, 9(12):1185–1188, Dec 2012.

[190] E. R. Mardis. Next-generation sequencing platforms. *Annu Rev Anal Chem (Palo Alto Calif)*, 6:287–303, 2013.

[191] J. C. Marioni, C. E. Mason, S. M. Mane, M. Stephens, and Y. Gilad. RNA-seq: An assessment of technical reproducibility and comparison with gene expression arrays. *Genome Research*, 18(9):1509–1517, Sep 2008.

[192] T. Marschall, I. G. Costa, S. Canzar, M. Bauer, G. W. Klau, A. Schliep, and A. Schönhuth. CLEVER: Clique-enumerating variant finder. *Bioinformatics*, 28(22):2875–2882, Nov 2012.

[193] J. Martin, V. M. Bruno, Z. Fang, X. Meng, M. Blow, T. Zhang, G. Sherlock, M. Snyder, and Z. Wang. Rnnotator: An automated de novo transcriptome assembly pipeline from stranded RNA-seq reads. *BMC Genomics*, 11:663, 2010.

[194] A. M. Maxam and W. Gilbert. A new method for sequencing DNA. *Proc Natl Acad Sci USA*, 74(2):560–564, Feb 1977.

[195] S. A. McCarroll, A. Huett, P. Kuballa, S. D. Chilewski, A. Landry, P. Goyette, M. C. Zody, J. L. Hall, S. R. Brant, J. H. Cho, R. H. Duerr, M. S. Silverberg, K. D. Taylor, J. D. Rioux, D. Altshuler, M. J. Daly, and R. J. Xavier. Deletion polymorphism upstream of IRGM associated with altered IRGM expression and Crohn's disease. *Nat Genet*, 40(9):1107–1112, Sep 2008.

[196] M. T. McCarthy and C. A. O'Callaghan. PeaKDEck: A kernel density estimator-based peak calling program for DNaseI-seq data. *Bioinformatics*, 30(9):1302–1304, May 2014.

[197] K. J. McKernan, H. E. Peckham, G. L. Costa, S. F. McLaughlin, Y. Fu, E. F. Tsung, C. R. Clouser, C. Duncan, J. K. Ichikawa, C. C. Lee, Z. Zhang, S. S. Ranade, E. T. Dimalanta, F. C. Hyland, T. D. Sokolsky, L. Zhang, A. Sheridan, H. Fu, C. L. Hendrickson, B. Li, L. Kotler, J. R. Stuart, J. A. Malek, J. M. Manning, A. A. Antipova, D. S. Perez, M. P. Moore, K. C. Hayashibara, M. R. Lyons, R. E. Beaudoin, B. E. Coleman, M. W. Laptewicz, A. E. Sannicandro, M. D. Rhodes, R. K. Gottimukkala, S. Yang, V. Bafna, A. Bashir, A. MacBride, C. Alkan, J. M. Kidd, E. E. Eichler, M. G. Reese, F. M. De La Vega, and A. P. Blanchard. Sequence and structural variation in a human genome uncovered by short-read, massively parallel ligation sequencing using two-base encoding. *Genome Research*, 19(9):1527–1541, Sep 2009.

[198] P. Medvedev, M. Fiume, M. Dzamba, T. Smith, and M. Brudno. Detecting copy number variation with mated short reads. *Genome Research*, 20(11):1613–1622, Nov 2010.

[199] P. Melsted and J. K. Pritchard. Efficient counting of k-mers in DNA sequences using a bloom filter. *BMC Bioinformatics*, 12:333, 2011.

[200] M. L. Metzker. Sequencing technologies: The next generation. *Nat Rev Genet*, 11(1):31–46, Jan 2010.

[201] K. H. Miga, C. Eisenhart, and W. J. Kent. Utilizing mapping targets of sequences underrepresented in the reference assembly to reduce false positive alignments. *Nucleic Acids Res*, 43(20):e133, Nov 2015.

[202] C. A. Miller, O. Hampton, C. Coarfa, and A. Milosavljevic. ReadDepth: A parallel R package for detecting copy number alterations from short sequencing reads. *PLoS One*, 6(1):e16327, 2011.

[203] R. E. Mills, K. Walter, C. Stewart, R. E. Handsaker, K. Chen, C. Alkan, A. Abyzov, S. C. Yoon, K. Ye, R. K. Cheetham, A. Chinwalla, D. F. Conrad, Y. Fu, F. Grubert, I. Hajirasouliha, F. Hormozdiari, L. M. Iakoucheva, Z. Iqbal, S. Kang, J. M. Kidd, M. K. Konkel, J. Korn, E. Khurana, D. Kural, H. Y. K. Lam, J. Leng, R. Li, Y. Li, C.-Y. Lin, R. Luo, X. J. Mu, J. Nemesh, H. E. Peckham, T. Rausch, A. Scally, X. Shi, M. P. Stromberg, A. M. Stütz, A. E. Urban, J. A. Walker, J. Wu, Y. Zhang, Z. D. Zhang, M. A. Batzer, L. Ding, G. T. Marth, G. McVean, J. Sebat, M. Snyder, J. Wang, K. Ye, E. E. Eichler, M. B. Gerstein, M. E. Hurles, C. Lee, S. A. McCarroll, J. O. Korbel, and . G. Project. Mapping copy number variation by population-scale genome sequencing. *Nature*, 470(7332):59–65, Feb 2011.

[204] M. Mohiyuddin, J. C. Mu, J. Li, N. Bani Asadi, M. B. Gerstein, A. Abyzov, W. H. Wong, and H. Y. K. Lam. MetaSV: An accurate and integrative structural-variant caller for next generation sequencing. *Bioinformatics*, 31(16):2741–2744, Aug 2015.

[205] V. Moncunill, S. Gonzalez, S. Beà, L. O. Andrieux, I. Salaverria, C. Royo, L. Martinez, M. Puiggròs, M. Segura-Wang, A. M. Stütz, A. Navarro, R. Royo, J. L. Gelpí, I. G. Gut, C. López-Otín, M. Orozco, J. O. Korbel, E. Campo, X. S. Puente, and D. Torrents. Comprehensive characterization of complex structural variations in cancer by directly comparing genome sequence reads. *Nat Biotechnol*, 32(11):1106–1112, Nov 2014.

[206] S. B. Montgomery, D. L. Goode, E. Kvikstad, C. A. Albers, Z. D. Zhang, X. J. Mu, G. Ananda, B. Howie, K. J. Karczewski, K. S. Smith, V. Anaya, R. Richardson, J. Davis, 1000 Genomes Project Consortium, D. G. MacArthur, A. Sidow, L. Duret, M. Gerstein, K. D. Makova,

J. Marchini, G. McVean, and G. Lunter. The origin, evolution, and functional impact of short insertion-deletion variants identified in 179 human genomes. *Genome Research*, 23(5):749–761, May 2013.

[207] A. Mortazavi, B. A. Williams, K. McCue, L. Schaeffer, and B. Wold. Mapping and quantifying mammalian transcriptomes by RNA-seq. *Nat Methods*, 5(7):621–628, Jul 2008.

[208] J. C. Mu, H. Jiang, A. Kiani, M. Mohiyuddin, N. Bani Asadi, and W. H. Wong. Fast and accurate read alignment for resequencing. *Bioinformatics*, 28(18):2366–2373, Sep 2012.

[209] J. M. Mullaney, R. E. Mills, W. S. Pittard, and S. E. Devine. Small insertions and deletions (INDELs) in human genomes. *Hum Mol Genet*, 19(R2):R131–R136, Oct 2010.

[210] O. Muralidharan, G. Natsoulis, J. Bell, D. Newburger, H. Xu, I. Kela, H. Ji, and N. Zhang. A cross-sample statistical model for SNP detection in short-read sequencing data. *Nucleic Acids Res*, 40(1):e5, Jan 2012.

[211] G. H. Murillo, N. You, X. Su, W. Cui, M. P. Reilly, M. Li, K. Ning, and X. Cui. MultiGeMS: Detection of SNVs from multiple samples using model selection on high-throughput sequencing data. *Bioinformatics*, 32(10):1486–1492, May 2016.

[212] E. W. Myers. The fragment assembly string graph. *Bioinformatics*, 21 Suppl 2:ii79–ii85, Sep 2005.

[213] E. W. Myers, G. G. Sutton, A. L. Delcher, I. M. Dew, D. P. Fasulo, M. J. Flanigan, S. A. Kravitz, C. M. Mobarry, K. H. Reinert, K. A. Remington, E. L. Anson, R. A. Bolanos, H. H. Chou, C. M. Jordan, A. L. Halpern, S. Lonardi, E. M. Beasley, R. C. Brandon, L. Chen, P. J. Dunn, Z. Lai, Y. Liang, D. R. Nusskern, M. Zhan, Q. Zhang, X. Zheng, G. M. Rubin, M. D. Adams, and J. C. Venter. A whole-genome assembly of drosophila. *Science*, 287(5461):2196–2204, Mar 2000.

[214] T. Namiki, T. Hachiya, H. Tanaka, and Y. Sakakibara. MetaVelvet: An extension of velvet assembler to de novo metagenome assembly from short sequence reads. *Nucleic Acids Res*, 40(20):e155, Nov 2012.

[215] G. Narzisi, J. A. O'Rawe, I. Iossifov, H. Fang, Y.-H. Lee, Z. Wang, Y. Wu, G. J. Lyon, M. Wigler, and M. C. Schatz. Accurate de novo and transmitted indel detection in exome-capture data using microassembly. *Nat Methods*, 11(10):1033–1036, Oct 2014.

[216] A. Natarajan, G. G. Yardımcı, N. C. Sheffield, G. E. Crawford, and U. Ohler. Predicting cell-type — specific gene expression from regions of open chromatin. *Genome Research*, 22(9):1711–1722, 2012.

[217] A. M. Newman, S. V. Bratman, H. Stehr, L. J. Lee, C. L. Liu, M. Diehn, and A. A. Alizadeh. FACTERA: A practical method for the discovery of genomic rearrangements at breakpoint resolution. *Bioinformatics*, 30(23):3390–3393, Dec 2014.

[218] K. P. Ng, A. M. Hillmer, C. T. H. Chuah, W. C. Juan, T. K. Ko, A. S. M. Teo, P. N. Ariyaratne, N. Takahashi, K. Sawada, Y. Fei, S. Soh, W. H. Lee, J. W. J. Huang, J. C. Allen, X. Y. Woo, N. Nagarajan, V. Kumar, A. Thalamuthu, W. T. Poh, A. L. Ang, H. T. Mya, G. F. How, L. Y. Yang, L. P. Koh, B. Chowbay, C.-T. Chang, V. S. Nadarajan, W. J. Chng, H. Than, L. C. Lim, Y. T. Goh, S. Zhang, D. Poh, P. Tan, J.-E. Seet, M.-K. Ang, N.-M. Chau, Q.-S. Ng, D. S. W. Tan, M. Soda, K. Isobe, M. M. Nöthen, T. Y. Wong, A. Shahab, X. Ruan, V. Cacheux-Rataboul, W.-K. Sung, E. H. Tan, Y. Yatabe, H. Mano, R. A. Soo, T. M. Chin, W.-T. Lim, Y. Ruan, and S. T. Ong. A common BIM deletion polymorphism mediates intrinsic resistance and inferior responses to tyrosine kinase inhibitors in cancer. *Nat Med*, 18(4):521–528, Apr 2012.

[219] P. Ng, C.-L. Wei, W.-K. Sung, K. P. Chiu, L. Lipovich, C. C. Ang, S. Gupta, A. Shahab, A. Ridwan, C. H. Wong, E. T. Liu, and Y. Ruan. Gene identification signature (GIS) analysis for transcriptome characterization and genome annotation. *Nat Methods*, 2(2):105–111, Feb 2005.

[220] A. Oshlack, M. D. Robinson, and M. D. Young. From RNA-seq reads to differential expression results. *Genome Biol*, 11(12):220, 2010.

[221] T. D. Otto, M. Sanders, M. Berriman, and C. Newbold. Iterative correction of reference nucleotides (iCORN) using second generation sequencing technology. *Bioinformatics*, 26(14):1704–1707, Jul 2010.

[222] R. Pagh and F. F. Rodler. Cuckoo hashing. *J. Algorithms*, 51(2):122–144, 2004.

[223] Q. Pan, O. Shai, L. J. Lee, B. J. Frey, and B. J. Blencowe. Deep surveying of alternative splicing complexity in the human transcriptome by high-throughput sequencing. *Nat Genet*, 40(12):1413–1415, Dec 2008.

[224] P. J. Park. ChIP-seq: Advantages and challenges of a maturing technology. *Nat Rev Genet*, 10(10):669–680, Oct 2009.

[225] S. Pathak and S. Rajasekaran. LFQC: A lossless compression algorithm for FASTQ files. *Bioinformatics*, 31(20):3276–3281, Oct 2014.

[226] C. E. Pearson, K. Nichol Edamura, and J. D. Cleary. Repeat instability: Mechanisms of dynamic mutations. *Nat Rev Genet*, 6(10):729–742, Oct 2005.

[227] Y. Peng, H. C. M. Leung, S.-M. Yiu, and F. Y. L. Chin. IDBA: A practical iterative de Bruijn graph de novo assembler. In *Annual Intl Conf on Comp Molecular Biology (RECOMB)*, pages 426–440, 2010.

[228] Y. Peng, H. C. M. Leung, S. M. Yiu, and F. Y. L. Chin. Meta-IDBA: A de novo assembler for metagenomic data. *Bioinformatics*, 27(13):i94–101, Jul 2011.

[229] Y. Peng, H. C. M. Leung, S.-M. Yiu, and F. Y. L. Chin. T-IDBA: A de novo iterative de Bruijn graph assembler for transcriptome. In *Annual Intl Conf on Comp Molecular Biology (RECOMB)*, pages 337–338, 2011.

[230] Y. Peng, H. C. M. Leung, S. M. Yiu, and F. Y. L. Chin. IDBA-UD: A de novo assembler for single-cell and metagenomic sequencing data with highly uneven depth. *Bioinformatics*, 28(11):1420–1428, Jun 2012.

[231] P. A. Pevzner, H. Tang, and M. S. Waterman. An Eulerian path approach to DNA fragment assembly. *Proc Natl Acad Sci USA*, 98(17):9748–9753, Aug 2001.

[232] J. K. Pickrell, D. J. Gaffney, Y. Gilad, and J. K. Pritchard. False positive peaks in ChIP-seq and other sequencing-based functional assays caused by unannotated high copy number regions. *Bioinformatics*, 27(15):2144–2146, Aug 2011.

[233] V. Plagnol, J. Curtis, M. Epstein, K. Y. Mok, E. Stebbings, S. Grigoriadou, N. W. Wood, S. Hambleton, S. O. Burns, A. J. Thrasher, D. Kumararatne, R. Doffinger, and S. Nejentsev. A robust model for read count data in exome sequencing experiments and implications for copy number variant calling. *Bioinformatics*, 28(21):2747–2754, Nov 2012.

[234] A. Pohl and M. Beato. bwtool: A tool for bigWig files. *Bioinformatics*, 30(11):1618–1619, Jun 2014.

[235] M. Pop, D. S. Kosack, and S. L. Salzberg. Hierarchical scaffolding with bambus. *Genome Research*, 14(1):149–159, Jan 2004.

[236] A. R. Quinlan, R. A. Clark, S. Sokolova, M. L. Leibowitz, Y. Zhang, M. E. Hurles, J. C. Mell, and I. M. Hall. Genome-wide mapping and assembly of structural variant breakpoints in the mouse genome. *Genome Research*, 20(5):623–635, May 2010.

[237] A. R. Quinlan and I. M. Hall. BEDTools: A flexible suite of utilities for comparing genomic features. *Bioinformatics*, 26(6):841–842, Mar 2010.

[238] A. R. Quinlan and I. M. Hall. Characterizing complex structural variation in germline and somatic genomes. *Trends Genet*, 28(1):43–53, Jan 2012.

[239] K. Ragunathan, G. Jih, and D. Moazed. Epigenetic inheritance uncoupled from sequence-specific recruitment. *Science*, 348(6230):1258699, Apr 2015.

[240] P. Ramachandran, G. A. Palidwor, C. J. Porter, and T. J. Perkins. MaSC: Mappability-sensitive cross-correlation for estimating mean fragment length of single-end short-read sequencing data. *Bioinformatics*, 29(4):444–450, Feb 2013.

[241] T. Rausch, T. Zichner, A. Schlattl, A. M. Stütz, V. Benes, and J. O. Korbel. DELLY: Structural variant discovery by integrated paired-end and split-read analysis. *Bioinformatics*, 28(18):i333–i339, Sep 2012.

[242] J. Reumers, P. De Rijk, H. Zhao, A. Liekens, D. Smeets, J. Cleary, P. Van Loo, M. Van Den Bossche, K. Catthoor, B. Sabbe, E. Despierre, I. Vergote, B. Hilbush, D. Lambrechts, and J. Del-Favero. Optimized filtering reduces the error rate in detecting genomic variants by short-read sequencing. *Nat Biotechnol*, 30(1):61–68, Jan 2012.

[243] H. S. Rhee and B. F. Pugh. Comprehensive genome-wide protein-DNA interactions detected at single-nucleotide resolution. *Cell*, 147(6):1408–1419, Dec 2011.

[244] F. J. Ribeiro, D. Przybylski, S. Yin, T. Sharpe, S. Gnerre, A. Abouelleil, A. M. Berlin, A. Montmayeur, T. P. Shea, B. J. Walker, S. K. Young, C. Russ, C. Nusbaum, I. MacCallum, and D. B. Jaffe. Finished bacterial genomes from shotgun sequence data. *Genome Research*, 22(11):2270–2277, Nov 2012.

[245] G. Rizk, D. Lavenier, and R. Chikhi. DSK: k-mer counting with very low memory usage. *Bioinformatics*, 29(5):652–653, Mar 2013.

[246] N. D. Roberts, R. D. Kortschak, W. T. Parker, A. W. Schreiber, S. Branford, H. S. Scott, G. Glonek, and D. L. Adelson. A comparative analysis of algorithms for somatic SNV detection in cancer. *Bioinformatics*, 29(18):2223–2230, Sep 2013.

[247] G. Robertson, M. Hirst, M. Bainbridge, M. Bilenky, Y. Zhao, T. Zeng, G. Euskirchen, B. Bernier, R. Varhol, A. Delaney, N. Thiessen, O. L. Griffith, A. He, M. Marra, M. Snyder, and S. Jones. Genome-wide profiles of STAT1 DNA association using chromatin immunoprecipitation and massively parallel sequencing. *Nat Methods*, 4(8):651–657, Aug 2007.

[248] G. Robertson, J. Schein, R. Chiu, R. Corbett, M. Field, S. D. Jackman, K. Mungall, S. Lee, H. M. Okada, J. Q. Qian, M. Griffith, A. Raymond, N. Thiessen, T. Cezard, Y. S. Butterfield, R. Newsome, S. K. Chan, R. She, R. Varhol, B. Kamoh, A.-L. Prabhu, A. Tam, Y. Zhao, R. A. Moore, M. Hirst, M. A. Marra, S. J. M. Jones, P. A. Hoodless, and I. Birol. De novo assembly and analysis of RNA-seq data. *Nat Methods*, 7(11):909–912, Nov 2010.

[249] L. Roguski and S. Deorowicz. DSRC 2–industry-oriented compression of FASTQ files. *Bioinformatics*, 30(15):2213–2215, Aug 2014.

[250] A. Roth, J. Ding, R. Morin, A. Crisan, G. Ha, R. Giuliany, A. Bashashati, M. Hirst, G. Turashvili, A. Oloumi, M. A. Marra, S. Aparicio, and S. P. Shah. JointSNVMix: A probabilistic model for accurate detection of somatic mutations in normal/tumour paired next-generation sequencing data. *Bioinformatics*, 28(7):907–913, Apr 2012.

[251] J. M. Rothberg, W. Hinz, T. M. Rearick, J. Schultz, W. Mileski, M. Davey, J. H. Leamon, K. Johnson, M. J. Milgrew, M. Edwards, J. Hoon, J. F. Simons, D. Marran, J. W. Myers, J. F. Davidson, A. Branting, J. R. Nobile, B. P. Puc, D. Light, T. A. Clark, M. Huber, J. T. Branciforte, I. B. Stoner, S. E. Cawley, M. Lyons, Y. Fu, N. Homer, M. Sedova, X. Miao, B. Reed, J. Sabina, E. Feierstein, M. Schorn, M. Alanjary, E. Dimalanta, D. Dressman, R. Kasinskas, T. Sokolsky, J. A. Fidanza, E. Namsaraev, K. J. McKernan, A. Williams, G. T. Roth, and J. Bustillo. An integrated semiconductor device enabling non-optical genome sequencing. *Nature*, 475(7356):348–352, Jul 2011.

[252] J. D. Rowley. Letter: A new consistent chromosomal abnormality in chronic myelogenous leukaemia identified by quinacrine fluorescence and giemsa staining. *Nature*, 243(5405):290–293, Jun 1973.

[253] J. Rozowsky, G. Euskirchen, R. K. Auerbach, Z. D. Zhang, T. Gibson, R. Bjornson, N. Carriero, M. Snyder, and M. B. Gerstein. PeakSeq enables systematic scoring of ChIP-seq experiments relative to controls. *Nat Biotechnol*, 27(1):66–75, Jan 2009.

[254] M. Ruffalo, T. LaFramboise, and M. Koyutürk. Comparative analysis of algorithms for next-generation sequencing read alignment. *Bioinformatics*, 27(20):2790–2796, Oct 2011.

[255] S. M. Rumble, P. Lacroute, A. V. Dalca, M. Fiume, A. Sidow, and M. Brudno. SHRiMP: Accurate mapping of short color-space reads. *PLoS Comput Biol*, 5(5):e1000386, May 2009.

[256] S. Saha, A. B. Sparks, C. Rago, V. Akmaev, C. J. Wang, B. Vogelstein, K. W. Kinzler, and V. E. Velculescu. Using the transcriptome to annotate the genome. *Nat Biotechnol*, 20(5):508–512, May 2002.

[257] M. K. Sakharkar, V. T. K. Chow, and P. Kangueane. Distributions of exons and introns in the human genome. *In Silico Biol*, 4(4):387–393, 2004.

[258] L. Salmela, V. Mäkinen, N. Välimäki, J. Ylinen, and E. Ukkonen. Fast scaffolding with small independent mixed integer programs. *Bioinformatics*, 27(23):3259–3265, Dec 2011.

[259] F. Sanger and A. R. Coulson. A rapid method for determining sequences in DNA by primed synthesis with DNA polymerase. *J Mol Biol*, 94(3):441–448, May 1975.

[260] F. Sanger, S. Nicklen, and A. R. Coulson. DNA sequencing with chain-terminating inhibitors. *Proc Natl Acad Sci USA*, 74(12):5463–5467, Dec 1977.

[261] J. F. Sathirapongsasuti, H. Lee, B. A. J. Horst, G. Brunner, A. J. Cochran, S. Binder, J. Quackenbush, and S. F. Nelson. Exome sequencing-based copy-number variation and loss of heterozygosity detection: ExomeCNV. *Bioinformatics*, 27(19):2648–2654, Oct 2011.

[262] A. Sboner, L. Habegger, D. Pflueger, S. Terry, D. Z. Chen, J. S. Rozowsky, A. K. Tewari, N. Kitabayashi, B. J. Moss, M. S. Chee, F. Demichelis, M. A. Rubin, and M. B. Gerstein. FusionSeq: A modular framework for finding gene fusions by analyzing paired-end RNA-sequencing data. *Genome Biol*, 11(10):R104, 2010.

[263] E. E. Schadt, S. Turner, and A. Kasarskis. A window into third-generation sequencing. *Hum Mol Genet*, 19(R2):R227–R240, Oct 2010.

[264] S. Schbath, V. Martin, M. Zytnicki, J. Fayolle, V. Loux, and J.-F. Gibrat. Mapping reads on a genomic sequence: An algorithmic overview and a practical comparative analysis. *J Comput Biol*, 19(6):796–813, Jun 2012.

[265] R. Schmieder and R. Edwards. Quality control and preprocessing of metagenomic datasets. *Bioinformatics*, 27(6):863–864, Mar 2011.

[266] J. P. Schouten, C. J. McElgunn, R. Waaijer, D. Zwijnenburg, F. Diepvens, and G. Pals. Relative quantification of 40 nucleic acid sequences by multiplex ligation-dependent probe amplification. *Nucleic Acids Res*, 30(12):e57, Jun 2002.

[267] J. Schröder, A. Hsu, S. E. Boyle, G. Macintyre, M. Cmero, R. W. Tothill, R. W. Johnstone, M. Shackleton, and A. T. Papenfuss. Socrates: Identification of genomic rearrangements in tumour genomes by re-aligning soft clipped reads. *Bioinformatics*, 30(8):1064–1072, Jan 2014.

[268] M. H. Schulz, D. R. Zerbino, M. Vingron, and E. Birney. Oases: Robust de novo RNA-seq assembly across the dynamic range of expression levels. *Bioinformatics*, 28(8):1086–1092, Apr 2012.

[269] Z. Shao, Y. Zhang, G.-C. Yuan, S. H. Orkin, and D. J. Waxman. MAnorm: A robust model for quantitative comparison of ChIP-seq data sets. *Genome Biol*, 13(3):R16, 2012.

[270] L. Shi, Y. Guo, C. Dong, J. Huddleston, H. Yang, X. Han, A. Fu, Q. Li, N. Li, S. Gong, K. Lintner, Q. Ding, Z. Wang, J. Hu, D. Wang, F. Wang, L. Wang, G. Lyon, Y. Guan, Y. Shen, O. Evgrafov, J. Knowles, F. Thibaud-Nissen, V. Schneider, C. Yu, L. Zhou, E. Eichler, K. So, and K. Wang. Long-read sequencing and de novo assembly of a Chinese genome. *Nat Comm*, 7:12065, Jun 2016.

[271] H. Shin, T. Liu, X. Duan, Y. Zhang, and X. S. Liu. Computational methodology for ChIP-seq analysis. *Quantitative Biology*, 1(1):54–70, Mar 2013.

[272] T. Shiraki, S. Kondo, S. Katayama, K. Waki, T. Kasukawa, H. Kawaji, R. Kodzius, A. Watahiki, M. Nakamura, T. Arakawa, S. Fukuda, D. Sasaki, A. Podhajska, M. Harbers, J. Kawai, P. Carninci, and Y. Hayashizaki. Cap analysis gene expression for high-throughput analysis of transcriptional starting point and identification of promoter usage. *Proc Natl Acad Sci USA*, 100(26):15776–15781, Dec 2003.

[273] J. T. Simpson and R. Durbin. Efficient construction of an assembly string graph using the FM-index. *Bioinformatics*, 26(12):i367–i373, Jun 2010.

[274] J. T. Simpson and R. Durbin. Efficient de novo assembly of large genomes using compressed data structures. *Genome Research*, 22(3):549–556, Mar 2012.

[275] J. T. Simpson, K. Wong, S. D. Jackman, J. E. Schein, S. J. M. Jones, and I. Birol. ABySS: A parallel assembler for short read sequence data. *Genome Research*, 19(6):1117–1123, Jun 2009.

[276] S. Sindi, E. Helman, A. Bashir, and B. J. Raphael. A geometric approach for classification and comparison of structural variants. *Bioinformatics*, 25(12):i222–i230, Jun 2009.

[277] S. Sindi, S. Onal, L. C. Peng, H.-T. Wu, and B. J. Raphael. An integrative probabilistic model for identification of structural variation in sequencing data. *Genome Biol*, 13(3):R22, 2012.

[278] E. Siragusa, D. Weese, and K. Reinert. Fast and accurate read mapping with approximate seeds and multiple backtracking. *Nucleic Acids Res*, 41(7):e78, Apr 2013.

[279] A. D. Smith, Z. Xuan, and M. Q. Zhang. Using quality scores and longer reads improves accuracy of Solexa read mapping. *BMC Bioinformatics*, 9:128, 2008.

[280] M. J. Solomon and A. Varshavsky. Formaldehyde-mediated DNA-protein crosslinking: A probe for in vivo chromatin structures. *Proc Natl Acad Sci USA*, 82(19):6470–6474, 1985.

[281] C. Soneson and M. Delorenzi. A comparison of methods for differential expression analysis of RNA-seq data. *BMC Bioinformatics*, 14:91, 2013.

[282] L. Song, Z. Zhang, L. L. Grasfeder, A. P. Boyle, P. G. Giresi, B.-K. Lee, N. C. Sheffield, S. Gräf, M. Huss, D. Keefe, et al. Open chromatin defined by DNaseI and FAIRE identifies regulatory elements that shape cell-type identity. *Genome Research*, 21(10):1757–1767, 2011.

[283] T. Souaiaia, Z. Frazier, and T. Chen. ComB: SNP calling and mapping analysis for color and nucleotide space platforms. *Journal of Computational Biology*, 18(6):795–807, Jun 2011.

[284] A. Srebrow and A. R. Kornblihtt. The connection between splicing and cancer. *J Cell Sci*, 119(Pt 13):2635–2641, Jul 2006.

[285] S. Stamm, J.-J. Riethoven, V. Le Texier, C. Gopalakrishnan, V. Kumanduri, Y. Tang, N. L. Barbosa-Morais, and T. A. Thanaraj. ASD: A bioinformatics resource on alternative splicing. *Nucleic Acids Res*, 34(Database issue):D46–D55, Jan 2006.

[286] P. Stankiewicz and J. R. Lupski. Structural variation in the human genome and its role in disease. *Annu Rev Med*, 61:437–455, 2010.

[287] P. J. Stephens, C. D. Greenman, B. Fu, F. Yang, G. R. Bignell, L. J. Mudie, E. D. Pleasance, K. W. Lau, D. Beare, L. A. Stebbings, S. McLaren, M.-L. Lin, D. J. McBride, I. Varela, S. Nik-Zainal, C. Leroy, M. Jia, A. Menzies, A. P. Butler, J. W. Teague, M. A. Quail, J. Burton, H. Swerdlow, N. P. Carter, L. A. Morsberger, C. Iacobuzio-Donahue, G. A. Follows, A. R. Green, A. M. Flanagan, M. R. Stratton, P. A. Futreal, and P. J. Campbell. Massive genomic rearrangement acquired in a single catastrophic event during cancer development. *Cell*, 144(1):27–40, Jan 2011.

[288] M. Sultan, M. H. Schulz, H. Richard, A. Magen, A. Klingenhoff, M. Scherf, M. Seifert, T. Borodina, A. Soldatov, D. Parkhomchuk, D. Schmidt, S. O'Keeffe, S. Haas, M. Vingron, H. Lehrach, and M.-L. Yaspo. A global view of gene activity and alternative splicing by deep sequencing of the human transcriptome. *Science*, 321(5891):956–960, Aug 2008.

[289] W.-K. Sung. *Algorithms in Bioinformatics: A Practical Introduction.* Chapman & Hall/CRC Mathematical and Computational Biology. CRC Press, 2009.

[290] W.-K. Sung, H. Zheng, S. Li, R. Chen, X. Liu, Y. Li, N. P. Lee, W. H. Lee, P. N. Ariyaratne, C. Tennakoon, F. H. Mulawadi, K. F. Wong, A. M. Liu, R. T. Poon, S. T. Fan, K. L. Chan, Z. Gong, Y. Hu, Z. Lin, G. Wang, Q. Zhang, T. D. Barber, W.-C. Chou, A. Aggarwal, K. Hao,

W. Zhou, C. Zhang, J. Hardwick, C. Buser, J. Xu, Z. Kan, H. Dai, M. Mao, C. Reinhard, J. Wang, and J. M. Luk. Genome-wide survey of recurrent HBV integration in hepatocellular carcinoma. *Nat Genet*, 44(7):765–769, May 2012.

[291] Y. Surget-Groba and J. I. Montoya-Burgos. Optimization of de novo transcriptome assembly from next-generation sequencing data. *Genome Research*, 20(10):1432–1440, Oct 2010.

[292] J. K. Teer and J. C. Mullikin. Exome sequencing: The sweet spot before whole genomes. *Hum Mol Genet*, 19(R2):R145–R151, Oct 2010.

[293] W. Tembe, J. Lowey, and E. Suh. G-SQZ: Compact encoding of genomic sequence and quality data. *Bioinformatics*, 26(17):2192–2194, Sep 2010.

[294] C. Tennakoon, R. W. Purbojati, and W.-K. Sung. BatMis: A fast algorithm for k-mismatch mapping. *Bioinformatics*, 28(16):2122–2128, Aug 2012.

[295] C. Trapnell, D. G. Hendrickson, M. Sauvageau, L. Goff, J. L. Rinn, and L. Pachter. Differential analysis of gene regulation at transcript resolution with RNA-seq. *Nat Biotechnol*, 31(1):46–53, Jan 2013.

[296] C. Trapnell, L. Pachter, and S. L. Salzberg. TopHat: Discovering splice junctions with RNA-seq. *Bioinformatics*, 25(9):1105–1111, May 2009.

[297] C. Trapnell, B. A. Williams, G. Pertea, A. Mortazavi, G. Kwan, M. J. van Baren, S. L. Salzberg, B. J. Wold, and L. Pachter. Transcript assembly and quantification by RNA-seq reveals unannotated transcripts and isoform switching during cell differentiation. *Nat Biotechnol*, 28(5):511–515, May 2010.

[298] T. J. Treangen, D. D. Sommer, F. E. Angly, S. Koren, and M. Pop. Next generation sequence assembly with AMOS. *Curr Protoc Bioinformatics*, Chapter 11:Unit 11.8, Mar 2011.

[299] A. Valouev, D. S. Johnson, A. Sundquist, C. Medina, E. Anton, S. Batzoglou, R. M. Myers, and A. Sidow. Genome-wide analysis of transcription factor binding sites based on ChIP-seq data. *Nat Methods*, 5(9):829–834, Sep 2008.

[300] V. B. Vega, E. Cheung, N. Palanisamy, and W.-K. Sung. Inherent signals in sequencing-based Chromatin-ImmunoPrecipitation control libraries. *PLoS One*, 4(4):e5241, 2009.

[301] V. E. Velculescu, L. Zhang, B. Vogelstein, and K. W. Kinzler. Serial analysis of gene expression. *Science*, 270(5235):484–487, Oct 1995.

[302] R. Wan, V. N. Anh, and K. Asai. Transformations for the compression of FASTQ quality scores of next-generation sequencing data. *Bioinformatics*, 28(5):628–635, Mar 2012.

[303] E. T. Wang, R. Sandberg, S. Luo, I. Khrebtukova, L. Zhang, C. Mayr, S. F. Kingsmore, G. P. Schroth, and C. B. Burge. Alternative isoform regulation in human tissue transcriptomes. *Nature*, 456(7221):470–476, Nov 2008.

[304] H. Wang, D. Nettleton, and K. Ying. Copy number variation detection using next generation sequencing read counts. *BMC Bioinformatics*, 15:109, 2014.

[305] J. Wang, C. G. Mullighan, J. Easton, S. Roberts, S. L. Heatley, J. Ma, M. C. Rusch, K. Chen, C. C. Harris, L. Ding, L. Holmfeldt, D. Payne-Turner, X. Fan, L. Wei, D. Zhao, J. C. Obenauer, C. Naeve, E. R. Mardis, R. K. Wilson, J. R. Downing, and J. Zhang. CREST maps somatic structural variation in cancer genomes with base-pair resolution. *Nat Methods*, 8(8):652–654, 2011.

[306] K. Wang, D. Singh, Z. Zeng, S. J. Coleman, Y. Huang, G. L. Savich, X. He, P. Mieczkowski, S. A. Grimm, C. M. Perou, J. N. MacLeod, D. Y. Chiang, J. F. Prins, and J. Liu. MapSplice: Accurate mapping of RNA-seq reads for splice junction discovery. *Nucleic Acids Res*, 38(18):e178, Oct 2010.

[307] X. Wang, Z. Wu, and X. Zhang. Isoform abundance inference provides a more accurate estimation of gene expression levels in RNA-seq. *J Bioinform Comput Biol*, 8 Suppl 1:177–192, Dec 2010.

[308] Z. Wang, M. Gerstein, and M. Snyder. RNA-seq: A revolutionary tool for transcriptomics. *Nat Rev Genet*, 10(1):57–63, Jan 2009.

[309] R. L. Warren, G. G. Sutton, S. J. M. Jones, and R. A. Holt. Assembling millions of short DNA sequences using SSAKE. *Bioinformatics*, 23(4):500–501, Feb 2007.

[310] J. D. Watson and F. H. Crick. Molecular structure of nucleic acids; a structure for deoxyribose nucleic acid. *Nature*, 171(4356):737–738, Apr 1953.

[311] D. Weese, A.-K. Emde, T. Rausch, A. Döring, and K. Reinert. RazerS: Fast read mapping with sensitivity control. *Genome Research*, 19(9):1646–1654, Sep 2009.

[312] C.-L. Wei, Q. Wu, V. B. Vega, K. P. Chiu, P. Ng, T. Zhang, A. Shahab, H. C. Yong, Y. Fu, Z. Weng, J. Liu, X. D. Zhao, J.-L. Chew, Y. L. Lee, V. A. Kuznetsov, W.-K. Sung, L. D. Miller, B. Lim, E. T. Liu, Q. Yu, H.-H. Ng, and Y. Ruan. A global map of p53 transcription-factor binding sites in the human genome. *Cell*, 124(1):207–219, Jan 2006.

[313] P. Weiner. Linear pattern matching algorithms. *Switching and Automata Theory*, pages 1–11, 1973.

[314] J. Weischenfeldt, O. Symmons, F. Spitz, and J. O. Korbel. Phenotypic impact of genomic structural variation: insights from and for human disease. *Nat Rev Genet*, 14(2):125–138, Feb 2013.

[315] D. A. Wheeler, M. Srinivasan, M. Egholm, Y. Shen, L. Chen, A. McGuire, W. He, Y.-J. Chen, V. Makhijani, G. T. Roth, X. Gomes, K. Tartaro, F. Niazi, C. L. Turcotte, G. P. Irzyk, J. R. Lupski, C. Chinault, X.-Z. Song, Y. Liu, Y. Yuan, L. Nazareth, X. Qin, D. M. Muzny, M. Margulies, G. M. Weinstock, R. A. Gibbs, and J. M. Rothberg. The complete genome of an individual by massively parallel DNA sequencing. *Nature*, 452(7189):872–876, Apr 2008.

[316] B. T. Wilhelm, S. Marguerat, S. Watt, F. Schubert, V. Wood, I. Goodhead, C. J. Penkett, J. Rogers, and J. Bähler. Dynamic repertoire of a eukaryotic transcriptome surveyed at single-nucleotide resolution. *Nature*, 453(7199):1239–1243, Jun 2008.

[317] A. Wilm, P. P. K. Aw, D. Bertrand, G. H. T. Yeo, S. H. Ong, C. H. Wong, C. C. Khor, R. Petric, M. L. Hibberd, and N. Nagarajan. LoFreq: A sequence-quality aware, ultra-sensitive variant caller for uncovering cell-population heterogeneity from high-throughput sequencing datasets. *Nucleic Acids Res*, 40(22):11189–11201, Dec 2012.

[318] K. Wong, T. M. Keane, J. Stalker, and D. J. Adams. Enhanced structural variant and breakpoint detection using SVMerge by integration of multiple detection methods and local assembly. *Genome Biol*, 11(12):R128, 2010.

[319] J. Wu, O. Anczuków, A. R. Krainer, M. Q. Zhang, and C. Zhang. OLego: Fast and sensitive mapping of spliced mRNA-seq reads using small seeds. *Nucleic Acids Res*, 41(10):5149–5163, Apr 2013.

[320] C. Xie and M. T. Tammi. CNV-seq, a new method to detect copy number variation using high-throughput sequencing. *BMC Bioinformatics*, 10:80, 2009.

[321] Y. Xie, G. Wu, J. Tang, R. Luo, J. Patterson, S. Liu, W. Huang, G. He, S. Gu, S. Li, X. Zhou, T.-W. Lam, Y. Li, X. Xu, G. K.-S. Wong, and J. Wang. SOAPdenovo-Trans: De novo transcriptome assembly with short RNA-seq reads. *Bioinformatics*, 30(12):1660–1666, Jun 2014.

[322] H. Xu, L. Handoko, X. Wei, C. Ye, J. Sheng, C.-L. Wei, F. Lin, and W.-K. Sung. A signal-noise model for significance analysis of ChIP-seq with negative control. *Bioinformatics*, 26(9):1199–1204, May 2010.

[323] H. Xu and W.-K. Sung. Identifying differential histone modification sites from ChIP-seq data. *Methods Mol Biol*, 802:293–303, 2012.

[324] H. Yang, Y. Zhong, C. Peng, J.-Q. Chen, and D. Tian. Important role of indels in somatic mutations of human cancer genes. *BMC Med Genet*, 11:128, 2010.

[325] L. Yang, L. J. Luquette, N. Gehlenborg, R. Xi, P. S. Haseley, C.-H. Hsieh, C. Zhang, X. Ren, A. Protopopov, L. Chin, R. Kucherlapati, C. Lee, and P. J. Park. Diverse mechanisms of somatic structural variations in human cancer genomes. *Cell*, 153(4):919–929, May 2013.

[326] L. R. Yates and P. J. Campbell. Evolution of the cancer genome. *Nat Rev Genet*, 13(11):795–806, Nov 2012.

[327] K. Ye, M. H. Schulz, Q. Long, R. Apweiler, and Z. Ning. Pindel: A pattern growth approach to detect break points of large deletions and medium sized insertions from paired-end short reads. *Bioinformatics*, 25(21):2865–2871, Nov 2009.

[328] S. Yoon, Z. Xuan, V. Makarov, K. Ye, and J. Sebat. Sensitive and accurate detection of copy number variants using read depth of coverage. *Genome Research*, 19(9):1586–1592, Sep 2009.

[329] D. R. Zerbino and E. Birney. Velvet: Algorithms for de novo short read assembly using de Bruijn graphs. *Genome Research*, 18(5):821–829, May 2008.

[330] Y. Zhang, E.-W. Lameijer, P. A. 't Hoen, Z. Ning, P. E. Slagboom, and K. Ye. PASSion: A pattern growth algorithm-based pipeline for splice junction detection in paired-end RNA-seq data. *Bioinformatics*, 28(4):479–486, Feb 2012.

[331] Y. Zhang, T. Liu, C. A. Meyer, J. Eeckhoute, D. S. Johnson, B. E. Bernstein, C. Nusbaum, R. M. Myers, M. Brown, W. Li, and X. S. Liu. Model-based analysis of ChIP-seq (MACS). *Genome Biol*, 9(9):R137, 2008.

[332] Z. D. Zhang, J. Du, H. Lam, A. Abyzov, A. E. Urban, M. Snyder, and M. Gerstein. Identification of genomic indels and structural variations using split reads. *BMC Genomics*, 12:375, 2011.

Index